Lecture Notes in Computer Science 8121

Commenced Publication in 1973
Founding and Former Series Editors:
Gerhard Goos, Juris Hartmanis, and Jan van Leeuwen

Sergey Balandin   Sergey Andreev
Yevgeni Koucheryavy (Eds.)

Internet of Things, Smart Spaces, and Next Generation Networking

13th International Conference, NEW2AN 2013
and 6th Conference, ruSMART 2013
St. Petersburg, Russia, August 28-30, 2013
Proceedings

 Springer

Volume Editors

Sergey Balandin
FRUCT Oy
Kissankellontie 20B
00930 Helsinki, Finland
E-mail: sergey.balandin@fruct.org

Sergey Andreev
Tampere University of Technology
Department of Electronics and Communications Engineering
Korkeakoulunkatu 1
33720 Tampere, Finland
E-mail: sergey.andreev@tut.fi

Yevgeni Koucheryavy
Tampere University of Technology
Department of Electronics and Communications Engineering
Korkeakoulunkatu 1
33720 Tampere, Finland
E-mail: yk@cs.tut.fi

ISSN 0302-9743   e-ISSN 1611-3349
ISBN 978-3-642-40315-6   e-ISBN 978-3-642-40316-3
DOI 10.1007/978-3-642-40316-3
Springer Heidelberg Dordrecht London New York

Library of Congress Control Number: 2013944980

CR Subject Classification (1998): C.2, C.4, K.6, D.2, G.3, I.2

LNCS Sublibrary: SL 5 – Computer Communication Networks
and Telecommunications

Typesetting: Camera-ready by author, data conversion by Scientific Publishing Services, Chennai, India

Printed on acid-free paper

Springer is part of Springer Science+Business Media (www.springer.com)

Preface

We welcome you to the joint proceedings of the 13th NEW2AN (Next Generation Teletraffic and Wired/Wireless Advanced Networking) and 6th conference on Internet of Things and Smart Spaces ruSMART (Are You Smart) held in St. Petersburg, Russia, during August 28–30, 2013.

Originally, the NEW2AN conference was launched by ITC (International Teletraffic Congress) in St. Petersburg in June 1993 as an ITC-Sponsored Regional International Teletraffic Seminar. The first edition was entitled "Traffic Management and Routing in SDH Networks" and held by R&D LONIIS. In 2002, the event received its current name, the NEW2AN. In 2008, NEW2AN acquired a new companion in Smart Spaces, ruSMART, hence boosting interaction between researchers, practitioners, and engineers across different areas of ICT. Presently, NEW2AN and ruSMART are well-established conferences with a unique cross-disciplinary mixture of telecommunications-related research and science. NEW2AN/ruSMART is accompanied by outstanding keynotes from universities and companies across Europe, USA, and Russia.

The 13th NEW2AN technical program addressed various aspects of next-generation data networks. This year, special attention was given to upper-layer networking protocols and applications as well as to cognitive radio networks. In particular, the authors demonstrated novel and innovative approaches to performance and efficiency analysis, employed game-theoretical formulations, Markov chain models, and advanced queuing theory. It is also worth mentioning the traditional emphasis on wireless technologies, including, but not limited to, ad hoc, cellular, satellite, sensor, and mesh networks.

The 6th conference on Internet of Things and Smart Spaces ruSMART 2013 provided a forum for academic and industrial researchers to discuss new ideas and trends in the emerging areas of Internet of things and smart spaces that create new opportunities for fully customized applications and services. The conference brought together leading experts from top affiliations around the world. This year, there was active participation by industrial world-leader companies and particularly strong interest from attendees representing Russian R&D centers, which have a good reputation for high-quality research and business in innovative service creation and applications development.

This year, the first day of NEW2AN/ruSMART Technical Program started with the keynote talk on "R&D Activities and Progress Around IoT in Finland" by Wilhelm Rauss, who is consortium director of Finland's national IoT Program, senior R&D engineer and technical coordinator of the Cloud Computing Center at Ericsson. Finland's IoT consortium consists of more than 250 scientists and international experts from more than 35 national and international companies, organizations, and universities. In this session, there was a unique opportunity to learn about the activities, achievements, and progress of IoT in Finland.

We would like to thank the Technical Program Committee members of both conferences, as well as the associated reviewers, for their hard work and important contribution to the conference. This year, the conference program met the highest quality criteria with the acceptance ratio of 30%.

The current edition of the conferences was organized in cooperation with Open Innovations Association FRUCT, ITC (International Teletraffic Congress), IEEE, Tampere University of Technology, St. Petersburg State University of Telecommunications, and Popov Society. The support of these organizations is gratefully acknowledged.

We also wish to thank all those who contributed to the organization of the conferences. In particular, we are grateful to Roman Florea for his substantial work on supporting the conference website and his excellent job on the compilation of camera-ready papers and interaction with Springer.

We believe that the 13th NEW2AN and 6th ruSMART conferences delivered an informative, high-quality, and up-to-date scientific program. We also hope that participants enjoyed both the technical and the social conference components, the Russian ways of hospitality, and the beautiful city of St. Petersburg.

August 2013

Sergey Balandin
Sergey Andreev
Yevgeni Koucheryavy

Organization

NEW2AN International Advisory Committee

Nina Bhatti	Hewlett Packard, USA
Igor Faynberg	Alcatel Lucent, USA
Jarmo Harju	Tampere University of Technology, Finland
Andrey Koucheryavy	Giprosviaz, St. Petersburg State University of Telecommunications, Russia
Villy B. Iversen	Technical University of Denmark, Denmark
Paul Kühn	University of Stuttgart, Germany
Kyu Ouk Lee	ETRI, Korea
Mohammad S. Obaidat	Monmouth University, USA
Michael Smirnov	Fraunhofer FOKUS, Germany
Manfred Sneps-Sneppe	Ventspils University College, Latvia
Ioannis Stavrakakis	University of Athens, Greece
Sergey Stepanov	Sistema Telecom, Russia
Phuoc Tran-Gia	University of Würzburg, Germany

NEW2AN Technical Program Committee

Ozgur B. Akan	Koc University, Turkey
Sergey Andreev	Tampere University of Technology, Finland
Konstantin Avrachenkov	MISTRAL project, INRIA Sophia Antipolis, France
Francisco Barcelo-Arroyo	Universitat Politecnica de Catalunya (UPC), Spain
Torsten Ingo Braun	University of Bern, Switzerland
Wei Koong Chai	University College London, UK
Chrysostomos Chrysostomou	Frederick University, Cyprus
Alexey Vinel	Tampere University of Technology, Finland
Sergey Gorinsky	IMDEA, Spain
Roman Dunaytsev	Democritus University of Thrace, Greece
Markus H Fidler	Leibniz Universität Hannover, Germany
Dieter Fiems	Ghent University, Belgium
Roman Florea	Tampere University of Technology, Finland
Ivan Ganchev	University of Limerick, Telecommunications Research Centre, Ireland
Visvasuresh Victor Govindaswamy	Texas A&M University-Texarkana, USA

Jarmo Harju	Tampere University of Technology, Finland
Yevgeni Koucheryavy	Tampere University of Technology, Finland
Jong-Hyouk Lee	RSM Department, TELECOM Bretagne, France
Tatiana Madsen	Aalborg University, Denmark
Maja Matijasevic	University of Zagreb, Croatia
Arturas Medeisis	Vilnius Gediminas Technical University, Lithuania
Paulo Mendes	SITILabs, University Lusofona, Portugal
Pedro Merino	University of Malaga, Spain
Edmundo Monteiro	University of Coimbra, Portugal
Dr. Nitin	Jaypee University of Information Technology, India
Stoyan Atanasov Poryazov	Institute of Mathematics and Informatics, Bulgarian Academy of Sciences, Bulgaria
Simon Pietro Romano	University of Naples Federico II, Italy
Zhefu Shi	University of Missouri - Kansas City, USA
Weilian Su	Naval Postgraduate School, USA
Paulo Carvalho	Universidade do Minho, Portugal
Zhenkai Zhu	UCLA, USA

ruSMART Executive Technical Program Committee

Sergey Boldyrev	Senior Manager, Nokia, Helsinki, Finland
Nikolai Nefedov	Principal Scientist, Nokia Research Center, Switzerland
Ian Oliver	Principle Architect, Nokia, Helsinki, Finland
Alexander Smirnov	Head of laboratory, SPIIRAS, St. Petersburg, Russia
Vladimir Gorodetsky	Head of laboratory, SPIIRAS, St. Petersburg, Russia
Michael Lawo	Professor, Center for Computing Technologies (TZI), University of Bremen, Germany
Michael Smirnov	Fraunhofer FOKUS, Germany
Dieter Uckelmann	LogDynamics Lab, University of Bremen, Germany
Cornel Klein	Program Manager, Siemens Corporate Technology, Germany

ruSMART Technical Program Committee

Sergey I. Balandin	FRUCT Oy, Finland
Michel Banâtre	INRIA, France
Mohamed Baqer	University of Bahrain, Bahrain
Gianpaolo Cugola	Politecnico di Milano, Italy
Didem Gozupek	University, Turkey

Table of Contents

Smart Systems

II NEW2AN

Performance and Efficiency Analysis I

Network and Transport Layer Issues

Cognitive Radio Networks

Sensor and Mesh Networks

Performance and Efficiency Analysis II

Upper Layer Protocols and Applications

Ad-Hoc, Cellular and Satellite Networks

Internet of Things: The Foundational Infrastructure for a Smarter Planet

Rob van den Dam

IBM Institute for Business Value,
Amsterdam, The Netherlands
rob_vandendam@nl.ibm.com

Abstract. Every day, our world is getting more instrumented and interconnected. Streams of data are continuously being generated by mobile devices,
personal computers, networks, sensors, RFID tags, web services, social media
and the like. IBM's newest study reveals how new technologies support the
development of the Internet of Things, and how the Internet of Things provides
the foundational infrastructure for a smarter planet. Key trends that relate to
the Internet of Things include Mobile, Big Data, Cloud Computing, and Smart
Networks. The paper describes the latest developments in Mobile, Big Data
(including cognitive systems capable to evaluate large amounts of both structured and unstructured data), Cloud Computing and Smarter networks (including software defined environments to cope with the ever increasing workloads
in the networks).

Keywords: IoT, Mobile, Big Data, Cloud Computing, Software Defined
Environment.

1 Introduction

The Internet of Things (IoT) is a technological revolution in the future of communication and computing that is based on the concept of any place, anytime connectivity for
anything [1]. Even in these early stages, the IoT has transformed the way consumers
and corporations interact with each other and the environment around them. IoT technologies have impacted solution domains, such as Smart Grid, Supply Chain Management, Smart Cities, and Smart Home. The IoT is a computing paradigm that will
change business consumer experiences, models, technology investments, and everyday life.

The IoT also represents a network of Internet-enabled, real-world objects, such as
consumer electronics, nanotechnology, home appliances, embedded systems, sensors
of all kinds, and personal mobile devices. It includes enabling network and communication technologies, such as web services, RFID, IPv6, and 4G networks. We are
already applying IoT solutions in practical ways by using mobile devices. For example, people can monitor their home security, lights, heating, and cooling from their
smartphone. They can purchase a refrigerator that monitors its processes and sends
reports to their smartphone.

S. Balandin et al. (Eds.): NEW2AN/ruSMART 2013, LNCS 8121, pp. 1–12, 2013.

It is critical for us to consider the challenges and approaches in an IoT-centric eco-system. The primary focus must be on critical operational considerations, such as scalability, availability, manageability, data management, and security.

• Scalability

An IoT environment contains two scalability issues, each of which poses unique chal-lenges. The first scalability issue is based on the number of connected devices and include the number of concurrent connections, or throughput, that a system can support and the quality of service (QoS) level that can be guaranteed. Here, Internet scalability is a critical factor. Currently, most Internet-connected devices use IPv4, which does not provide sufficient unique addresses for the IoT, and optimum scalability would require migrating to IPv6 [2]. The second issue is based on the volume of generated data, and highlight performance issues that are associated with data collection, processing, sto-rage, query, and display. IoT systems need to handle both device and data scalabilities.

• Availability

IoT availability involves reliability and recoverability. One architecture implication to availability is driven by the increased demand around cloud computing and x-as-a-service, such as software as a service. Corporations must closely look at the implica-tions to the services and capabilities that are required in an IoT environment. An innovative solution addresses fault avoidance/intolerance in ways that will facilitate a business to meet enterprise needs and customer expectations.

• Manageability

Currently, only IT-related systems, such as computers, servers, and storage devices, are managed under a governance model. Most other IoT devices are not managed systematically as part of a larger ecosystem. Many devices operate remotely without direct human interaction, which requires management of such devices in the same way, that is, remotely and without human intervention. New approaches are required to develop an IoT architecture and to manage its lifecycle.

• Managing data

Big data and the IoT are computing paradigms that, together, fundamentally change the nature of how we work, and interact with our environment. Where big data is all about volume, velocity, verity, and veracity, the IoT is about using that data in mea-ningful ways to improve productivity and quality of life. For example, the IoT can collect temporospatial information, which is both temporal (time) and spatial (loca-tion) data. This information, when combined with analytic technology, provides new insight into when, where, and how devices and humans can or should interact. The key issue is how corporations handle storing, managing, and manipulating this data.

• Security

Traditional IT security establishes secure boundaries and firewalls around internal IT systems. But with the Internet of Things, the concept of *controlled access* has changed to one of *controlled trust* that offers the widest range of possible solutions. Security challenges require IoT implementations to effectively deal with authentication, authorization, access control, trust and privacy requirements without negatively impacting usability.

The IoT represents the logical evolution in mobile, big data, cloud and smarter networks:

- *Mobile*: Mobile is becoming a part of everything we do
- *Big data*: The increasing number of 'things' will produce a tsunami of data, taxing our already complex information management systems
- *Clouds*: IoT systems require systems to scale quickly and autonomously
- *Smart Networks*: Networks must enable trillions of devices and objects to 'talk'

Most of the elements in the IoT are *mobile* devices whether they are smart phones, tablets, automobiles or smart grids. They are connected wirelessly, and they each have an IP address. It is estimated that by 2015 we will have 1 trillion of such devices. These devices are not isolated instruments. The whole purpose of the device is to connect to networks, applications, data and people, that is to be interconnected. Mobile can be leveraged to gather information, often in real time, and push out insights for better, faster decision making.

Mobile extends to Machine-to-Machine (M2M) as well. Advances in technology are enabling M2M connections that are creating new operating models and opportunities to provide business value. We are able to identify things through tagging and sensing them. Advances in nanotechnology are helping us infuse intelligence and processing power into mobile objects to create thinking things. Advances in power technology allow us to power things more efficiently and for longer periods of time and in increasingly remote location. The ability to tag, sense, power and shrink things has extended mobility beyond people to nearly every type of object on the world. Mobile is right at the center of the IoT story.

In the following sections, we will focus on three other key trends that relate to the IoT: Big Data (analytics), Cloud computing and Smart networks.

2   Big Data Is Getting Bigger

Ten to fifteen years ago data was coming in from just a few sources, mainly customer transactions and supply chain transactions. Today, data is coming in from everywhere, ranging from the consumer space (personal devices like smartphones, gadgets, implants, etc.) to the service space (Internet services, local environment services, etc.). In addition, the huge growth of pictures, audio, video, social media and other unstructured data is taxing the storage systems and information databases of many data centers.

Our appetite for creating, gathering and storing data continues to grow and grow and grow. IDC forecasts 15 billion devices will be communicating over the network in 2015 [3]. Ericsson estimates 50 billion devices will be connected to the Web by the year 2020, producing enormous streams of data [4].

To more clearly understand how organizations – and in particular communications service providers (CSPs) – view big data in the Internet of Things, the IBM Institute for Business Value conducted a global big data study, surveying 1144 businesses and IT professionals in 95 countries [5]. The CSPs in this survey responded quite differently on a number of questions as compared to the other industries. For instance, more than any other industry we studied, CSP respondents define big data as the capabilities needed to perform '*real-time*' information analysis (see Fig 1). In fact, 40% percent of CSPs defined big data as such, in contrast to only 15 percent in the total sample.

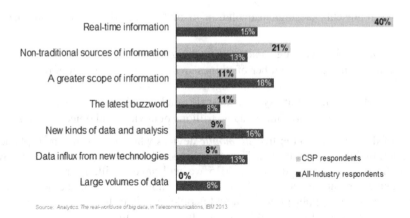

Source: Analytics: The real-world use of big data, in Telecommunications, IBM 2013

Fig. 1. For CSPs, big data is best described by the emerging requirements for real-time information

While large volumes of data are not new to CSPs – collecting millions of call detail records per day has become routine – the level of complexity of data today is a significant challenge. Analyzing the contextual data provided today by smartphones, tablets, personal computers, networks, sensors, RFID tags, web services, social media and the like, in near real-time is becoming increasingly complex yet crucial. Moreover, with the advent of smart phones, tablets and other devices that are application dependent, the volume of signaling data – i.e. non-message information about the device, its location and updates – has also increased significantly.

The real-time aspect of big data is extremely important for the IoT. For consumers, IoT and Contextual Services will make their lives fully connected. Today, mass production is no longer good enough; smarter consumers expect unique products and services customized for themselves. Interconnected devices are enabling mass personalization through contextual-awareness by utilizing continuous processing to make meaningful inferences that can benefit the end user.

For businesses and government organizations the real-time aspect is very important in cases that require quick decision-making, such as logistic problems or spread of infectious diseases. Big data approaches can help cities to leverage near-real time city information, anticipate incidents and coordinate resources to give support in the event of an emergency. The Operations Center implanted in Rio de Janeiro, for example, is using a forecasting system that synthesizes data from rivers, historical rainfall logs and radar feeds in order to anticipate heavy rains, flash floods, landslides, power outages and traffic hazards [6].

Big data itself does not create value in the IoT, until it is put to use to solve important issues or challenges. This requires access to more - and different kinds of - data, as well as adequate analytics capabilities.

Examining the responses in our global big data survey, more than 75 percent of all types of organizations reported to have started with a strong core of analytics capabilities - such as data mining - to analyze big data to support key decision processes. This can transform data into insight by delivering relevant, integrated, timely and actionable information. Two-thirds reported using predictive modeling, enabling them to start the transition to an optimized, 'outcomes-focused' environment. Predictive capabilities can create foreknowledge and deep awareness of consumer, operations and network behaviors [7].

Big data increasingly creates the need to analyze multiple data types, including location data, social media, data from sensors and natural language text. In more than half of the active big data efforts, respondents reported using advanced capabilities designed to analyze text in its natural state. These analytics include the ability to interpret and understand the nuances of language, such as sentiment, slang and intentions.

Intelligent analytics and 'autonomics' can help create a highly dynamic and efficient information-centric environment. A system that is automated and aware of real-time events can provide input to promote solutions for immediate execution. As contextual information from the IoT and all other types of environments – including data produced by social platforms – becomes more important to support effective decision making, organizations should increasingly focus on acquiring the capabilities needed to wield prescriptive analytics designed to automate actions.

New powerful technologies – including cognitive systems [8] – are capable of evaluating large amounts of both structured and unstructured data in or near real-time. Whereas in today's programmable era, computers essentially process a series of "if then what" equations, cognitive systems learn, adapt, and ultimately hypothesize and suggest answers.

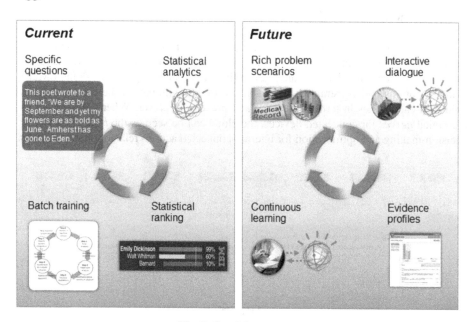

Fig. 2. Cognitive systems

Cognitive computing aims to break the conventional programmable machine paradigm and to evolve to entirely new computing architectures and programming paradigms. The end goal: ubiquitously deployed computers imbued with a new intelligence that can integrate information from a variety of sensors and sources, deal with ambiguity, respond in a context-dependent way, learn over time and carry out pattern recognition to solve difficult problems based on perception, action and cognition in complex, real-world environments (Fig 2).

Watson technology, for example, applies advanced natural language processing, information retrieval, knowledge representation and reasoning, and machine learning technologies to answering questions. It can sift through an equivalent of about 1 million books or roughly 200 million pages of data, and analyze this information and provide precise answers in less than three seconds [9].

Cognitive computers are expected to learn through experiences, find correlations, create hypotheses, and remember - and learn from - the outcomes, mimicking the brains structural and synaptic plasticity. Next steps to more advanced cognitive computing includes:

- development of chips that enable brain's abilities to perception, action and cognition
- computer simulation of the human brain
- exascale computing that will be more than 80% faster than today's fastest computers while consuming just a trickle of energy through designs inspired by human brains

3   Clouds That Scale

For the full power of the IoT to be realized, the utilization of cloud computing is fundamental. The 'things' in the IoT - such as sensors in cars, cameras, food packages, refrigerators and the like - are generally very small, primary doing M2M communications and constrained in capacity. Cloud computing on the other hand has unlimited capabilities in terms of storage and processing power. Where sensors act as the digital *nerves* for connected devices, the cloud can be seen as the *brain* to improve decision-making and optimization for internet-connected actions related to these devices.

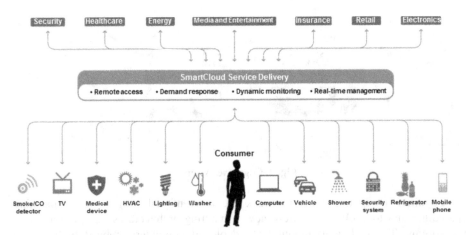

Fig. 3. Devices connect to the Cloud, reducing complexity and further empowering the consumer

Cloud-based M2M technology is becoming increasingly important in smarter healthcare, smarter utilities, smarter cities and smarter homes, to mention a view (see Fig 3). The next wave of Internet applications promise to have a massive impact on the society as a whole, and cloud computing – though IoT may change the overall architecture of the cloud – will provide the virtual infrastructure for all these applications.

Home automation is the "top of mind" application for many when discussing IoT. The market for cloud-based smarter home services is maturing in four key segments:

- *Entertainment and convenience:* create an open platform for new lines of televisions and devices that feature a portal which personalizes entertainment content from numerous content providers
- *Energy management*: automatically synchronize lighting, home appliances, climate control sensors and other electronics to minimize energy use based on changing exterior conditions and usage patterns in the home
- *Safety and security*: deploy home sensors that can instantly notify the homeowner, police and fire departments and selected neighbors – enhancing home security and providing peace of mind
- *Health and wellness*: continually monitor the health and fitness of patients using implanted or other at-home medical devices, avoiding the need to hospitalization or office visits.

Some CSPs are already exploring the potential to combine the pervasiveness of mobile technology with the ubiquity of the cloud platform. This milestone will pave the way for consumers to better control 'their' IoT, such as managing heating, lighting, laundry and the like via the CSP's mobile wireless network (see Fig 4).

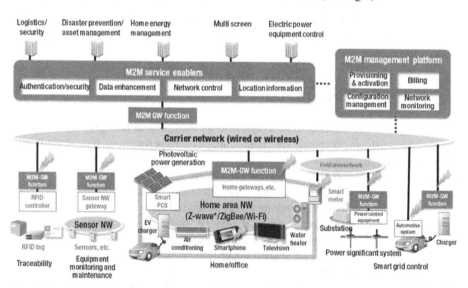

Fig. 4. In a M2M platform a variety of different devices are connected through fixed and wireless networks [10]

Vodafone, for example, combines mobile communications and cloud computing for the remote management of smart home appliances. The system combines Vodafone's Global M2M platform with a Smart Cloud platform to enable connected "smart" appliances to feed useful data to either service provider or vendor [11]. The system allows customers to control appliances through smartphones, with remote activities including the viewing of utility consumptions, security control, heating and lighting systems and the activation of appliances. It also provides vendors with a scaleable appliance management cloud-based platform, as well as the means to quickly introduce new consumer services.

This Vodafone smarter home initiative is a peek into a larger vision for mobile + Cloud-based innovation that extends into cities, buildings, banking, healthcare, government and more. Strong examples of how cloud can enable IoT connectivity exist in the healthcare industry. It includes a CSP's ability to offer health-specific solutions such as cloud-based remote monitoring to manage chronic diseases or sophisticated remote diagnostic capabilities. A&T, for instance, set up its ForHealth business unit in 2010, with a vision to accelerate the delivery of innovative, wireless, networked and cloud solutions specifically for the healthcare industry [12]

CSPs have a crucial role to play and they are well-suited to take a central position in the cloud-based IoT [13]. They fully control QoS at every point in the network. In addition, a cloud solution requires an infrastructure that is reliable and secure, and CSPs have a good reputation for managing large-scale infrastructures, dealing with personal and business-sensitive data in a confidential way.

They also have access to a wealth of information about their customers' behavior, preferences and movements; applying analytics to produce the insights and context that are necessary to optimize decision making. And they are uniquely positioned to group and structure a wide variety of their own and third-party applications and services relevant in the IoT.

4 Networks That Are Smart

Ultra-fast broadband continue to be a high priority for CSPs, as they prepare to meet a considerable rise in demand for data capacity. Experimental fixed-broadband programs plans to offer 1000/1000 Mbit/s symmetrical connections directly to consumer homes. The race for ultra-fast broadband appears to have been decided in favor of LTE, in combination with WiFi mobile traffic offload and femtocells. LTE technology is to handle the tremendous surge in data traffic.

High speed broadband, as important it may be, doesn't make a network 'smart'. We need the network to be multidirectional instead of point-to-point. Smart networks must be infused with advanced analytics and intelligence, so they can identify connected instrumented things and collect relevant data from them. They'll have to be built on a foundation of standards and software that allows trillions of devices and objects to 'talk'.

The realization of the Internet of things have placed requirements upon mobile networks that the infrastructure was not originally designed to accommodate. Adding more intelligence at the *edge* of mobile networks is one of the options CSPs have to optimize their infrastructure to deal with unprecedented amounts of traffic, enhance latency-critical applications and to provide a distribution computing environment that can analyze massive amounts of information [14].

A more structural way of coping with the increasing workloads is the approach of Software Defined Environments (SDE) [15]. In the past (traditional) workload scenarios, there were few, stable and well known workloads (see Fig 5). The software was manually stood up on a fixed set of devices with minimum requirements for configuration. Often the hardware and software were tightly integrated.

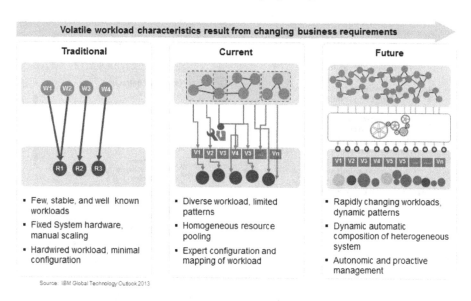

Fig. 5. Today's environment are making workloads (and Networks) more volatile

With the emergence of cloud, traditional workloads were virtualized and moved to the cloud, but at large, they were continued to be manually tuned and managed. In addition, new workloads started to emerge. Though limited workloads are available, expert configurations and mapping are still required. Cloud infrastructures are largely based on pooling of *homogeneous* resources.

Today's environment with mobile first, social technologies and scalable service ecosystems is making workloads (and networks) extremely volatile. The environment becomes dramatically *heterogeneous*. The exploding number and increased volatility of workloads and applications coupled with the heterogeneity leads to a situation where standing and tuning applications can no further be done in a manual fashion. A need to automate the deployment and to continuously and optimally manage these workloads is evident. This includes both the software and infrastructure of workload fit systems. This is what SDE is about.

SDEs provide abstractions of workloads, services and infrastructure and end-to-end mappings (see Fig 6). Being able to abstract workloads and the infrastructure, the workloads, the network, the storage environment and the computing environment can be virtualized, as well as common services. As part of the deployment the abstract workload is mapped to the best suited sources at the time. The system is dynamically constructed and configured and the workload is mapped to this system. As a result SDE provides an end-to-end orchestration at the workload, services and system (resource) level.

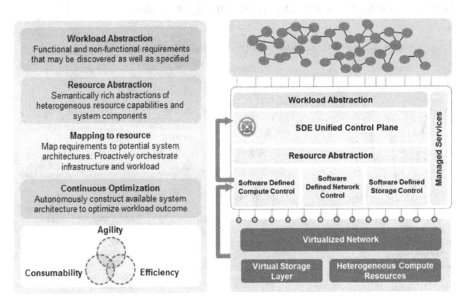

Fig. 6. Software Defined Environments provide abstractions of workloads, services and infrastructure and end-to-end mappings

The mapping/orchestration is autonomous, continuously monitoring and assessing the state of the system and if necessary reconfiguring hardware and software top optimize the outcome. SDEs bring together three key requirement of workload deployment in cloud environments: agility, efficiency and consumability. The vision of the SDE is one of an intelligent data-driven ecosystem that is easily managed and scaled to meet the requirements of the IoT.

5   Concluding Remarks

The IoT is a technological revolution in the future of computing and communications that is based on the concept of any time, any place connectivity for anything. The Internet of Things provides the foundational infrastructure for a smarter planet. It comprises billions of sensors and actuators embedded in physical objects, which are linked through fixed and wireless networks parallel to the hundreds of millions of people who have access to the Internet. When the objects in the IoT are able to interpret the continuous flow of data, to sense what is happening and to communicate with each other, the IoT will enable applications and uses in people's live that were previously unimaginable. The IoT vision is to provide a dynamic and global infrastructure which is characterized by intelligent and self-configuring capabilities.

A number of trends will be fundamental for realization of IoT. Communications and connectivity of IoT is enabled by IPv6, which is replacing IPv4. Sensors are getting smarter, smaller and cheaper, and there will be billions of them. Sensors and systems of sensors will be increasingly talking to each other and data centers via wireless communications. IoT will act as a nervous system, enabling an automated sense and respond system for any business process or application. All these 'things' will produce an ever-increasing amount of structured and unstructured data that requires advanced analytics to provide insights from all the 'things'. And IoT systems may have to scale quickly and autonomously. All these trends will result in new innovative and smart applications and services we only can dream of today. The IoT will impact every single industry and become a part of people's lives.

References

1. IERC. The Internet of Things, New Horizons, Halifax, UK (2012)
   ISBN 978-0-9553707-9-3
2. IoT6 project, Universal Integration of the Internet of Things through an IPv6-based service oriented architecture, http://www.iot.eu
3. Rise of the embedded Internet. Intel (2010),
   http://download.intel.com/embedded/15billion/
   applications/pdf/322202.pdf
4. More than 50 billion connected devices. Ericsson (2012),
   http://www.ericsson.com/res/docs/whitepapers/
   wp-50-billions.pdf
5. Analytics: The real-world use of big data in Telecommunications, IBM Institute for Business Value (2013), http://www-935.ibm.com/services/us/gbs/
   thoughtleadership/communications.html
6. Mission Control, built for cities. New York Times (2012),
   http://www.nytimes.com/2012/03/04/business/
   ibm-takes-smarter-cities-concept-to-rio-de-
   janeiro.html?pagewanted=all&_r=0
7. Analytics: The real world use of big data, IBM Institute for Business Value (2012),
   http://www-935.ibm.com/services/us/gbs/
   thoughtleadership/ibv-big-data-at-work.html
8. New ways of Thinking. IBM Research (2012),
   http://www.ibm.com/smarterplanet/us/en/business_analytics/
   article/cognitive_computing.html
9. IBM's Watson Computer heads to Wall Street for post-Jeopardy Gig. Time Magazine (2012), http://business.time.com/2012/03/07/
   ibms-watson-supercomputer-heads-to-wall-street/

10. Telecommunications systems for realizing a smart city. Hitachi (2012),
    http://www.hitachi.com/products/smartcity/
    smart-infrastructure/communication/life.html
11. IBM, Vodafone team up for advanced smarter home initiative. Telecompaper (2012),
    http://www.telecompaper.com/news/
    ibm-vodafone-team-up-for-advance-smarter-home-
    initiative-893408
12. Extend your reach. AT&T press release (2010), http://www.att.com/
    gen/press-room?pid=18711&cdvn=news&newsarticleid=31334
13. The natural fit of Cloud with Telecommunications, IBM Institute for Business Value
    (2012), http://www-935.ibm.com/services/
    us/gbs/thoughtleadership/ibv-telecom-cloud.html
14. NSN, IBM push intelligence to the edge. Telecoms.com (2013),
    http://www.telecoms.com/109332/
    nsn-ibm-push-intelligence-to-the-edge/
15. IBM advances the Software Defined Environment with solution that boosts overall net-
    work performance, scalability and management. MarketWatch, Wall Street Journal (2013),
    http://www.marketwatch.com/story/ibm-advances-the-software-
    defined-environment-with-solution-that-boosts-overall-
    network-performance-scalability-and-management-2013-03-26

Leveraging Multi-domain Links
via the Internet of Things

Towards Horizontal Integration of Vertical Pilots

Aliaksei Andrushevich[1], Bertrand Copigneaux[2], Rolf Kistler[1], Alexander Kurbatski[3],
Franck Le Gall[2], and Alexander Klapproth[1]

[1] CEESAR-iHomeLab, Lucerne University of Applied Sciences
Technikumstrasse 21, Horw, 6048, Switzerland
{aliaksei.andrushevich,rolf.kistler,alexander.klapproth}@hslu.ch
[2] Inno TSD, Place Joseph Bermond Ophira 1 – BP 63, Sophia Antipolis, 06902, Cedex France
{b.copigneaux,f.le-gall}@inno-group.com
[3] Belarusian State University, Nezavisimosti Prospekt 4, Minsk, 220000, Belarus
kurbatski@bsu.by

Abstract. This article describes the work on validation and measurement while dealing with integration of heterogeneous IoT systems being done within BUTLER research project. First, we give the brief motivation for integration work and reference the visionary Smart Life scenario of hypothetic personas' day from the year 2020. Presentation of identified security and privacy concerns of IoT deployments follows Smart Life scenario because of inevitable direct impact on large number of human end-users. Development objectives towards integration of heterogeneous IoT systems follow the defined scenario and are accompanied by IoT technical feasibility and user feedback validation targets and measurement values for the four to-be-deployed IoT applications. The architectural system overview of technological enablers of interoperability and interconnection between and within IoT applications are also important parts of the on-going research. The IoT application example gives an understanding on applicability of previously described considerations particularly in Smart Home domain. We round off with discussion on our developments at the end.

Keywords: Internet of Things, experimental deployments, heterogeneity, integration, validation.

1 Introduction

Active digitization and automation in all the life areas starting from the end of nineties has led to a number of specialized systems from different domains, from partially to totally incompatible with each other. In opposite, the vision of ubiquitous Internet of Things [1] implies the interdisciplinary system scalable to all areas of human activity. However, specific area impedes scalability because of the differences of logic processes in various fields and international differences. That is why the exemplary

S. Balandin et al. (Eds.): NEW2AN/ruSMART 2013, LNCS 8121, pp. 13–24, 2013.
© Springer-Verlag Berlin Heidelberg 2013

focusing on different vertical domains like Smart Transport, Smart Health, Smart Shopping, Smart Home and Smart City is essential to show the ways for seamless integration of heterogeneous domain specific IoT technologies into one coherent technical solution facilitating the context-aware 24/7 information support for end-users while preserving privacy and anonymity. In order to achieve this ambitious goal for interconnection and integration of IoT applications BUTLER project is designing and demonstrating the first prototype of a comprehensive, pervasive and effective context-aware information system, which will operate transparently and seamlessly across various scenarios.

2   Defined Horizontal Scenario

Horizontality is technically understood as complementary data exchange between different application domains generating the added value through availability of the non-core (also called contextual) information for a given application domain. Horizontality is a key feature of BUTLER's Smart Life Vision for ubiquitous context-aware and secure IoT. Concretely, after several iterations the following multi-domain or horizontal scenario has been defined within BUTLER project based on the users' stories [2] and the initial IoT applications available [3]. This horizontal scenario serves as a conceptual target to be reached by the end of the project and the final IoT platform is meant to implement this level of horizontality.

Table 1. Horizontal Scenario

Donald is 51 years old, works at a bank, married with Daisy. They have 2 children. They live in a house and own a chalet in the mountains. He enjoys eating out, despite food allergies and is interested in energy efficiency.

Daisy is 45 years old, a housewife and part-time shop assistant. She is married to Donald and has 2 children. Daisy suffers from diabetes, tries to do fitness workouts regularly and likes shopping with her girlfriends.

Story: Donald lives together with his wife Daisy and the two kids in its own one-family house in a suburb of a bigger city. While he enjoys living and working near the city, he loves to spend his holidays in the mountains. They are just about to go there for a family skiing week…

Donald is at home; he uses his tablet to check on the chalet and recognises that the temperature is just 12°C. Is the heating broken? BUTLER should have started heating the chalet up. He turns on the heating manually over his tablet and a second later the system comes up with notification: The alternative wind energy he had chosen for the operation of the electric heating system is very expensive at the moment as there is not much wind to drive the wind farms during the next 5 hours. So that was the reason why BUTLER hasn't yet turned on the heating. However, the system predicts low energy prices for the night and still enough time to heat up the chalet to 20°C by tomorrow morning. It also gives Donald the opportunity to switch the energy source and/or turn on the heating anyway. No, that is not necessary.

Table 1. (*Continued.*)

By the way, Donald is curious about the real energy consumption of his new TV. Using his tablet, he points the camera at the TV screen and sees the current watts augmented over the cam image. 20W, that's ok.

But for now Donald decides to watch the news magazine which is just about to start. After a while, his wife enters the living room. She has finished packing and wants to look at the "Ice Age"-Movie she recorded yesterday. No problem. Donald decides to go to the TV in the sleeping room and also start packing there for tomorrow. So he presses a button on his phone and Ice Age starts in the living room. He first goes to the kitchen to drink a glass of water and when he finally enters the sleeping room, the news show continues exactly at the point where he stopped watching them in the living room.

During the movie, her smart phone reminds Daisy of checking her blood values. As it's very easy, she stops the movie for a moment. Then she takes the small testing kit and starts the app on her phone. A few moments later she sees the actual amount of sugar in her blood. She also takes a look at the measurement history of the last week and the data of her step counter. The BUTLER App suggests more sports and fewer sweets. It was a stressful week and she ate too much of them. So it is about time to go to the mountains for few days of skiing.

After the movie, Daisy gets another notification from BUTLER. Moon boots on sale at xyz-Store. It is one of those spark deal offers where you can register as a customer and get notified if they have your article. It is a very good offer! Those are exactly the sort of boots she wanted to have and they are on sale. She could use them well in the mountains. But she had to buy them today as they are leaving tomorrow morning. There would be even half an hour time to buy them. But finding an available parking space at this time is a nightmare. Let's ask BUTLER. Thank god, there is a free parking space right down the corner of the shop. She immediately reserves it and heads into the town where her boots already wait for her. When Daisy arrives to the shop, she utilizes her NFC-enabled mobile to pay for the usage of the reserved parking space.

During the night, the heating turns on. BUTLER reminds that the old electric heating system is very inefficient and a nasty waste of energy. It context-sensitively calculates what he could save by replacing it.

When leaving the home for their holidays the next morning, the house automatically switches to the holiday mode.

On Monday morning, Donald's tablet suddenly buzzes. Someone stands at the door of his home back in the city. Donald uses the app on the tablet to look who this could be. The front door is lit (BUTLER turned on the lights) and the postman stands there. Donald shortly opens the door and lets him enter the home with the packet. A minute later, he sees the postman leaving the house and BUTLER notifies "door locked" again.

Initial high level view on components part of horizontal IoT system for context-aware 24/7 user information support, described in horizontal scenario, is shown on Figure 1.

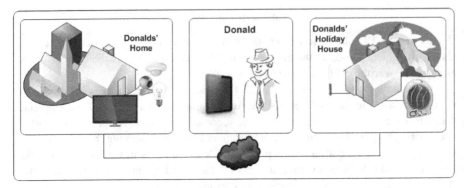

Fig. 1. Scenario setup

3   Security and Privacy Considerations in IoT

The scenario described before as well as the full "One Day in 2020" [1] inevitably causes numerous issues about security and privacy [4] because they gather and analyze personal and behavioral data before, during and after the IoT pilot operation. The following main concerns have been identified by IoT community [5] and are actively addressed within our work:

- Sensibility of user data: The IoT applications gather lots of information about the user (willingly or even without notice), that, despite a harmless appearance, can turn out to become sensitive by revealing daily behavior if analyzed on a large scale [6]
- Security of user data: The user data must therefore be protected against unauthorized access, and this security should be ensured at each level of communication. The plethora of communicating devices in the IoT increases the complexity as the number of to be protected links grows exponentially
- Management of data: Even when the security of the user data can be guaranteed against unauthorized access, the question of the actual management and storage of the information by the service provider remains
- Ownership and communication of data: The ownership question of the data collected is also central to the IoT applications. The monetization of user data can raise several issues on sharing the control and revenue, user awareness of monetization or third parties access
- Captivity of data: what happens to the user data if the user leaves the service? And how feasible is it for a user or consumer to change service provider once he has been engaged with one for a significant time? These questions are important to avoid consumer captivity through data that would result in an unfair advantage
- Availability of information: the quality of the information available to the user is key to the management of the ethical issues: the service provider must ensure not only that the information is available, but that it is presented in a way that ensure it is correctly understood by the user

These ethical concerns lead to the so called "Privacy Paradox": By collecting personal data, better "personalized' services can be offered to the users; but on the other side, these personal data can be collated, linked with transactional data, processed by powerful data mining techniques and assembled into user profiles that may become so detailed, that identification becomes possible. BUTLER project addresses this paradox using "Privacy by Design" approach [7] doing the systematic questioning of security and privacy issues at every step of new service design. Privacy is seen and treated as an ethics of knowledge:

- Transparency data usage: User shall give explicit consent of data usage
- Collected data shall be adequate, relevant and not excessive: The data shall be collected on "need to know" or "Data Minimization" principle that helps to setup user contract, to fulfill the data storage regulation / enhance the "Trust" paradigm
- Collector shall use data for explicit purpose: Data shall be collected for legitimate reasons and shall be deleted (or anonymized) as soon as data is no longer relevant
- Collector shall protect data at communication level: The required level of protection depends on the data to be protected according the cost of the protection and the consequence of data disclosure to unauthorized systems
- Collector shall protect collected data at data storage: User has accepted to disclose information to a specific system, not all the systems. It also could be mandatory to get infrastructure certification. Security keys at device side and server side are very exposed and shall be properly protected against hardware attacks
- Collector shall allow user to access / remove personal data: Personal data may be considered as a property of the user. User shall be able to verify correctness of the data and ask – if necessary – correction. Dynamic Personal Data – for instance home electricity consumption – shall also be available to the user for consultation

4   Development Objectives

The horizontal scenario, implementing it IoT applications as well as the full "One Day in 2020" [1] shows a fully cross domain (horizontal) exploitation of the IoT. Horizontality as a main objective is achieved through interconnection and integration of heterogeneous systems [8]. The final goal is to prove that horizontality is technically possible and partially attainable. The approach to implement this horizontality consists of the following activities:

- Improving and / or creating enabling technologies to implement a well-defined vision of secure, pervasive and context-aware IoT, where links are inherently secure (from PHY to APP layers) applications cut across different scenarios (Home, Office, Transportation, Health, etc.), and the network reactions to users are adjusted to their needs, learned and monitored in real time
- Integrating / developing a new flexible device-centric network architecture where platforms (devices) function according to three well-defined roles: smart Object (sensors, actuators, gateways), smart Mobile (user's personal device) and smart Server (providers of contents and services), interconnected over IPv6
- Building a series of field trials, which progressively integrate and enhance state-of-the-art technologies to showcase BUTLER's secure, pervasive and context-aware vision of IoT

From the methodology viewpoint, the first step on the way to horizontality is to gather, evaluate and validate the technical requirements of the different IoT application domains that may share some technologies but generally are not based on a similar architecture or platform [9]. This has been done in the following IoT application domains:

- Smart Home - Monitoring and controlling home appliances demonstrates the monitoring of the energy consumption of home appliances and how a context aware application can give advice on when to execute certain tasks to use renewable energies as much as possible
- Smart Home - Media everywhere demonstrates how a context aware media application can follow the users when they switch activities and places
- Smart Health - Personalized diabetes self-healthcare demonstrates how a context aware health application can follow the user lifestyle and recommend medication
- Smart City - Smart parking space management demonstrates how a smart city could monitor the use of parking spaces and provide reservation services
- Smart Shopping - Context-aware spark deals (product discounts) demonstrates a context-aware shopping application enabling users to benefit of specific and personalized deals
- Smart Transport - Safe transportation of school kids demonstrates several scenarios based on the precise localization of a bus driver, school kids and their teacher on a school trip

The second methodological step on the way to horizontality and also the the main technical objective of on-going work is focused on the integration of the heterogeneous IoT solutions within and between IoT application domains of smart home, smart health, smart city, smart shopping and smart transport.

5   Validation Targets and Measurement Values for IoT Pilots

Considering real deployments resulting in user- and business relevant applications of IoT technologies, the BUTLER project understands the need of setting added-value-driven and solution-oriented objectives as well as adequate metrics for progress monitoring and success measurement within IoT application pilot deployments. In this regard a very promising approach for modeling the adoption process of consumer wireless initiatives has been recently presented by S. Roebuck and S. A. Snyder in [10]. According to this approach the IoT application adoptions are often less a challenge of technical implementation and are more a question of identifying / validating use cases that drive real business benefits. Having this in mind, the proposed score framework recognizes four key driving factors for the adoption of IoT solutions:

- Strategic purpose: the offered solution has to be a part of the general strategy followed by the company implementing it.
- Solution integration: the entire value chain of interconnected solutions has to be able to benefit from the existing frameworks, technologies and services.

- Consumer value proposition: only appropriate quantitative and / or qualitative rewards motivate the end-users to be a part of a solution ecosystem in a long-term.
- User experience: minimal intrusion and seamless integration into existing processes of end-users are success factors for adoption

The full explanation can be found in [10] but we would like to briefly list all the parts of the driving IoT factors here. Strategic purpose factors include customer retention, revenue generating, new channels and brand enhancement. Solution integration factors include ease of integration, internal device capabilities, connectivity standards (wired/wireless) and data analytics. Consumer value proposition factors include convenience, fiscal incentive, personalization and effective results. Finally, user experience factors include engagement, simplicity, security/privacy and reliability.

Inspired by this approach [10] but realizing the differences between business product adoption and experimental IoT pilot execution in a research project, the BUTLER consortium has modified and adapted the original approach. The resulting list of pursued IoT pilot validation targets has been defined in regard to the BUTLER specific objectives of technological horizontality and ubiquitous context-awareness, Table 2.

Table 2. IoT pilot validation targets and measurement values

Validation target	Values	Description
Technical feasibility	Feasible now Feasible in less than 3 years Feasible in more than 3 years Not feasible until 2020	Field Trial aims to validate if the proposed IoT solution, service or application can be technically implemented using a state-of-the-art technology set today or within the next years to come
Technology integration	Open for integration Open data formats only Standard communication protocols only Integration hardly possible	Field Trial aims to validate if the proposed IoT solution, service or application is ready for integration into the IoT ecosystem including data formats and communication protocols
Deployment efforts	Low – Remote automatic Feasible – Remote semiautomatic Considerable – Onsite semiautomatic High – Onsite manual	Field Trial aims to validate if the proposed IoT solution, service or application can be easily deployed
Maintenance efforts	Low – Remote automatic Feasible – Remote semiautomatic Considerable – Onsite semiautomatic High – Onsite manual	Field Trial aims to validate if the proposed IoT solution, service or application can be easily maintained

Table 2. (*Continued.*)

Validation target	Values	Description
Scalability	> 1'000'000 connected nodes 100'000 – 1'000'000 connected nodes 10'000 – 100'000 connected nodes 1'000 – 10'000 connected nodes < 1'000 connected nodes	Field Trial aims to validate if the proposed IoT solution, service or application can be scaled to a large amount of nodes
User experience / comfort / perception	User-friendly, simple, secure, reliable, privacy-driven User-affine - Any three properties Limited - Any two properties Very limited - Any property	Field Trial aims to validate if the proposed IoT solution, service or application attracts with a well-designed user experience
User acceptance	High – Daily usage Considerable – Weekly usage Borderline – Monthly usage Low – Less than monthly usage	Field Trial aims to validate if the proposed IoT solution, service or application is widely accepted

As one can also see from Table 2, the requirement gathering through IoT applications continues through involvement of real users and stakeholders to gather not only technical feedbacks but more importantly user experience and acceptance data. The following IoT applications are being tested now based on access to the end-users as well as possibility to measure its technical feasibility:

• Smart Office scenario – the main goal here is to optimize the information flows between all the office users and this way to increase flexibility / reaction times and to eliminate information losses inside of organization. Technically, smart office IoT application is aiming to develop, deploy and get the user feedback on united informational environment enabled by interconnection of common work tools (domain login/logoff times, calendar, phone and VoIP service, etc.), sensing information (energy consumption, temperature, humidity, air quality, luminosity) with personalized context information (activity, presence, location) within one organization. The expected feedbacks will come from office staff and system maintenance staff. Main users of this IoT application are the colleagues from BUTLER team involving in total around 100 users at different European locations

- Smart Shopping scenario - the main goal here is to enable the customers to benefit from personalized special product or service offers while saving time and money spent on search of the best preferences – quality – price ratio. For the shop owners the Smart Shopping IoT application opens a new advertisement channel thus enhancing their access to potential customers. This IoT application will be deployed in Europe over 2013, with a first identified preliminary set of 500 users. The expected feedback will come from customers and shop owners including quantitative (based on usage statistics and feedback forms) and qualitative (customers and shop owners interviews) feedbacks
- Smart Transport scenario - the main goal here is to provide serving train staff with possibility to improve their service quality while serving their train passengers. This is being done by indoor localization system allowing releasing the staff from the duty to operate the internal doors by arms that are often busy with dishes or train operation materials. The initial deployment consists of a 6 week running test starting from January 2013 on the Glacier Express line in Switzerland. The expected feedback will come from train staff, passengers and system maintenance staff
- Smart Parking scenario – the main goal here is to eliminate the effort of citizens for parking search in European urbanized areas. This IoT application is being deployed as part of the Smart Santander initiative. As a side effect of this deployment we expect the improvement of general traffic situation since fewer cars will circulate on the roads in a parking search process. The expected feedbacks will come from the end-users of reservation system, city authorities and system maintenance staff

The next logical step is to deploy the multi-domain IoT platform into real IoT applications enabling communication and interoperability between them. This will not only extend the user basis and the associated users' feedbacks but also test and demonstrate the readiness of the platform to be used by new stakeholders. This horizontality requires that the different IoT applications share a common basis, a common technical infrastructure.

6   IoT System Overview

The implementation of horizontal technological IoT solution within BUTLER is organized in three flows, each one dedicated to the implementation of one platform. The 3 platforms together constitute the IoT Horizontal Platform that is being integrated into interconnected heterogeneous IoT system, offering common functionalities to application developers. In particular, Figure 2 gives schematic view on platform parts and relations between them:

- Smart Server platform: an integration of different servers to provide API building blocks on the IoT cloud for application and service integration. It provides to developers components that realize server-side functionalities horizontal to all application, for example: complex event processing, localization engine, context management framework, user profile and user behavior model, security policies

- Smart Mobile platform: a mobile application reference framework that provides mobile-side functionalities horizontal to all applications. Common mobile client functionalities are: uniform GUI, security (privacy), communication with server components, "local" data-mining algorithms
- Smart Object platform: integration of different Smart Object Gateway and related object technologies through a uniform "Object API", so that applications and mobile clients will be able to access objects through the same interfaces. Again common functionalities are made available like: object discovery, security and management

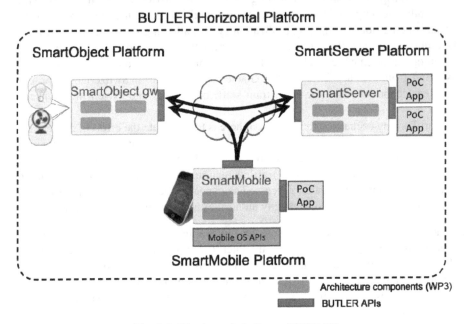

Fig. 2. IoT horizontal platform of BUTLER

7   IoT Application Example

Considering presented objectives, methodology, IoT system overview and ethical issues from end-user perspective the BUTLER team has developed and deployed the following IoT application example or pilot in Smart Home domain [11]. The basic idea is practical integration of different platforms from different vendors into one show case covering a more or less fluent scenario. The authors don't yet claim the full usage of a multi-domain IoT horizontal BUTLER platform but point the direction where the development is heading through a certain degree of interoperability, integration and interconnection as shown on the Figure 3.

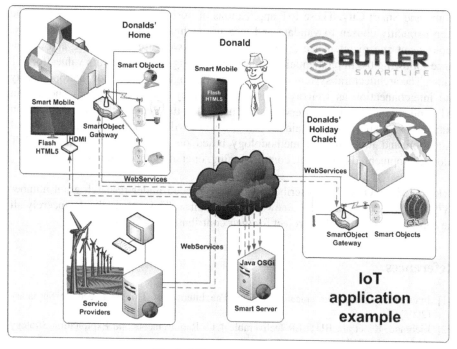

Fig. 3. IoT application example

As a common option, the particular external services from different providers are connected to the cloud for publishing their data and control API to the central control and visualization server. This server accesses the gateways and web services in order to fetch data coming from the attached sensors as well as from the prediction algorithms (green lines). It processes the data and finally generates the information to be visualized and sent to the TV as well as the tablet (red lines). Interaction with the user happens over the tablet.

Although all platforms come with their own visualization and input variants, this approach over the central server and its connected Flash Clients running on different target hardware was chosen in order to provide this IoT application example with a homogenous user interface (red lines). The data interfaces to get to the data and switch the loads are specifically defined between service providers and central server (green lines).

8   Concluding Discussion

After short introduction into the IoT world challenges of isolated domain specific applications we have presented the vision on integration and interconnection of heterogeneous IoT subsystems. The main goal of this integration is a generation the added value in the context-aware 24/7 information support for end-users through availability of the non-core information for a given application domain IoT applications from different vertical domains like Smart Transport, Smart Health, Smart Shopping, Smart

Home and Smart City. These IoT applications of horizontal BUTLER scenario have been explicitly chosen to validate and measure technical and integration feasibility, deployment and maintenance efforts, scalability as well user experience and acceptance following the implementation of previously suggested by authors values for IoT pilots. The architectural system overview of technological enablers of interoperability and interconnection as well as security and privacy concerns between and within IoT applications have been afterwards instantiated in the first concrete IoT application example in Smart Home domain. The future work will be focused on finalization of validation and measurement methodology based on feedback collection from IoT pilot development, deployment, commissioning, operation and maintenance users.

Acknowledgements. The described work is co-funded under research grant number 287901 of the European 7th Framework Programme. The authors thank sincerely all the consortium of BUTLER project for their contributions.

References

[1] International Telecommunication Union. The Internet of Things. Executive Summary (2005)

[2] Liebrand, K., et al.: BUTLER Deliverable 1.1 - Requirements and Exploitation Strategy (January 31, 2012)

[3] Gurgen, L., et al.: BUTLER Deliverable 3.1 Architectures of BUTLER Platforms and Initial Proofs of Concept (October 31, 2012)

[4] Raij, A., et al.: Privacy Risks Emerging from the Adoption of Innocuous Wearable Sensors in the Mobile Environment. In: Proc. Conf. Human Factors in Computing Systems (CHI 11), pp. 11–20. ACM (2011)

[5] Massimi, M., et al.: Understanding Recording Technologies in Everyday Life. In: IEEE Pevasive Computing, July-September, pp. 64–71 (2010)

[6] Narayanan, A., et al.: De-Anonymizing Social Networks. In: Proc. 30th Symp. Security and Privacy (S&P 2009), pp. 173–187. IEEE (2009)

[7] Langheinrich, M.: Privacy by design - principles of privacy-aware ubiquitous systems. In: Abowd, G.D., Brumitt, B., Shafer, S. (eds.) UbiComp 2001. LNCS, vol. 2201, pp. 273–291. Springer, Heidelberg (2001)

[8] Petcu, D.: Portability and interoperability between clouds: Challenges and case study. In: Abramowicz, W., Llorente, I.M., Surridge, M., Zisman, A., Vayssière, J. (eds.) ServiceWave 2011. LNCS, vol. 6994, pp. 62–74. Springer, Heidelberg (2011)

[9] Miller, J.H., et al.: Complex Adaptive Systems: An Introduction to Computational Models of Social Life. Princeton Univ. Press (2007)

[10] Roebuck, S., Snyder, S.A.: An Adoption Model for Consumer Wireless Sensor Networks. IEEE Cons. Electronics Magazine 2(2), 34–41 (2013)

[11] Helal, S., et al.: The Gator Tech Smart House: A Programmable Pervasive Space. Computer, 64–74 (2005)

On the Relevance of Using Interference and Service Differentiation Routing in the Internet-of-Things

Antoine Bigomokero Bagula[1], Djamel Djenouri[2], and Elmouatezbillah Karbab[2]

[1] Department of Computer Science, University of Cape Town, South Africa
[2] CERIST Research Center, Algiers, Algeria
bagula@cs.uct.ac.za, {ddjenouri,m.karbab}@mail.cerist.dz

Abstract. Next generation sensor networks are predicted to be deployed in the *Internet-of-the-Things (IoT)* with a high level of heterogeneity. They will be using sensor motes which are equipped with different sensing and communication devices and tasked to deliver different services leading to different energy consumption patterns. The application of traditional wireless sensor routing algorithms designed for sensor motes expanding the same energy to such heterogeneous networks may lead to energy unbalance and subsequent short-lived sensor networks resulting from routing the sensor readings over the most overworked sensor nodes while leaving the least used nodes idle. Building upon node interference awareness and sensor devices service identification, we assess the relevance of using a routing protocol that combines these two key features to achieve efficient traffic engineering in IoT settings and its relative efficiency compared to traditional sensor routing. Performance evaluation with simulation reveals clear improvement of the proposed protocol vs. state of the art solutions in terms of load balancing, notably for critical nodes that cover more services. Results show that the proposed protocol considerably reduce the number of packets routed by critical nodes, where the difference with the compared protocol becomes more and more important as the number of nodes increases. Results also reveal clear reduction in the average energy consumption.

1 Introduction

1.1 Motivations

The recent advances in Radio Frequency Identification (RFID) and Wireless Sensor/Actuator Networks (WSANs) have led to a new information technology (IT) era where devices built around these technologies are deployed in our daily living environments to provide services that range from the most common, such as weather forecasting, to most unusual such as body area monitoring. While RFID systems are used in such environments to accurately identify objects in a number of applications such as asset tracking, telemetry-based remote monitoring, and real time supply chain management, they usually fail short to accurately locate these objects and sense what is happening in their surrounding. On the other hand, while being good in the localization and recognition of the physical parameters of the environment in applications such as precision agriculture, fire detection, weather and pollution monitoring and many others, sensor devices are unable to identify objects. The integration of both technologies

S. Balandin et al. (Eds.): NEW2AN/ruSMART 2013, LNCS 8121, pp. 25–35, 2013.

into hybrid sensor devices capable of both sensing and identifying objects present a great advantage compared to using a single technology or deploying these technologies separately. When deployed in a hospital setting, for example, to monitor babies in a maternity ward, hybrid sensors can both localize the movement of each baby during daily care, e.g., what treatment stations the baby has been through, and report on the environmental conditions he has been exposed to, e.g, temperature, humidity, light exposure, etc. Separate deployment of these technologies may lead to a duplication of resources both hardware and software, complex and costly system management and difficult software trouble shooting and maintenance. The relevance of using hybrid sensors compared to single or separate technology deployment can also be demonstrated in underground mine monitoring where the placement of such devices in different locations of a mine may enable both localization of miners and identification of the environmental parameters they are exposed to in order to trigger early warning in case of high exposure to high levels of gazes and danger of explosion.

Ubiquitous Sensor Networks (USNs) [1] are emerging as a family of networks that build upon the integration and networking of RFID, WSAN and hybrid devices into a common communication platform capable of identifying the objects in our living environment and sense what is happening in such environment to enable pervasive access to the information carried by a multitude of user applications and produced by a multitude of objects that surround us. When endowed with an IP address (or any global ID), USN devices may transform the objects and things we use in our daily environment into *"smart objects"* capable of using the Internet and web services to communicate among themselves, and with humans in the *"Internet-of-the-Things (IoT)"* [2]. Born between 2008 and 2009 when the number of objects/things connected to the Internet exceeded the number of people connected, the IoT is raising a great interest by both the research and practitioner's communities as a network of the future that is predicted to connect by 2020, billions of objects outfitted with sensor, actuator and RFID devices. It is also expected to provide access to the information not only *"anytime"* and *"anywhere"*, but also by *"anyone"* and using *"anything"* with projected high impact in the development of innovative technologies that will lead the near future. Based on their scientific, economic and engineering benefits, these technologies are opening tremendous opportunities for a large number of novel applications that promise to revolutionize and improve the quality of our lives.

Traditional WSN routing protocols have been designed on a routing model that route sensor readings from nodes to a gateway by assuming that the sensor nodes are of the same fabric and expected to deliver the same service: sensing and forwarding the sensor readings towards the sink node. The application of these routing protocols in the heterogeneous IoT settings may lead to performance degradation as different nodes might exhibit different levels of service heterogeneity: e.g some nodes might be sensing their environment and using their GPRS modem to send SMSs in fire-fighting applications, other nodes might be tasked to achieve both sensing, identification and forwarding as illustrated by the underground mining example above.

1.2   Related Work

Integration of sensors and RFID devices have been largely investigated in the literature [3–6]. In [3] for example, a two-tiered RFID sensor network where readers collect data from tags and forward it to the base station is proposed. The authors identified energy unbalance in the network caused by an increase in the amount of traffic as the distance to the base station gets shorter. Consequently, readers closer to the base station die quicker. To solve the problem, they propose a scheme that balances load among readers by adding more readers in areas near the base station. The results obtained from the simulation show that the network lifetime increases as the number of readers close to the base station increases. The solution is very expensive considering the current cost of RFID readers. Furthermore, an increase in the number of reader nodes may lead to an increase in the number of collisions in the network.

In [4–6], different techniques for integrating sensor nodes with RFIDs are discussed. The objective of the different integrations is to achieve an ad-hoc network similar to WSNs. The integrated readers collect data from the environment and share the data among themselves. This type of integrated network has similar energy limitations to WSNs because all the nodes have the same properties. In order to save energy in the network, the authors in [4] decreased energy consumption of the network by proposing an on-demand wakeup capability that eliminates idle listening. This approach saves power, but it is a Medium Access Control (MAC) protocol and not a routing protocol. Multi-objective routing solutions [7–9] have also been proposed to improve the QoS delivery in sensor networks. While [7] uses an energy constrained multipath routing approach, the works in [8, 9] are based on geographic routing but using service differentiation with respect to the traffic classes and requirements in a homogeneous environment. This differs from the solution proposed in this paper where service differentiation is related to the delivered services of the sensor nodes in a heterogeneous environment.

Data collection protocols such as collection tree protocol (CTP) [10], TinyOS beaconing (TOB) [11] and RPL [13] are closely related to the routing solution proposed in this paper. They are designed around a collection tree structure where minimum- cost trees for nodes that advertise themselves as tree roots are built and maintained to forward the sensor readings from nodes to the base-station. CTP and RPL use the trickle algorithm to enable data traffic to quickly discover and fix routing inconsistencies by relying on the *Collection tree* and *adaptive beaconing* features to reduce route repair latency and beacon messages. It has been credited to the TOB protocol the attractive feature of node simplicity and the advantage of not having to maintain large routing tables or other complicated data structures. However, this attractive feature has to be weighted against some of the inefficiencies of the beaconing protocol. These include 1) the lack of resilience to node failures, leading to an entire sub-tree being cut off from the base-station during the current epoch when a parent node fails, 2) the tree-like m-to-1 sensor readings dissemination model leading to uneven power consumption across network nodes as the nodes surrounding the base-station tasked to forward packets from all the nodes in their sub-tree consume a lot of power, whereas the leaf nodes in the spanning tree, which do not perform any forwarding, consume least power. These shortcomings are addressed in this paper.

1.3   Contributions Overview

This paper tackles the issue of energy efficiency for USNs to evaluate the impact of using role-based service differentiation on USN efficiency in IoT settings. We propose the LIBP protocol that combines node interference with role-aware service differentiation to enable USN devices of different predefined roles to provide different routing services and thus avoid to over-stretching the most over-worked sensor nodes. Our simulation results obtained using TOSSIM [12] reveal the relative scalability and efficiency of the traffic engineering scheme resulting from LIBP compared to state of the art collection protocols TOB and CTP. The remainder of this paper is organized as follows: Section 2 presents the proposed model and protocol. The experimental results obtained through comparative simulation study are presented in Section 3, and finally Section 4 draws the conclusions.

2   Proposed Solution

2.1   Path Finding Scenario

Fig 1 depicts a USN as a trap topology graph with the sink located at node 0 and the edges showing potential wireless links that can be used to route the sensor readings from nodes to sink. The application of any of collection protocol to the USN illustrated by Fig 1 may lead to two sensor network routing configurations, depending on how the parent nodes are selected at each epoch: A path multiplexing configuration illustrated by Figure Fig 2 (a) and a path separated configuration revealed by Fig 2 (b). The path separated configuration is a load balanced configuration which can be useful in 1) interference-aware routing schemes 2) service-aware routing schemes and

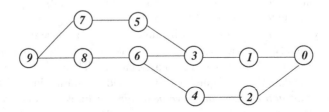

Fig. 1. Network Topology Connectivity

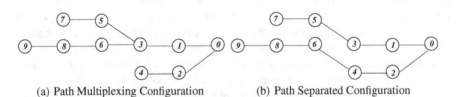

(a) Path Multiplexing Configuration   (b) Path Separated Configuration

Fig. 2. Path Discovery

3) heterogeneous routing situations combining both schemes which we predict to be common in the IoT. Interference-aware routing aims to minimize traffic flows interference on nodes with the expectation of reducing energy usage as each node will route less traffic. Interference-aware routing also protects the network against the impact of node failures by having less branches cut from the network upon failure. Service-aware routing protects critical nodes from being overworked by the routing process while leaving the less critical nodes idle. The *"least interference beaconing (LIB)"* model proposed in this paper is a scheme where a weighted combination of interference and service-aware routing is piggy-backed on the beaconing process applied to collection protocols to achieve efficient and scalable USN management. As applied in this paper, path separation can 1) protect node 3 in interference-aware routing from becoming a single point of interference consuming high energy and leading to the high traffic loss under failure and 2) protect critical node 3 in service-aware routing from being overworked while less critical nodes are idle.

2.2 The Routing Problem

The routing in USNs can be formulated as a zero-one linear problem consisting of finding for each node n, the subset $\mathbf{N_0} \subseteq \mathbf{N}[n]$ of its neighbours that solves the following zero-one linear problem

$$\min \quad \sum_{j \in \mathbf{N}[n]} x_j \tag{1}$$
$$subject\ to$$

$$\begin{cases} w(n) & = \alpha w_i(n) + \beta w_s(n) & (2) \\ parent(j) = n \mid w(n) = \min_{x \in \mathcal{N}(j)}\{w(x)\} & (3) \\ x_j & = 0 \text{ or } 1, \forall j \in \mathbf{N}[n] & (4) \end{cases}$$

where $\beta = 1 - \alpha$ while $parent(j)$ is a function that returns the preferred parent for a given node n. $w(n)$ is the routing metric associated with node n. It is a weighted expression of its interference in the number of children that it is carrying $w_i(n) = \sum_{j \in \mathbf{N}[n]} x_j$ and the penalty related to the role played by the node in the network expressed by $w_s(n)$. It can be set high to protect critical nodes from battery depletion or low to steer the traffic flows towards less critical nodes. Note that as expressed above, the problem formulation expresses the node interference minimization and role-based differentiation of services and how they are mapped into i) *a routing metric/cost* expressed by equation (2), ii) *a parent selection* expressed by equation (3) and iii) *the zero-one linearity model* expressed by equation (4). The routing model formulated above is a local optimization problem that may be solved using a heuristic solution described in subsection 2.3, and then implemented as a protocol in subsection 2.4. The β value and consequently $\alpha = 1 - \beta$ is an important parameter that defines the routing model. As expressed below

$$\beta = \begin{cases} 0 & \text{Interference-aware routing} \\ 1 & \text{Service-aware routing} \\ x \in]0 \ldots 1[& \text{Hybrid routing.} \end{cases}$$

It expresses the network administration preference for a given routing model.

2.3   LIBA: An Algorithmic Solution

Least Interference Beaconing Algorithm (LIBA) is an algorithmic solution to the routing problem formulated by (1)-(4). It uses a time-bound by "epoch" distributed breadth-first search model to find the routing paths for the traffic flows carrying the sensor readings from nodes to the sink/gateway node. A high-level description of LIBA is presented in Figure 3 (a) for the sensor nodes, where T_e is the duration of an epoch, and *mod* is the modulo operation used in our case to compute the beginning of a new epoch. Its gateway version is depicted by Figure 3 (b).

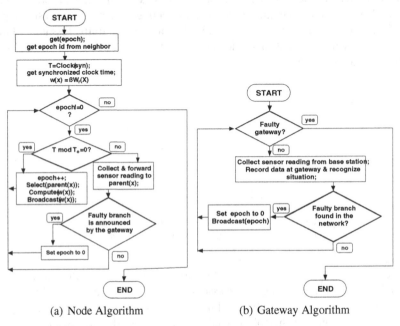

(a) Node Algorithm   (b) Gateway Algorithm

Fig. 3. Least Interference Beaconing Algorithm

As presented in Figure 3 (a), LIBA provides a heuristic solution to the least interference routing problem expressed by (1) by using a similar scheme to TinyOS beaconing, but with a slight modification to the beaconing process in order to meet the routing constraints (2), (3) and (4) as follows:

- When broadcasting the beacon after the initial step, the parent computes its weight specifying a weighted average of the number of children it is supporting (interference) and the role played by the node (service delivery), as expressed by the routing constraint (2). It then includes the calculated weight in the beacon that is being broadcasted.
- Upon reception of the beacons from potential parents, the children nodes select their preferences for the least weighted parent and update their forwarding tables based on the expression of the routing constraint (3).

- The zero-one linearity routing constraint (4) can also be expressed by

$$x_j = \begin{cases} 1 & \text{parent}(j) = i \\ 0 & \text{otherwise.} \end{cases}$$

It suggests the creation of a breadth-first spanning tree rooted at the sink through recursive broadcasting of routing update beacon messages and recording of parents.

Figure 3 (b) presents a high level description of the algorithm implemented by the sink/gateway. It involves a situation recognition process that triggers recovery mechanisms, by reinitializing the epoch counter, $epoch = 0$, upon detection of a node failure. However, in this paper, situation recognition has been limited to ensuring that as a protocol implementation of the zero-one linear formulation, LIBA leads to a connected network. The study of the recovery processes under failure conditions are beyond the scope of this current work. It should be noted that the LIBA algorithm depicted in Fig 3 (a) might (i) lead to a path multiplexing configuration such as illustrated in Fig 2 (a) during an epoch where all weights are equal, and it might (ii) converge to a path separated configuration as depicted in Fig 2 (b) after computation and broadcasting of weights. In the illustration provided in Fig 2, the convergence to a path separated configuration happens after weight allocation and broadcasting in a given epoch where from a path multiplexing, node 3 informs nodes 5 and 6 that it has a $weight = 2$. In this case, during the parent selection process that follows the weight allocation and broadcasting, node 5 will select node 3 as parent while node 6 will prefer node 4 as parent.

2.4 LIBP: A Protocol Implementation

LIBP is a protocol implementation of the LIBA algorithm described above. It builds upon an ad hoc routing model similar to TOB in terms of simplicity, and to the emerging RPL protocol in terms of structure. It uses beacons and acknowledgements as main messages and weight updating, weight broadcasting, and parent selection as its main operations. The beacon messages carry the sender's identity and weight, and they are broadcasted to potential children by senders. Parent selection is performed at reception of the beacon messages but acknowledged to only the selected parents, which subsequently increase their weights only after receiving the acknowledgement message. LIBP is based on the following key features:

- Use of a simple ad hoc routing protocol, which creates a breadth-first spanning tree rooted at the sink through recursive broadcasting of routing update beacon messages and recording of parents.
- The beacon messages are (1) broadcasted periodically at intervals called epochs, (2) propagated progressively to neighbours and (3) received by a few nodes located in the vicinity of the source of the beacon message.
- The transmission of the beacon is built around a source marking progressive propagation to neighbours and rebroadcasting progress, which sets up a breadth-first spanning tree rooted at the sink.
- The least interference paradigm is integrated into the process through selection of a parent node that has the least weight. It is thus a point of least burden in terms of node interference and service delivery.

– While the LIBP protocol leads to the same number of messages exchanged as TOB, it implements a different parent selection model where instead of selecting the first parent node they heard from, the sensor nodes hear from a set of neighbours and select the least burdened (in number of children, task, or trading-off both depending on the values of β and α) as the parent node.

Note that by piggy-backing the parent identification into the beacon broadcasting process and adding parent identification to the packet header, our model may avoid the signalling overheads related to the addition of an acknowledgement into the routing process. However, as LIBP acknowledgements are sent to only the selected parents, they are bound by the maximum number of nodes in the network, thus reducing tremendously the signalling overheads during an epoch.

3   Simulation Study

To evaluate the performance of the proposed protocol and compare it with CTP [10] and TinyOs Beaconing (TOB) [11], extensive simulations have been conducted with TOSSIM [12]. The number of nodes have been varied from 20 to 200, and β, from 0.2 to 1. In each scenario, 10% of nodes where set to be critical (hybrid) nodes whose energy resource management is of high importance due to the high loads they are required to perform. These node should route as few packets as possible to ensure a long network lifetime. The number of packets forwarded by these nodes is thus the key performance metric that should be optimized (minimized) in this hybrid environment. Table 1 sketches the most relevant simulation parameters. Each point of the plots is the average of several runs, and results are presented with 99% confidence interval. The number of packets forwarded by critical nodes has been measured. Fig. 4 depicts the number of packets forwarded by critical nodes in LIBP vs. β. The plots show averaged values of the minimum number, the maximum number, and the mean number of the forwarded packets by the 10 critical nodes in the 100 nodes scenario. We can see that there is a sharp decrease from $\beta = 0$ to $\beta = 0.4$, then all the numbers become more or less stable with some but insignificant fluctuation. We conclude that setting β to 0.4 is sufficient enough– in the simulated scenarios– to enable relaxing routing load at critical nodes . β is thus fixed to 0.4 for LIBP in what follows. The number forwarded by critical nodes is presented in Fig. 5. Fig. 5 a) depicts the mean values of packets forwarded by critical nodes for both LIBP and TOB vs. the number of nodes. CTP has also been simulated, but its mean values are very fluctuating with very high error bars. It has been removed to make the figure legible. It is clear from the figure that LIBP reduces the routing load on critical nodes compared to TOB. The inevitable increase vs. the number of nodes is much smoother for LIBP, and the difference between the protocol becomes more important as the number of nodes rises. This is justified by the fact that the more nodes are n

Table 1. Simulation Setup

Traffic	every node sends a 28-byte packet every 5 sec
Number of nodes	20: 200
Topology	random
Simulation duration	900 sec
beacon interval	20 s

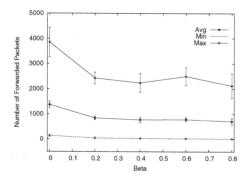

Fig. 4. Impact of β on packet forwarding by critical nodes in LIBP

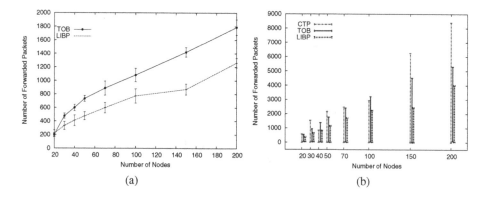

Fig. 5. Packet forwarding of critical nodes vs. number of nodes a) Average, b) min/max dispersal

the network, the more choices will be available to permit routing around critical nodes. Fig. 5 b) shows the interval of the number of forwarded packets by critical nodes (the minimum/maximum dispersal), where CTP is also depicted. Here, it is clear how the difference between the minimum and the maximum values is huge for CTP that does not apply any load balancing, and that the CTP tree construction strategy resulted in some bottleneck nodes amongst the critical ones. On contrary, LIBP demonstrated the best performance owing to its strategic load balancing.

Finally, Fig.6 a) and b) plot the total instantaneous number of data packets received by the sink and those sent by the nodes, respectively, vs. time in 100 nodes scenario. From these plots, it can be seen that CTP implementation results in higher latency owing to the spanning tree construction that takes a long time compared to the other protocols. This explains non-transmission (and accordingly no reception) of packets at the beginning, and peaks in a later stage of the experimentation. Using Avrora, we measured the average energy consumption of all nodes in the network for the tree protocols. Fig. 7 depicts the obtained results vs. the number of nodes. It is clear from the figure taht CTP leads to a drastic rise of energy consumption when the number of nodes reaches 70, while both TOB and LIBP scale with the increase in the number of nodes. LIBP reveals the lowest energy consumption with the increase of number of USN nodes.

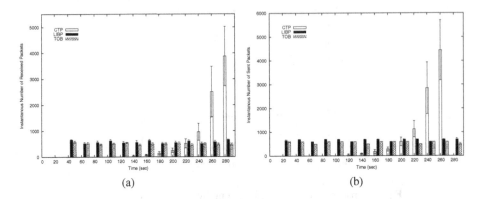

Fig. 6. Instantaneous number of packets a) received, b) sent

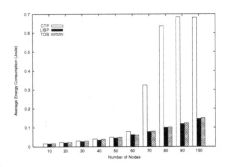

Fig. 7. Average radio energy consumption vs. number of nodes

4   Conclusion and Future Work

This paper presents LIBP, a new routing protocol that builds upon routing simplicity, minimization of the interference among competing traffic flows and service differentiation to achieve efficient traffic engineering of the emerging islands of USNs that form the IoT. Preliminary experimental results using TOSSIM reveal the relative efficiency of LIBP compared to CTP and TOB protocols. These results reveal that the "path separation" principle behind the "least interference beaconing" paradigm embedded into LIBP and the "least interference optimization" paradigm proposed in [14, 15] translates into network efficiency.

There is room for further investigation of the LIBP protocol in terms of its fault tolerance capabilities upon failure, its dependability in terms of protection against jamming attacks, and its relative performance compared to recently standardized protocols such as RPL. When deployed to support sensing operations in intermittent power supply environments, a flexible and robust gateway such as proposed in [16] may be augmented with situation recognition capabilities to improve USN security and efficiency. This is another avenue for future research.

References

1. Bagula, A., et al.: Ubiquitous Sensor Networking for Development (USN4D): An Application to Pollution Monitoring. MDPI Sensors 12(1), 391–414 (2012)
2. Vasseur, J., Dunkels, A.: Interconnecting Smart Objects with IP, The Next Internet. Morgan Kaufmann (July 2010) ISBN: 9780123751652
3. Yang, G., Xiao, M., Chen, C.: A simple energy-balancing method in RFID sensor networks. In: Proceedings of 2007 IEEE International Workshop on Anti-Counterfeiting, Security, Identification, Xiamen China, April 16-18, pp. 306–310 (2007)
4. Ruzzelli, A.G., Jurdak, R., O'Hare, G.M.P.: On the RFID wake-up impulse for multi-hop sensor networks. In: Proceedings of 1st ACM Workshop on Convergence of RFID and Wireless Sensor Networks and their Applications (SenseID) at the 5th ACM Conference on Embedded Networked Sensor Systems (ACM SenSys 2007), Sydney, Australia (November 2007)
5. Englund, H.W.: RFID in wireless sensor network, Tech. Report, Dept. of Signals & systems, Chalmers Univ. of Technology, Sweden, pp. 1-69 (2004)
6. Dyo, V., Ellwood, S.A., Macdonald, D.W., Markham, A., Mascolo, C., Pasztor, B., Scellato, S., Trigoni, N., Wohlers, R., Yousef, K.: Wildlife and Environmental Monitoring using RFID and WSN Technology. In: Proc. of the 7th ACM Conference on Embedded Networked Sensor Systems (SenSys 2009), Berkeley CA, USA (2009)
7. Bagula, A.B.: Modelling and Implementation of QoS in Wireless Sensor Networks: A Multi-constrained Traffic Engineering Model. Eurasip Journal on Wireless Communications and Networking 2010, Article ID 468737 (2010), doi:10.1155/2010/468737
8. Djenouri, D., Balasingham, I.: Traffic-Differentiation-Based Modular QoS Localized Routing for Wireless Sensor Networks. IEEE Transactions on Mobile Computing 10(6), 797–809 (2011)
9. Felemban, E., Lee, C.-G., Ekici, E.: MMSPEED: Multipath Multi-Speed Protocol for QoS Guarantee of Reliability and Timeliness in Wireless Sensor Networks. IEEE Trans. Mobile Computing 5(6), 738–754 (2006)
10. Gnawali, O., et al.: Collection Tree Protocol. In: Proc. of ACM SenSys 2009, Berkeley, CA, USA, November 4-6 (2009)
11. Hill, J., et al.: System architecture directions for networked sensors. In: Proc. of ACM Architectural Support for Programming Languages and Operating Systems (ASPLOS IX) (2000)
12. Levis, P., Lee, N., Welsh, M., Culler, D.: TOSSIM: Simulating large wireless sensor networks of tinyos motes. In: Proc. of ACM SenSys 2003, Los Angeles, CA, pp. 126–137 (November 2003)
13. Winter, T., et al.: RPL: IPv6 Routing protocol for Low-Power and Lossy Networks. In: RFC 6550 (March 2012)
14. Bagula, A.: Hybrid traffic engineering: the least path interference algorithm. In: Proc. of ACM Annual Research Conference of the South African Institute of Computer Scientists and Information Technologists on IT Research in Developing Countries, pp. 89–96 (2004)
15. Bagula, A.: On Achieving Bandwidth-aware LSP/LambdaSP Multiplexing/Separation in Multi-layer Networks. IEEE Journal on Selected Areas in Communications (JSAC): Special issue on Traffic Engineering for Multi-Layer Networks 25(5) (June 2007)
16. Zennaro, M., Bagula, A.B.: Design of a flexible and robust gateway to collect sensor data in intermittent power environments. International Journal of Sensor Networks 8(3/4) (2010)

Adaptive and Context-Aware Service Discovery for the Internet of Things

Talal Ashraf Butt[1], Iain Phillips[1], Lin Guan[1], and George Oikonomou[2]

[1] Department of Computer Science, Loughborough University,
Loughborough, Leicestershire, LE11 3TU, UK
{T.A.Butt,I.W.Phillips,L.Guan}@lboro.ac.uk
[2] University of Bristol, Faculty of Engineering, Merchant Venturers Building,
Woodland Road, Clifton BS8 1UB, UK
g.oikonomou@bristol.ac.uk

Abstract. The Internet of Things (IoT) vision foresees a future Internet encompassing the realm of smart physical objects, which offer hosted functionality as services. The role of service discovery is crucial when providing application-level, end-to-end integration. In this paper, we propose TRENDY: a RESTful web services based Service Discovery protocol to tackle the challenges posed by constrained domains while offering the required interoperability. It provides a service selection technique to offer the appropriate service to the user application depending on the available context information of user and services. Furthermore, it employs a demand-based adaptive timer and caching mechanism to reduce the communication overhead and to decrease the service invocation delay. TRENDY's grouping technique creates location-based teams of nodes to offer service composition. Our simulation results show that the employed techniques reduce the control packet overhead, service invocation delay and energy consumption. In addition, the grouping technique provides the foundation for group-based service mash-ups and localises control traffic to improve scalability.

Keywords: Adaptive, Context-aware, Service Discovery, 6LoWPAN, Internet of Things, CoAP, RESTful, Web of Things.

1 Introduction

The Internet of Things (IoT) concept has revolutionised the vision of the future Internet with the advent of standards such as 6LoWPAN making it feasible to extend the Internet into previously unreachable environments, e.g. Wireless Sensor Networks (WSN). The abstraction of resources as services, has opened WSNs to a new plethora of potential applications. Moreover, the web service paradigm can be used to provide interoperability by offering a standard interface to interact with these services. However, these networks pose many challenges in terms of limited resources. Consequently, the adaptability of existing IP-based solutions is not feasible. As traditional service discovery and selection solutions demand heavy communication and use bulky formats, which are unsuitable for these

S. Balandin et al. (Eds.): NEW2AN/ruSMART 2013, LNCS 8121, pp. 36–47, 2013.

resource-constrained devices incorporating sleep cycles to save energy. Even a registry-based approach exhibits burdensome traffic in maintaining the availability status of the devices. The feasible solution for service discovery and selection is instrumental in enabling wide application coverage of these networks in the future [1].

The contribution of this paper is a compact and optimise registry based service discovery solution with context awareness for the IoT, which is more focused on constrained domains such as 6LoWPAN. It uses CoAP-based [14] RESTful web services to provide a standard interoperable interface which can be easily inter-worked with HTTP. The modular design of protocol features allows its implementation on the constrained devices. High capability devices can benefit by implementing profiles to share the load of other devices. Thus, it allows the productive usage of resources in the network. TRENDY intelligently uses the context information to provide optimal service selection, which make sure that more superior hosts will be suggested to an enquirer. This paper extends our previous work [3] by introducing new adaptive timer and caching techniques. Adaptive timer minimises the protocol's control overhead and energy consumption. Its grouping mechanism is based on location tags to localise status maintenance traffic and to compose and offer new group based services. The APPUB (Adaptive Piggybacked Publish) technique balances the trade-off between service invocation delay and packet overhead by adaptively making cache available for highly requested resources. We have performed simulations to demonstrate the benefit of using TRENDY techniques in terms of energy consumption, packet overhead, scalability (packets towards the sink and cache hits) and service invocation time.

This paper covers the related work in Section 2, before describing protocol and its architecture, entity interactions and techniques in Section 3. In the end, Section 4 discusses the performed experiments and generated results.

2   Related Work

The service discovery protocols are generally classified into three broad categories on an architectural basis: *centralised, distributed* and *hierarchical* [2]. Centralised architectures have a directory, where Service Agents (SA) register their services. Subsequently, User Agents (UA) discover the services by sending unicast queries to the directory. On the other hand, distributed architectures demand that nodes collaborate using broadcast or multicast to discover a service. An example of a distributed service discovery mechanism is ADDER [12]. Hierarchical architectures employ some nodes with high capabilities, to represent a cluster of nodes in their vicinity.

The industry-standard, IP-based Service Discovery Protocols (SDP) including SLP, UPnP, JINI and Salutation are not directly applicable to 6LoWPAN because of the employed complex formats and high communication demand. uBonjour [8] is bonjour's compact variant, based on mDNS and DNS-SD. Even though mDNS/DNS-SD message sizes were recently optimised for 6LoWPANs [9], uBonjour still relies on the availability of IP multicast and entails more communication

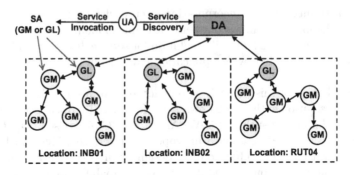

Fig. 1. Hybrid architecture of TRENDY

overhead. A SLP adaptation approach [4] employs SSLP inside the 6LoWPAN and provides interoperability with SLP by using a Translation agent (TA). However, this solution involves complexity and delay of translation; each time message is translated to or from SLP. An industry focused SOA (Service Oriented Architecture) based middle-ware solution [7] uses WS-* and RESTful web services. A recent approach [10] provides RESTful web services using HTTP based service discovery with existing or injected strategies. The IETF Resource directory [15] uses CoAP as an underlying communication protocol for service discovery. However, most of these existing directory-based solutions do not address the service discovery requirements of IoT environments. This paper proposes TRENDY service discovery solution that provides context-aware discovery, efficient service management, service selection, caching and service composition.

3   TRENDY: Trend-Based Service Discovery for the IoT

This section describes the various design aspects of the TRENDY service discovery protocol, including architecture, interaction between entities in context of protocol features, adaptive timer and caching techniques.

3.1   Architecture

TRENDY maintains a registry: the DA (Directory Agent), where SAs (Service Agents) register services. UAs (User Agents) query the DA, to find the location of a service. The grouping mechanism further categorises SAs into GLs (Group Leaders) and GMs (Group Members). Fig. 1 presents TRENDY's architecture.

Directory Agent (DA): The DA has a backbone role in TRENDY's architecture maintaining the registry and using a demand-based adaptive timer (Section 3.3) to increase or decrease the interval between status maintenance updates. The DA responds to service discovery requests and uses collected context information to provide optimal service selection. In case of a constrained network, e.g. 6LoWPAN, it can be at the root of RPL (IPv6 Routing

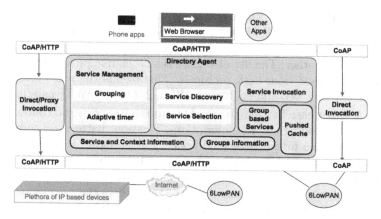

Fig. 2. Interoperable framework of TRENDY: The DA collects service and context information to provide service discovery and selection, and uses grouping and adaptive timer mechanisms to reduce status maintenance traffic

Protocol for Low power and Lossy Networks) routing protocol. However, any other routing protocol can be used for this purpose. In our work, the DA role is embedded in an edge router that acts as a bridge between the WSN and IP networks using adaptation layer. However, the only requirement is that the DA is on a resource-rich node with IP reachability to the SAs.

Group Member (GM): This is the most basic entity of TRENDY, and represents a service host who registers its services with the DA. Furthermore, it periodically sends status updates to the DA using TRENDY's UPD (Update) message by selecting a random interval of 50% to 100% of the DA's time window between messages.

Group Leader (GL): The GL plays a key role in the grouping mechanism. TRENDY's modular design allows a GL to choose a different feature-set depending on its available resources and application needs. The responsibilities of a GL depend on its implemented resources; it can just collect the status updates, forwards the query to group, aggregate the results for a query or can act like a local registry or proxy.

User Agent (UA): The UA is a client that dicovers services available in a network by sending queries to the DA. It can be either part of the sensor network or can send a request from elsewhere in the Internet.

3.2   Entity Interaction

TRENDY introduces an open and interoperable framework to deal with the diversity of networks that can be the IoT. The challenge posed by the IoT's requirements is managed by employing a layered architecture, intelligent DA and enabling a RESTful web service paradigm to deal with various formats and protocols as shown in Figure 2.

This section covers the detail of TRENDY's features from the perspective of different entities.

Web Service Paradigm: TRENDY uses a RESTful web service paradigm. Entities use either CoAP (default) or HTTP (in case the targeted host understands it or DA is acting as a proxy) to define their services and to communicate with each other. The use of CoAP/HTTP simple proxy can seamlessly translate requests from both protocols. This blends the real-world devices into the existing web and enables the Web of Things (WoT) paradigm.

Context-Awareness: In TRENDY, the DA stores all service and contextual information, including service descriptions, location, battery consumed, and registration time for all registered nodes. Furthermore, it maintains a hit counter for each service, which is incremented whenever a service is discovered and selected. The grouping, optimal GL and service selection is based on the available context information. TRENDY allows the use of any context attributes in an attribute-value pair format separated by "=". These context attributes can be defined in the service description by adding "," to separate them e.g. "l=INB01,b=10" describes a host's location and battery attribute.

Grouping: Context-based grouping serves several purposes, including simple localisation of status maintenance, execution of group-based queries to offer an optional local service repository. It costs in terms of some packet overhead. However, networks can get the benefit in the form of localised communication, which conserves energy. In addition, this enables a DA to compose and offer group-based services, e.g. to actuate a command in a certain area.

The DA periodically analyses its registry for grouping and for every ungrouped GM, it sends a YGM (Your Group Member) message (with GM's IP) to a GL in the same location. In case of multiple GLs, it selects one based upon available context information. The GL then responds with an acknowledgement and completes grouping process by sending a YGL (Your Group Leader) message to GMs. This shifts the status maintenance burden to the GL, which reports the DA about each unresponsive GM. A GL can inform the DA about a missing GM using NRP (Not reported) message and its depleting battery using a GLD (Group Leader Done) message.

Hybrid Architecture: Basically, TRENDY has a centralised architecture that converges to a distributed one when the DA uses context information to group GMs as shown in Fig. 1.

Service Descriptions: There are diverse requirements for service descriptions posed by the IoT. TRENDY defines a default compact format for service description consisting of only semi-colon separated URLs of resources offered by a device. This simplistic format of resource description is a compact and efficient choice for constrained environments. However, TRENDY considers the requirement of extra semantic information, and recommends the IETF Core Link Format[1] while allowing any other format depending on the application.

Service Management: The DA maintains all service records with soft states, which need to be updated regularly by SAs. TRENDY introduces an adaptive timer that considerably reduces the number of update messages. Timer

---

[1] https://datatracker.ietf.org/doc/draft-ietf-core-link-format/

Table 1. A UA specifies *trendy/server* URL with *GET* method and some URL queries for a service discovery request to get the appropriate service

URL queries for appropriate discovery	Matching criteria
?location=INB01	Location based
?location=INB01&type=temperature	Location and type based
?location=INB01&type=temperature&info=sensor	Location, type and relevant information based

adapts to the demand of a service to increase or decrease the status update interval.

Service Discovery: The DA determines the matching service from the registry using the attributes of the UA request described by the URL queries (examples shown in Table 1). Subsequently it responds back to the UA by appending the service information (resource's URL and IP address of the host) of one or more matching services in the payload.

Service Selection: A UA can specify best in a service discovery request's payload, to seek the DA's assistance in selecting the best matching host if multiple prospective hosts are found. In this case, the DA determines the most appropriate service (if multiple services have been discovered) using available user and network context information, e.g. battery, hops count, UA location, etc.

Service Invocation: Service discovery is completed when an application gets the response with a service identifier and address of its host. TRENDY, however, enables service invocation using a RESTful web service interface and takes a step ahead by defining an adaptive publishing protocol (Section 3.4).

3.3   TRENDY Timer

The DA maintains soft state for each service description, so host devices send status updates within a time period given by the DA. TRENDY uses an adaptive timer to vary this time window length by maintaining a TRENDY counter for individual nodes. The algorithm senses the demand of services to adaptively increase the status maintenance interval for individual nodes. This significantly decreases the number of packets required for status maintenance by the nodes.

The DA is configured with global attributes including, hit count threshold value, timer step, retain threshold and the maximum TRENDY counter value, which can be changed dynamically. It also maintains individual maximum counter values for all registered SAs, which are changed in response to high demand of services hosted by a node. Figure 3a describes the adaptive timer in a scenario. Whenever the DA receives a status update message from a new node, it acknowledges the registration with a time window for the next status maintenance update. The subsequent update messages from the SA are responded with a TRENDY counter value which is incremented every time a new update message

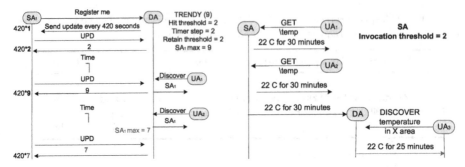

(a) Adaptive timer: The DA keeps on increasing the status update interval for a SA until its hosted services become popular.

(b) APPUB: The SA pushed the cache to the DA after two service invocations, which is used by the DA to serve a UA.

Fig. 3. Scenarios demonstrating adaptability of TRENDY

is received. The SA multiplies the acknowledged TRENDY counter with the basic time window period to determine when the next update message is expected by the DA. The DA keeps on increasing the TRENDY counter up to the maximum value for the SA.

This maximum counter value for the SA is decremented by timer step, when hit count (number of times discovered and selected) of its hosted services surpasses the hit count threshold value during the passage of a time window. Figure 3a shows how the maximum counter value of a SA decreased to 7 from 9 after two discoveries. If the hit count of a SA remained below the retain threshold value, then the counter is increased by the timer step. In case of grouping, all GMs follow the GL's TRENDY counter value.

3.4   Adaptive Piggybacked Publishing (APPUB)

TRENDY devises a demand-based caching technique the APPUB (Adaptive Piggybacked Publishing) as an alternative to balance the trade-off between service invocation delay and network efficiency. It adapts to the demand of a resource for caching rather than blindly maintaining cache of all resources in the network.

The DA maintains cached values with cached time and cache lifetime for each service. The SA implements APPUB algorithm by sending cached values and corresponding lifetime values to the DA, when the number of invocations exceeds the hit count threshold. This enables a SA to get the help from the DA to share the burden by acting as a proxy in busy times. The DA does not pass the node's IP address to a UA, if the fresh (not expired) cached value of the resource is available. Figure 3b shows how the cache is pushed by a SA and then used by the DA to serve a UA.

4 Experiments and Results

Our simulations use the CONTIKI with RPL as a routing protocol and employed COOJA [13] to simulate all SAs (GMs and GLs). ContikiMAC [5] is used as the Radio Duty Cycling (RDC) scheme and Carrier Sense Multiple Access (CSMA) as MAC protocol. Two implementations of CoAP are used in experiments. Erbium [8] is a Contiki based CoAP implementation, which is used inside the 6LoWPAN for SAs; and JAVA based Californium[2] is used to implement the DA and UA. All simulations consist of 36 Tmote Sky nodes where one node acts as a border router to connect the COOJA-based 6LoWPAN to the DA running as Linux process via the Serial Line Internet Protocol (SLIP). All nodes are placed randomly in a $190m \times 180m$ wide field. Each node hosts three resources: temperature, humidity and light. All SAs also share their location and current state of battery with the DA. The nodes are given 5 different location tags with 7 nodes (for grouping: 1 GL and 6 GMs) for each location. The implemented GL nodes in grouping scenarios are only capable of maintaining the status of their GMs. Contiki's ENERGEST [6] module is used to measure the energy consumed at each node. All simulations are executed for 20 DA time windows each of 7 minutes long. Each experiment was repeated 10 times using a different random seed for each iteration. All UA queries are stateless (each is sent as if from a new UA) and randomly selected to send a *GET* request for a resource value in one of the five locations after a random interval between 0 and 10 seconds. The number of queries are varied (100 and 1000) for following Scenarios:

Case 1 - *Basic TRENDY Service discovery (SD)*: This scenario only enables the basic functionality of TRENDY. The UA gets an appropriately selected resource's URL and IP address of a SA hosting the matching service.

Case 2 - *Basic TRENDY SD with adaptive timer*: This scenario has a TRENDY timer with global maximum counter fixed at 9 and hit count threshold at 2 on top of case 1's functionality.

Case 3 - *Basic TRENDY SD with adaptive timer and grouping*: In this scenario, grouping technique is employed with the functionality of case 2.

Case 4 - *TRENDY APPUB and timer*: This scenario employs TRENDY's APPUB technique with the threshold for service invocations fixed at 2 on top of case 2. Therefore, SAs send the cached value of a resource to the DA after two service invocations.

Case 5 - *TRENDY APPUB with timer and grouping*: In this scenario, grouping is also enabled on top of the case 4.

The service invocation delay, number of control packets, number of packets at the DA and energy consumption are measured in all experiments.

4.1 Measurements

Service Invocation Delay: Service Invocation (SI) delay is defined as the time interval between issuing an invocation request and the reception of a response.

---

[2] http://people.inf.ethz.ch/mkovatsc/californium.php

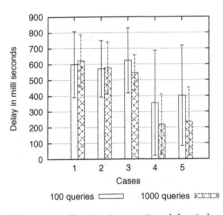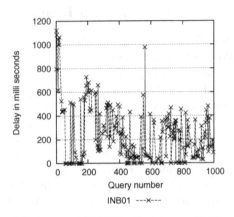

(a) Average Service invocation delay is less in cases 4 and 5 which have employed TRENDY's APPUB.

(b) TRENDY APPUB's effect in case 5: The service invocation delay falls after few service invocations.

Fig. 4. Service Invocation delay for queries (small is better)

The network traffic load, mean path length and message processing time are the factors which affect the SI delay. Figure 4a shows the advantage of using TRENDY's APPUB technique in cases 4 and 5, which have the lowest average service invocation delay. SAs push the cache value to the DA after two service invocations, which is used to serve the next UA queries resulting in cache hits. Figure 4b depicts this trend for case 5 in one of the five locations.

Control Overhead: A protocol's control overhead is measured as the number of control packets used by the protocol to complete its operational and management tasks. Figure 5a shows the control packet overhead as the sum of all registrations, grouping and reporting messages. The TRENDY adaptive timer used in cases 2 to 5 has reduced control overhead. Subsequently, in Figure 5b we additionally display the number of service invocation messages directly served by nodes in the control overhead equation. This figure illustrates the benefit of using TRENDY's APPUB and adaptive timer in cases 4 and 5.

Scalability Factor: Packets Received at the DA: With TRENDY, all services are stored and maintained by the DA. This requires messages for registration and status maintenance to be sent to the DA from all the SAs. Consequently, the number of messages generated by nodes can overwhelm the network, as most of the messages need to pass through multiple hops to reach the destination. Thus, we consider the number of packets received by the DA from 6LoWPAN network as an important scalability factor of a service discovery solution. Figures 6a and 6b show that case 3 and 5 using grouping mechanism reduced the flood of messages towards the DA by localising status maintenance.

Energy Consumption and Network Lifetime: To estimate network lifetime, we have considered top nodes in terms of energy consumption from each of

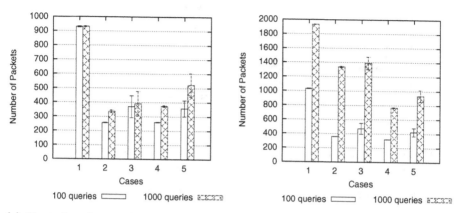

(a) Control packet overhead is far less in cases 2-5 using adaptive timer.

(b) Cases 4-5 with APPUB performed best when service invocations are included.

Fig. 5. Control packet overhead of 35 nodes after 8400 seconds (small is better)

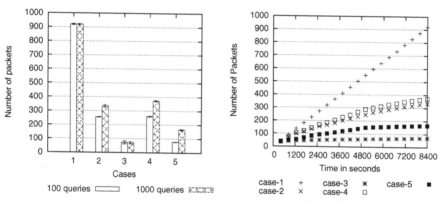

(a) After 8400 seconds: Cases 3 and 5 with grouping have performed best because of localised status maintenance.

(b) Over the time for 1000 queries: Case 5 has sent more packets to the DA compare to case 3 because of APPUB messages.

Fig. 6. Total Packets received at the DA (small is better)

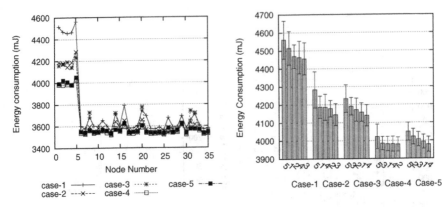

(a) All 35 nodes: case 4-5 with TRENDY techniques are the best.

(b) Top nodes in each location: Network will last longer in Cases 4-5.

Fig. 7. Energy Consumption after 8400 seconds of simulation and 1000 queries

the five locations. Figure 7a shows the individual energy consumption of all 35 nodes, whereas only top nodes are considered in Fig 7b. Both figures show that cases 4 and 5 using TRENDY's APPUB and adaptive timer will increase the energy efficiency and network lifetime.

5   Conclusion and Future Work

This paper presents TRENDY: an adaptive and context-aware Service Discovery Protocol for the IoT. This protocol employs CoAP based RESTful web services, which enable application-layer integration of constrained domains and the Internet. TRENDY's resource directory provides service discovery with a context-aware service selection using user- and network-based context. The trade-off between status maintenance load and reliability is managed by TRENDY's adaptive timer based on demand. TRENDY's APPUB technique has the following benefits: it allows the service hosts to share their load with the resource directory and also decreases the service invocation delay. Furthermore, TRENDY introduces a context-based grouping technique where the resource directory divides the network at the application layer, by creating location-based groups. This grouping of nodes localizes the control overhead and provides the base for service composition, localized aggregation and processing of data. Our simulation results show that TRENDY's techniques decrease the control overhead, energy consumption and service invocation delay. Additionally, the grouping technique considerably decreases the number of packets towards the sink and thus improves scalability in a multi-hop network. In future work, we intend to experiment with service composition by employing more appropriate group leaders for the groups. In addition, the experiments with multiple heterogeneous networks and physical hardware testbeds [11] are also in the pipeline.

References

1. Vermesan, O., Friess, P., Guillemin, P., Gusmeroli, S., Sundmaeker, H., Bassi, A., Jubert, I.S., Mazura, M., Harrison, M., Eisenhauer, M., et al.: Internet of Things Strategic Research Roadmap. In: Internet of Things-Global Technological and Societal Trends, pp. 9–52 (2011)
2. Zhu, F., Mutka, M.W., Ni, L.M.: Service Discovery in Pervasive Computing Environments. IEEE Pervasive Computing 4(4), 81–90 (2005)
3. Butt, T.A., Phillips, I., Guan, L., Oikonomou, G.: TRENDY: An Adaptive and Context-Aware Service Discovery Protocol for 6LoWPANs. In: Proc. Third International Workshop on the Web of Things, p. 2. ACM (2012)
4. Chaudhry, S.A., Jung, W.-D., Hussain, C.S., Akbar, A.H., Kim, K.-H.: A proxy-enabled service discovery architecture to find proximity-based services in 6LoW-PAN. In: Sha, E., Han, S.-K., Xu, C.-Z., Kim, M.-H., Yang, L.T., Xiao, B. (eds.) EUC 2006. LNCS, vol. 4096, pp. 956–965. Springer, Heidelberg (2006)
5. Dunkels, A.: The ContikiMAC Radio Duty Cycling Protocol. Swedish Institute of Computer Science (2011)
6. Dunkels, A., Österlind, F., Tsiftes, N., He, Z.: Software-based Sensor Node Energy Estimation. In: Proceedings of the 5th International Conference on Embedded Networked Sensor Systems, pp. 409–410. ACM (2007)
7. Guinard, D., Trifa, V., Karnouskos, S., Spiess, P., Savio, D.: Interacting with the SOA-Based Internet of Things: Discovery, Query, Selection, and On-Demand Provisioning of Web Services. IEEE Trans. on Services Computing 3(3), 223–235 (2010)
8. Klauck, R., Kirsche, M.: Bonjour contiki: A case study of a DNS-based discovery service for the internet of things. In: Li, X.-Y., Papavassiliou, S., Ruehrup, S. (eds.) ADHOC-NOW 2012. LNCS, vol. 7363, pp. 316–329. Springer, Heidelberg (2012)
9. Klauck, R., Kirsche, M.: Enhanced DNS Message Compression - Optimizing mDNS/DNS-SD for the Use in 6LoWPANs. In: Proc. 9th International Workshop on Sensor Networks and Systems for Pervasive Computing (PerSeNS 2013) (March 2013)
10. Mayer, S., Guinard, D.: An Extensible Discovery Service for Smart Things. In: Proc. Second International Workshop on Web of Things, p. 7. ACM (2011)
11. Oikonomou, G., Phillips, I.: Experiences from Porting the Contiki Operating System to a Popular Hardware Platform. In: Proc. 2011 International Conference on Distributed Computing in Sensor Systems and Workshops (DCOSS) (June 2011)
12. Oikonomou, G., Phillips, I., Guan, L., Grigg, A.: ADDER: Probabilistic, Application Layer Service Discovery for MANETs and Hybrid Wired-Wireless Networks. In: Proc. 9th Annual Communication Networks and Services Research Conference (CNSR 2011), Ottawa, Canada, pp. 33–40 (May 2011)
13. Österlind, F., Dunkels, A., Eriksson, J., Finne, N., Voigt, T.: Cross-Level Sensor Network Simulation with COOJA. In: Proceedings 2006 31st IEEE Conference on Local Computer Networks, pp. 641–648. IEEE Computer Society Press, Los Alamitos (2006)
14. Shelby, Z.: Embedded Web Services. IEEE Wireless Communications 17(6), 52–57 (2010)
15. Shelby, Z., Krco, S., Bormann, C.: CoRE Resource Directory. draft-shelby-core-resource-directory-05, IETF (2013)

Deployment of Smart Spaces in Internet of Things: Overview of the Design Challenges*

Dmitry G. Korzun[1,2], Sergey I. Balandin[3], and Andrei V. Gurtov[1]

[1] Helsinki Institute for Information Technology (HIIT), Aalto University
P.O. Box 19800, 00076 Aalto, Finland
{dkorzun,gurtov}@hiit.fi
[2] Department of Computer Science, Petrozavodsk State University (PetrSU)
33, Lenin Ave., Petrozavodsk, 185910, Russia
dkorzun@cs.karelia.ru
[3] FRUCT Oy, Helsinki, Finland
sergey.balandin@fruct.org

Abstract. The smart spaces paradigm and the M3 concept have already showed their potential for constructing advanced service infrastructures. The Internet of Things (IoT) provides the possibility to make any "thing" a user or component of such a service infrastructure. In this paper, we consider the crucial design challenges that smart spaces meet for deploying in IoT: (1) interoperability, (2) information processing, (3) security and privacy. The paper makes a step toward a systematized view on smart spaces as a computing paradigm for IoT applications. We summarize the groundwork from pilot M3 implementations and discuss solutions to cope with the challenges. The considered solutions can be already used in advanced service infrastructures.

Keywords: Interoperability, Semantic web, Security.

1 Introduction

The amount of information and services is growing so fast that users cannot efficiently utilize the existing Internet service infrastructure. Low communication between services results in high fragmentation of information. The huge opportunity is in analysis and efficient use of all information by all applications, involvement of many surrounding physical/digital objects into the service provision chain and enabling proactive delivery of the services.

The smart spaces paradigm aims at constructing advanced service infrastructures that follow the ubiquitous computing vision: smart objects are executed on a variety of digital devices and services are constructed as interaction of agents in information sharing environment [1, 2]—smart space. Its users connect new devices flexibly to the space and consume information from any of the services.

The M3 concept further considers the Multidevice, Multidomain, and Multivendor properties of smart spaces [3,4], resulting in M3 spaces. Smart-M3 [5–7]

---

* This joint research is supported by Academy of Finland project SEMOHealth, TiViT IoT SHOK program and ENPI CBC Karelia grant KA-179.

S. Balandin et al. (Eds.): NEW2AN/ruSMART 2013, LNCS 8121, pp. 48–59, 2013.

implements a pilot open-source interoperability platform of M3 spaces. The runtime information and majority of the underlying mechanisms are visible and manageable via a common knowledge base, which exploits Resource Description Framework (RDF) of the Semantic Web.

In contrast to Giant Global Graph of the Semantic Web, M3 spaces are of local and dynamic nature [3]. This property suits well for the Internet of Things (IoT) with its ubiquitous interconnections of highly heterogeneous networked entities and networks. IoT becomes a feasible internetworking substrate on top of which M3 spaces can be deployed. Autonomous everyday objects, being augmented with sensing, processing, and network capabilities, are transformed into smart objects that understand and react to their environment [8,9]. It has led recently to revision of application programming techniques and met with new design challenges for development of IoT service infrastructures.

This paper sorts out the following design challenges, which smart space deployment and in particular M3-based service infrastructures meet in IoT.

Interoperability: How to manipulate with information in an open dynamic multi-device environment and to offer services to the users.

Information Processing: How to reason over the information and to construct the services, despite of environment heterogeneity, volatility, and ad-hoc nature.

Security and Privacy: How to provide integrity and confidentiality of processed data and communication as well as authentication of services and users.

We expect that these challenges are most crucial on the recent phase of M3 concept realization. Other challenges are their instances to certain extent. That is, seamless device integration is connected to interoperability and security, knowledge exchange between services and understanding of the current situation are related to interoperability and information processing.

This overview continues our work [10] on the M3 concept. We analyze its IoT-related challenges and their impact on service infrastructure development and deployment. The analysis considers the latest achievements from recent pilot implementations of M3 spaces and services, including results from regular discussions on ruSmart and FRUCT conferences. We systematize potential responses to the challenges as well as existing M3-based solutions.

The rest of the paper is organized as follows. Section 2 introduces the smart spaces paradigm, M3 concept, and Smart-M3 platform. Section 3 shows an example of M3 space for illustrating the challenges. Sections 4, 5, and 6 sequentially consider the design challenges of smart spaces deployment in IoT and overview existing solutions and research directions. Section 7 summarizes the paper.

2 M3 Spaces

Smart space is an ecosystem of interacting computational objects on shared knowledge base. The key goal is seamless provision of users with information using the best available resources for all kinds of devices that the users can use in the ecosystem [1,3]. M3 spaces focus further on dynamic mash-up and integration of many users, devices, applications, where domains span from embedded digital equipment and consumer electronics to Web [4,5,10]. The fusion of physical and information worlds is not bound to any device type, device vendor, or application domain. Provision of end-users is localized within the situational environment and users' needs, including personalized and context-aware services.

Smart-M3 platform [6] can be used to deploy an M3 space, providing a space-based communication and synchronization substrate to independent agents—knowledge processors (KPs). They run on devices available in the environment and communicate by inserting information to the space and querying the information in the space. The space is represented by one or more semantic information brokers (SIBs); they maintain a knowledge base—a named search extent of information. Smart-M3 employs term "knowledge": a space keeps habitual data, relations between them, and even such information as computations. Existing SIB software implementations include original Smart-M3 SIB [6], RedSIB [7], OSGi SIB [11], RIBS [12], and ADK [13].

Instant content is stored as an RDF graph, adopting the low-level triple-based approach of the Semantic Web. The RDF model allows easy linking and semantic-level interoperability when there are many content producers and consumers. Each SIB performs RDF triples governance in possible cooperation with other SIBs of the same space. Transactions between SIB and KP follow Smart Space Access Protocol (SSAP), which supports the basic space primitives: join/leave, insert/update/remove, query, (un)subscribe.

M3 space application is an ad-hoc assembly of KPs implementing collaboratively a service scenario to meet users' goal. The high-level view is illustrated in Fig. 1. Scenario steps emerge from actions taken by the KPs and observable in the application M3 space. Access to global knowledge is possible via a gateway KP to the external world. A scenario can be composed from multiple applications. The key point is the loose coupling between the participating KPs.

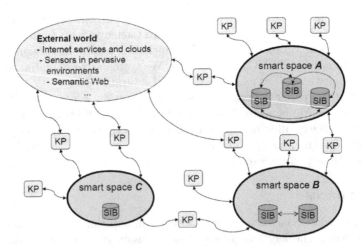

Fig. 1. Each space is maintained by own SIBs to host applications. An application is formed by KPs that publish and query shared content. Some KPs are wrappers for external services; some are used for knowledge exchange between spaces.

The impact of each KP to others is limited by the knowledge the KP provides into the space. Within the same application its space can conform to a common ontology [14, 15], though it can be modular, multi-domain, composed from multiple ontologies. Each KP applies own "sub-ontology" to interpret the accessible part of the shared content. The application is not fixed since its KPs may join and leave the space. Several demo pilots have been already developed and showed the feasibility of the M3 concept, see [2, 5, 10, 16–18] and references therein.

Consider the M3 ontology-driven computing formalism based on the generic smart space model [19]: a space is a knowledge base $S = (n, I, \rho)$, where n is its name, I is information content, and ρ is a rule set to deduce knowledge. Content I is represented as an RDF graph. When ρ is OWL ontology O then I is further structured with classes and properties from O. Deduction in S can be performed on the ontology instance graph using techniques of Semantic Web [3, 19, 20]. Such a graph is formed by individuals (nodes) that are interlinked with object properties (links) and have data properties (attributes). That is, we can treat an application M3 space as $S = (I, O)$, where I is an RDF graph and O is an OWL ontology. A portion of knowledge $x \in S$ is an ontology instance graph that is a part of the deductive closure calculated from I according to O. We also refer to S as to the space unique name.

3   Explanatory Example: Smart Room

Let us consider an example M3 space—Smart Room [18] to explain the reasons and importance of challenges of smart spaces deployment in an IoT environment. Smart room system scope is shown in Fig. 2.

Communication in a smart room uses a wireless local area network (WLAN) attached to the Internet. Participants are chairman, active speaker (in turn relay manner), and spectators (including inactive speakers). Two public screens are available: (1) Agenda shows the event timetable and (2) Presentation shows material that each speaker presents.

The participants access services in the smart room using personal mobile computers (e.g., smartphones, tablets, laptops). The room is equipped with sensor

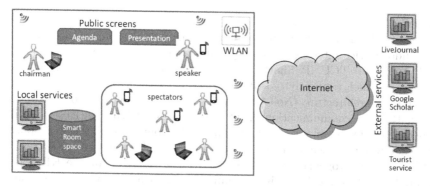

Fig. 2. Equipment and service environment for participants in Smart Room

devices that sense the physical parameters of the environment and participant activity. All knowledge is collected, organized, shared, and searched in a common smart space. Local services run on local computers or nearby servers. The system accesses the external world for appropriate Internet services. Service outcome is visible online on the public screens or personalized on mobile clients.

The heterogeneity of participating devices and information sources immediately faces with the interoperability challenge. Smart room service set is formed from multiple sources of heterogeneous information and requires intensive processing, including service discovery and the provision to a particular participant or a group of them. Personal information is essential for smart room operation, but it must support rigorous security and privacy defense mechanisms.

Examples of particular problems are the following. Integration of joining devices (e.g., personal devices) is seamless. Service management is adaptive, e.g., when some services become temporarily unavailable. Knowledge exchange is supported: one service utilizes knowledge deduced by another service, e.g., discussion of participants in the blog leads to updates in the agenda.

4   Interoperability

Information available on some devices may be interesting to other devices of the same environment. Furthermore, some devices should communicate with the external world. Currently, the standards for interoperability have been mostly created for single domains or are controlled by a single company. Such domain specific standards pose considerable challenges for IoT devices. The traditional standardization approach cannot achieve the basic IoT property: a device can interoperate with whatever devices accessible at the given time.

The smart spaces concept makes clear separation between device, service, and information level interoperability [5, 21]. Device interoperability covers technologies for devices to discover and network with each other. Service interoperability covers technologies for space participants to discover services and use of them. Information interoperability covers technologies and processes for making information available without a need to know interfacing methods of the entity creating or consuming the information.

Application developer uses KP Interface (KPI) for programming KP logic and its interaction with the space by SSAP primitives [6, 22]. The M3 concept requires that SIB supports a number of solutions for network connectivity, yielding multivendor device interoperability. For Internet communication, SIB supports HTTP and plain TCP/IP. Short-range wireless communications of mobile devices can use such connectivity solutions as Bluetooth or 6LoWPAN. Network on Terminal Architecture (NoTA) provides a possible solution for embedded devices. Reliable communication on top of IPv4 and IPv6 uses Host Identity Protocol [23], which supports mobility and multi-homing, see Sect. 6. Application code developer selects a connectivity mechanism for a device family.

The device heterogeneity introduces additional difficulty for the KP development. If the hosting device is a computer (i.e., relatively powerful OS and ability to run non-trivial programs), then KP can run directly on the device.

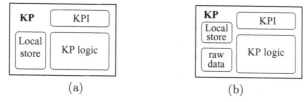

Fig. 3. Typical KP architecture: (a) KP running on a computer, (b) KP serving as a gateway for low-capacity devices

The computational resources have to be allocated for KPI (SSAP operations, XML processing, networking), KP logic (written by the developer), and local store (knowledge that KP directly processes), see Fig. 3 (a).

Techniques for efficient KP programming for mid-capacity devices (laptops, smartphones, tablets, etc.) and certain embedded devices (with embedded Linux, Contiki, etc.) exist [14]. If a device is very primitive (even no operating system, like in a sensor), then hosting a KP on the device is unreasonable or even impossible. In this case, such a device can be attached to the space via a dedicated computer running a gateway KP, see Fig. 3 (b). The KP has to transform data between the given data format of satellite device and the ontological representation in the KP local store. For instance, this approach is used for constructing personal smart space in healthcare applicationss [24].

Many of primitive devices are pure data producers in the M3 space. Thus the required processing at the KP side is small due to the machine-oriented property of RDF. It includes transformation of the raw data into triples and construction of simple SSAP packets in XML or binary format [12,22]. The above processing can be implemented on the hardware level.

Information sharing in M3 spaces is based on the same mechanisms as in the Semantic Web, thus allowing multidomain applications, where the RDF representation allows easy exchange and linkage of data between different ontologies. It makes cross-domain interoperability straightforward [6]. Application domains are localized, limiting the search extent and ontology governance. That is, for each application its space $S = (I, O)$ is relatively small, allowing computationally reasonable knowledge maintenance at SIB and moderate performance expenses at KP. The interoperability is due to the locally agreed unification of semantics when accessing the same part of the space content I. That is, the space-wide ontology O is a virtual application-level component.

The space content I is organized into an RDF graph. Although explicit use of a specific ontology is not demanded, additional semantics are provided by an ontology O, usually defined in OWL. For example, a group of KPs can agree an aligned ontology for interpretation of a certain part of the space. The consistency of stored information is not guaranteed; KPs are free to interpret information in whatever way they want. This RDF-based low-level model requires KP code to operate with triples following the SSAP operations directly; the triples are basic exchange elements in communication with the M3 space.

For development efficiency, the high-level ontology-driven KP programming is supported, e.g., SmartSlog SDK [14]. The approach is based on an ontology library, which is automatically generated mapping the ontology to code in a given programming language. The KP logic then is written using high-level ontology entities (classes, relations, individuals). They are implemented with predefined data structures and methods. It essentially simplifies the KP code; the developer has the programming language-like tools to manipulate with the concepts defined in the ontology. The number of domain elements is reduced since an ontology entity consists of many triples. The library API is generic: its syntax does not depend on a particular ontology, ontology-related names do not appear in names of API methods, and ontology entities are used only as arguments.

Notably that ontology library is less machine-dependent than low-level KPI. The same high-level KP code is suitable for different devices since the ontology library can wraps the appropriate KPI.

5   Information Processing

The SIB side of M3 space provides mechanisms for knowledge discovery and first-order logic reasoning. Each space may contain its own set of reasoning capabilities. The most important mechanisms are semantic queries and subscription.

To find appropriate knowledge in S the KP constructs a query using semantic query languages as SPARQL. The SIB resolves the query [7] and returns an ontology instance graph $x \in S$. The KP interprets the result locally and then can insert new knowledge to S or update some previous instances. The appropriate deduction (e.g., deductive closure) is performed by SIB dynamically—at query-time (also at insert-time in some spaces).

A subscription operation is a special case of query—a persistent query, realizing the publish/subscribe communication model in smart spaces [25]. Changes in the space content trigger actions from participating KPs. Subscription is used (i) for synchronizing KP's local knowledge storage with the shared space, as well as (ii) for receiving notifications about recent changes. The latter is a way for a KP to detect events happening in the system.

The RDF-based semi-structured knowledge representation with no strict ontology conformance shifts the responsibility of knowledge interpretation and truth maintenance to the agents. Each KP u manages a non-exclusive part I_u of knowledge and applies own expertise for reasoning over I_u. A KP u uses own ontology O_u as an assistance tool that helps to achieve a common understanding with other KPs, see Fig. 4. Each KP publishing its shared knowledge provides meta-information to indicate intention for interpretation. Thus, an application is aware of the unification of semantics, which can be done in a localized manner (between a group of KPs) and even runtime. It may result in information inconsistency in the space and misinterpretation on the reader side. The supporting mechanisms to deal with this problem are under development [14, 26, 27].

The M3 concept supports multi-space applications when a KP needs the information from several spaces (Fig. 1 in Sect. 2). It provides an opportunity for

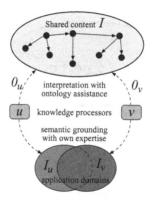

Fig. 4. KPs u and v make own interpretations of the content. The ontologies and semantic grounding are agreed to ensure a common interpretation and aligned domains.

applications integration in an ad-hoc manner. Notably that the coupling between the participating KPs is loose and KP granularity can be extremely fine (e.g., a KP can implement a function outside of the hosting UI concept). When an application needs a service from another Smart-M3 application, a KP mediator can be used to connect these spaces [16]. The mediator should properly interpret corresponding knowledge in the source and target spaces, construct a mapping between them, and execute the exchange. Ontology accompanied with logic programming rules can define constraints to which exchanged instances satisfy as well as specify mappings between the spaces.

6   Security

The security challenge includes traditional issues of open distributed systems, such as key exchange and resource restrictions, and specific problems caused by the dynamicity and heterogeneity of smart spaces [28]. We classify smart space security components onto (a) share level, (b) space access control, and (c) communication. Let u and v be KPs in space S.

Security of the share level is based on a sharing function $\sigma_u(S) \subset I$ that defines which locally available knowledge to publish in S for sharing with others. Each KP makes own decisions on its share level, keeping essentially private knowledge at the local storage only. It does not prevent u to combine private and shared knowledge in local reasoning.

In the space access control, an access function ϕ_u limits other KPs in access u's shared content; $\phi_u \subset I$ is the knowledge that u allows for v. Hence u and v collaborate in the content $\phi_u \cup \phi_v$ (Fig. 5). Since SIB enforces access control over brokered information, application-specific policies need additional support.

Access control benefits from meta-information published in the space. Exclusive access to the content can be on RDF level. The method of [29] allows restricting the access for an arbitrary set of triples. Meta-information is additional triples that specify which data are protected and which KP is their owner.

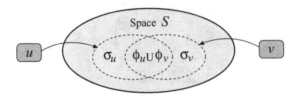

Fig. 5. KPs u and v provide restricted content to share in S. They collaborate accessing only allowed parts of content of each other.

The method may be embedded in middleware data access primitives of a standard KPI, so becoming hidden to the KP developer. Although the method allows extension for more security attributes beside synchronization, any KP is able to see what has been protected in the space.

IoT applications need context-dependent and fine-grained access control [28, 30]. Smart space access control policies define which KPs are allowed to access which objects. Security level of joined devices is measured. An access control ontology allows representing meta-information about the context and granularity. SIB utilizes this information to authorize the access to the space content. The approach enables devices to share knowledge with the same security level even when these devices do not have interoperable security protocols for direct confidential communication. Note that combining security policy rules with reasoning allows exploiting further the advantages of logic programming and description logic [29, 31]. Unfortunately, many reasoning problems in general form require exponential time in the worst case.

Accessing the space is session-based by join/leave operations [32]. It forms a base for mechanisms of access control and secure communication. For instance, KP identity and cryptographic keys can be implemented with Host Identity Protocol (HIP) [23], which is standardized by IETF. The HIP exchange authenticates a KP-to-SIB communication session based on robust identities. They can be used for access control to different parts of the space, implementing the access function ϕ. All transferred data are encrypted, so providing confidentiality and message integrity in communication with SIB. The same approach was discussed in [30] for Transport Layer Security (TLS) protocol. However, TLS does not support mobility and multi-homing as well as causes significant overhead.

Many IoT devices are of low capacity (memory, CPU, battery, etc.), and they cannot use the full scale of security capabilities that the basic HIP or other Internet protocols provide [9]. A HIP-based extension for secure transfer of private data in M3 spaces is proposed in [24] for healthcare applications with wearable and implantable medical devices. The proposal employs HIP Diet Exchange (DEX) to establish secure associations between KP and SIB. HIP DEX requires rather limited computation capabilities from the devices since it uses elliptic curve cryptography to distribute the shared secret. Although HIP DEX is designed for resource-restricted devices it still provides possibility to control performance level by adjusting cryptographic computation difficulty.

Table 1. Summary of challenges and their solutions

Challenge	Provided solutions and feasible directions
Interoperability: device, service, information	Many network protocols. RDF-based operation of SSAP. Multiplatform KPIs and reusable code. Ontology libraries and code generation. Development tools for mid- and low- capacity devices.
Information processing	SPARQL queries and first-order reasoning. Subscription and proactive services. Ontology-driven development and runtime mechanisms. Multi-space operation and mediator-based synchronization.
Security and Privacy	RDF-based knowledge access control and mutual exclusion. HIP-based network communication. Ontology-based control policies and context-aware security.

7 Conclusion

This paper considered the smart spaces paradigm and its potential for service infrastructure development in IoT environment. The discussion focused on the design challenges of smart spaces deployment: (1) interoperability, (2) information processing and (3) security and privacy. Table 1 lists the corresponding solutions. Although full-valued solutions are still under development, the presented summary shows the overall feasibility and applicability of the smart spaces computing paradigm for IoT settings.

Knowledge processors running on IoT devices and cooperating in various service scenarios are loosely coupled. The shared content conforms an ontological knowledge representation, supporting also localized agreements and personalization. These properties provide a base to tackle the interoperability challenge.

The semantic reasoning mechanisms and their distributed nature support effective processing within huge multi-source information collections. The M3 concept states localized ad-hoc spaces and integrates the Semantic Web with other information on surrounding electronic devices. This groundwork feeds and catalyzes solutions to the information processing challenge.

Progress in Internet security protocols provides promising solutions for confidential communications and authentication of the participants with strong cryptographic identities. The computation overhead can be made low, and even low-capacity devices are involved into the service infrastructure. Additionally, advanced semantic models equipped with logic programming techniques support fine-grained context-dependent access control to the shared content.

References

1. Cook, D.J., Das, S.K.: How smart are our environments? an updated look at the state of the art. Pervasive and Mobile Computing 3(2), 53–73 (2007)
2. Smirnov, A., Kashevnik, A., Shilov, N., Oliver, I., Balandin, S., Boldyrev, S.: Anonymous agent coordination in smart spaces: State-of-the-art. In: Balandin, S., Moltchanov, D., Koucheryavy, Y. (eds.) ruSMART 2009. LNCS, vol. 5764, pp. 42–51. Springer, Heidelberg (2009)

3. Oliver, I.: Information spaces as a basis for personalising the semantic web. In: Proc. 11th Int'l Conf. Enterprise Information Systems (ICEIS 2009), pp. 179–184 (May 2009)

4. Balandin, S., Waris, H.: Key properties in the development of smart spaces. In: Stephanidis, C. (ed.) UAHCI 2009, Part II. LNCS, vol. 5615, pp. 3–12. Springer, Heidelberg (2009)

5. Liuha, P., Lappeteläinen, A., Soininen, J.P.: Smart objects for intelligent applications - first results made open. ARTEMIS Magazine (5), 27–29 (2009)

6. Honkola, J., Laine, H., Brown, R., Tyrkkö, O.: Smart-M3 information sharing platform. In: Proc. IEEE Symp. Computers and Communications (ISCC 2010), pp. 1041–1046. IEEE Computer Society Press (June 2010)

7. Morandi, F., Roffia, L., D'Elia, A., Vergari, F., Cinotti, T.S.: RedSib: A Smart-M3 semantic information broker implementation. In: Balandin, S., Ovchinnikov, A. (eds.) Proc. 12th Conf. of Open Innovations Association FRUCT and Seminar on e-Tourism, SUAI, pp. 86–98 (November 2012)

8. Kortuem, G., Kawsar, F., Sundramoorthy, V., Fitton, D.: Smart objects as building blocks for the internet of things. IEEE Internet Computing 14(1), 44–51 (2010)

9. Bonetto, R., Bui, N., Lakkundi, V., Olivereau, A., Serbanati, A., Rossi, M.: Secure communication for smart IoT objects: Protocol stacks, use cases and practical examples. In: Proc. IEEE Int'l Symposium on a World of Wireless, Mobile and Multimedia Networks (WoWMoM 2012), pp. 1–7. IEEE Computer Society (2012)

10. Korzun, D.G., Balandin, S.I., Luukkala, V., Liuha, P., Gurtov, A.V.: Overview of Smart-M3 principles for application development. In: Proc. Congress on Information Systems and Technologies (IS&IT 2011), Conf. Artificial Intelligence and Systems (AIS 2011), vol. 4, pp. 64–71. Physmathlit, Moscow (2011)

11. Manzaroli, D., Roffia, L., Cinotti, T.S., Ovaska, E., Azzoni, P., Nannini, V., Mattarozzi, S.: Smart-M3 and OSGi: The interoperability platform. In: Proc. IEEE Symp. Computers and Communications (ISCC 2010), pp. 1053–1058. IEEE Computer Society (2010)

12. Suomalainen, J., Hyttinen, P., Tarvainen, P.: Secure information sharing between heterogeneous embedded devices. In: Proc. 4th European Conf. Software Architecture (ECSA 2010), Companion Volume, pp. 205–212. ACM (2010)

13. Gómez-Pimpollo, J.F., Otaolea, R.: Smart objects for intelligent applications – ADK. In: Proc. 2010 IEEE Symp. Visual Languages and Human-Centric Computing (VL/HCC), pp. 267–268 (September 2010)

14. Korzun, D.G., Lomov, A.A., Vanag, P.I., Honkola, J., Balandin, S.I.: Multilingual ontology library generator for Smart-M3 information sharing platform. International Journal on Advances in Intelligent Systems 4(3&4), 68–81 (2011)

15. Lomov, A.A.: Ontology-based KP development for Smart-M3 applications. In: Balandin, S., Trifonova, U. (eds.) Proc. 13th Conf. of Open Innovations Association FRUCT and 2nd Seminar on e-Tourism for Karelia and Oulu Region, SUAI, pp. 94–10 (April 2013)

16. Korolev, Y., Korzun, D., Galov, I.: Smart space applications integration: A mediation formalism and design for smart-M3. In: Andreev, S., Balandin, S., Koucheryavy, Y. (eds.) NEW2AN/ruSMART 2012. LNCS, vol. 7469, pp. 128–139. Springer, Heidelberg (2012)

17. Kiljander, J., Ylisaukko-oja, A., Takalo-Mattila, J., Eteläperä, M., Soininen, J.P.: Enabling semantic technology empowered smart spaces. Journal of Computer Networks and Communications (2012)

18. Galov, I., Korzun, D.: Smart room service set at Petrozavodsk State University: Initial state. In: Balandin, S., Ovchinnikov, A. (eds.) Proc. 12th Conf. of Open Innovations Association FRUCT and Seminar on e-Tourism, SUAI, pp. 239–240 (November 2012)
19. Oliver, I., Boldyrev, S.: Operations on spaces of information. In: Proc. IEEE Int'l Conf. Semantic Computing (ICSC 2009), pp. 267–274. IEEE Computer Society (September 2009)
20. Gutierrez, C., Hurtado, C.A., Mendelzon, A.O., Pérez, J.: Foundations of semantic web databases. J. Comput. Syst. Sci. 77(3), 520–541 (2011)
21. Ovaska, E., Cinotti, T.S., Toninelli, A.: The design principles and practices of interoperable smart spaces. In: Liu, X., Li, Y. (eds.) Advanced Design Approaches to Emerging Software Systems: Principles, Methodology and Tools, pp. 18–47. IGI Global (2011)
22. Kiljander, J., Morandi, F., Soininen, J.-P.: Knowledge sharing protocol for smart spaces. International Journal of Advanced Computer Science and Applications (IJACSA) 3, 100–110 (2012)
23. Gurtov, A., Komu, M., Moskowitz, R.: Host Identity Protocol (HIP): Identifier/Locator Split for Host Mobility and Multihoming. Internet Protocol Journal 12(1), 27–32 (2009)
24. Gurtov, A., Nikolaevskiy, I., Lukyanenko, A.: Using HIP DEX for key management and access control in smart objects. In: Proc. of Workshop on Smart Object Security (March 2012) (Position paper)
25. Lomov, A.A., Korzun, D.G.: Subscription operation in Smart-M3. In: Balandin, S., Ovchinnikov, A. (eds.) Proc. 10th Conf. of Open Innovations Association FRUCT and 2nd Finnish–Russian Mobile Linux Summit, SUAI, pp. 83–94 (November 2011)
26. Smirnov, A., Kashevnik, A., Shilov, N., Balandin, S., Oliver, I., Boldyrev, S.: On-the-fly ontology matching in smart spaces: A multi-model approach. In: Balandin, S., Dunaytsev, R., Koucheryavy, Y. (eds.) ruSMART 2010. LNCS, vol. 6294, pp. 72–83. Springer, Heidelberg (2010)
27. Janhunen, T., Luukkala, V.: Meta programming with answer sets for smart spaces. In: Krötzsch, M., Straccia, U. (eds.) RR 2012. LNCS, vol. 7497, pp. 106–121. Springer, Heidelberg (2012)
28. Evesti, A., Suomalainen, J., Ovaska, E.: Architecture and knowledge-driven self-adaptive security in smart space. Computers 2(1), 34–66 (2013)
29. D'Elia, A., Honkola, J., Manzaroli, D., Cinotti, T.S.: Access control at triple level: Specification and enforcement of a simple RDF model to support concurrent applications in smart environments. In: Balandin, S., Koucheryavy, Y., Hu, H. (eds.) NEW2AN 2011 and ruSMART 2011. LNCS, vol. 6869, pp. 63–74. Springer, Heidelberg (2011)
30. Suomalainen, J., Hyttinen, P.: Security solutions for smart spaces. In: Proc. 2011 IEEE/IPSJ Int'l Symposium on Applications and the Internet (SAINT 2011), pp. 297–302. IEEE Computer Society, Washington, DC (2011)
31. Aziz, R.A., Janhunen, T., Luukkala, V.: Distributed deadlock handling for resource allocation in smart spaces. In: Balandin, S., Koucheryavy, Y., Hu, H. (eds.) NEW2AN 2011 and ruSMART 2011. LNCS, vol. 6869, pp. 87–98. Springer, Heidelberg (2011)
32. Lomov, A.A.: SmartSlog session in Smart-M3. In: Balandin, S., Ovchinnikov, A. (eds.) Proc. 12th Conf. of Open Innovations Association FRUCT and Seminar on e-Tourism, SUAI, pp. 66–71 (November 2012)

Agent Substitution Mechanism for Dataflow Networks: Case Study and Implementation in Smart-M3

Ilya Paramonov, Andrey Vasilev, Denis Laure, and Ivan Timofeev

P.G. Demidov Yaroslavl State University, Yaroslavl, Russia
ilya.paramonov@fruct.org,
{vamonster,den.a.laure,skat.set}@gmail.com

Abstract. The paper continues the study of dataflow networks based on top of Smart-M3 platform. The goal is to provide support for implementation of reliable ubiquitous services based on the dataflow model. We propose agent substitution as a way to make services robust and describe behaviour and implementation aspects of the agent substitution mechanism in semantic information broker. The mechanism allows to safely transfer computational context from one agent to another preventing long service downtime when an agent unanticipatedly disconnects from the network. The potential benefit of using such a mechanism is discussed for the medical telemonitoring service.

Index Terms: RedSib, Smart-M3, Agent substitution, dataflow network, Reliability.

1 Introduction

Dataflow network is one of available architectures for distributed computing systems. It consists of computing units that operate in different information processing levels: each unit receives information from the lower level units or information sources (e.g., sensors), processes the information and sends results to the higher level units for further processing.

Sometimes devices that host processing units may lose connection and thereby violate the whole network operation. One way to solve the issue is to replace the disconnected unit with another one. In [11] we proposed the idea of substitution mechanism for dataflow networks that allows to manage the substitution process by detecting the disconnected unit, sending the context to a substitute one and restoring the original unit operation in case of its returning to the network. In [12] we presented the refined version of the substitution mechanism and identified key situations for data flows tempering.

This paper provides a case study of the health monitoring service based on dataflow network and discusses details of the substitution mechanism implementation on Smart-M3 platform.

S. Balandin et al. (Eds.): NEW2AN/ruSMART 2013, LNCS 8121, pp. 60–71, 2013.

The object of the case study is the health monitoring system that allows to automatically monitor the patient's vital signs and inform a doctor about patient's condition in case of emergency. The case study demonstrates applicability of the dataflow network model for modern mobile services and reveals a kind of motivation for the agent substitution.

Implementation of the proposed mechanism is based on Smart-M3—open-source platform for smart space applications development [4]. To implement the mechanism we modified the latest version of Semantic Information Broker (SIB)–RedSib–that provides access to shared information in the smart space. The modifications include integration of several new modules to SIB architecture and implementation of special operations allowing agents of the platform—knowledge processors (KPs)—to function as dataflow network nodes and to be substituted with other agents in case of failure.

The paper is structured as follows. In Section 2 we present the model of dataflow network used in our research. Section 3 provides the case study. Description of the proposed substitution mechanism and its implementation are presented in Sections 4 and 5 respectively. In section 6 we consider our work in context of related papers. Conclusion summarizes the main results of the work.

2   Dataflow Network on Smart-M3 Platform

In this section we give coherent description of dataflow network implementation on Smart-M3 platform, which was presented in previous publications. Dataflow network is a network of parallel executed and cooperated processes [6]. The network consists of computation nodes that processes information.

In our research we treat nodes as finite-state machines. Each node has inputs, outputs and an internal state. Node outputs can be connected to inputs of other nodes, thereby transferring data tokens from one node to another for further processing.

When a node receives a data token from one of its inputs, this token becomes the current value of the input. Then the node generates an output token and changes the internal state. The both one are computed from all current input values and the internal state according to some functional relation.

The input tokens are allowed to come from outside of the network. They represent the data from information sources, such as sensors or network services. Similarly, the output value of any node can be used from outside the network, for example, to control some kind of device. Such inputs and outputs make up input and output of the whole network.

To implement the network we use Smart-M3 platform [4] based on the smart spaces concept [1]. The main components of the platform are semantic information broker (SIB) and knowledge processors (KPs). SIB manages information storage of the smart space and provides access to it; KP is a client that queries and modifies information in the smart space.

The data storage contains information in the form of triples, each of which consists of three components: subject, predicate and object. To operate with data from the storage, SIB provides the following operations: "insert", "update", "delete", "query" and "subscription". The first three operations perform the corresponding changes of the data in the smart space. The "query" operation allows KP to retrieve triples according to a template. Subscription is the most interesting operation, as it allows KP to track data changes and receive notifications once they get modified.

In the implementation of dataflow network on Smart-M3 platform each node of the network corresponds to a KP. For each input of the node we define a triple that stores current value of the input. For example, for the input value from a luminosity sensor can be stored in the triple of form: *("luminosity", "is", actual value of luminosity)*. The same assumption is true for outputs. In our implementation the internal state of the node is also represented as a set of triples in the data storage. These triples are considered as private for the particular node and cannot be accessed by the others.

The node operation is subscription-driven. On start it subscribes for all input triples and then performs the cycle that includes the following steps:

1. Wait until a subscription fires.
2. Read current values of inputs and the state from the corresponding triples in the data storage.
3. Compute output values and new state from inputs and current state.
4. Update the triples in the data storage that correspond to output values and the internal state.

On finishing the cycle KP is ready for the next one.

The using of the Smart-M3 platform allows to implement end-user services on the base of dataflow network. The next section provides an example of such a service.

3   Case Study: Health Monitoring System

The goal of this case study is to show that the substitution mechanism for dataflow network can improve existing and future services based on this model by making them more robust. The benefits of the substitution mechanism can be seen clearly on systems and applications in medical care field, because a failure of such systems and applications can cost a patient's life.

The object of the case study is a monitoring system for people from risk groups (suffering from diabetes, vascular or heart diseases etc.). These people need a real time monitoring and quick action in case of emergency. There are several studies on systems that can solve this problem (for example, [8], [3]). Unfortunately, these studies often do not consider cases of failure of the system components despite the fact that such failure can affect the whole system, thus threatening the patient's health and life.

The system under consideration of our case study monitors patient's electrocardiogram (ECG) and blood pressure using wearable or stationary sensors and informs the doctor in case of emergency. The dataflow network of the monitoring system is shown in Fig. 1. The system includes several additional agents that are not shown in the figure. These agents use output of the network to display information or to inform the doctor.

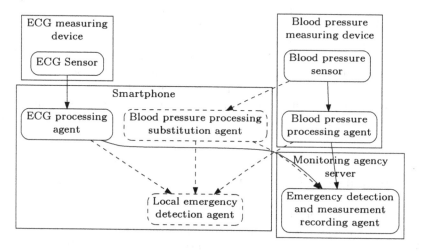

Fig. 1. Dataflow network structure of the health monitoring system

The ECG measuring device is a wearable sensor that provides real-time measurements (e.g., Alive Heart and Activity Monitor). The ECG processing agent running on patient's smartphone uses this data to calculate heart rate and display it on the mobile phone screen. Both ECG data and heart rate are sent further to the remote monitoring server of the hospital or agency in charge of the medical observation.

The blood pressure sensor and processing agent run on the stationary blood pressure measuring device (e.g., Numera Home Hub). Blood pressure processing agent classifies measurements, checks whether pressure is too low or too hight and sends this information to the monitoring server and to the external visualization agent running on the patient's mobile.

The monitoring server uses gathered ECG, blood pressure and heart rate measurements to inform the doctor in case of strong deviations in received values. Collected data are also added to the patient's record for further analysis and creation of the personalized health model.

In the situation, when the patient's mobile phone looses connection with the monitoring server, the doctor will not be informed in case of emergency and the patient's life may be threatened. To lower this risk the local emergency detection agent starts to track measurements instead of the remote monitoring server. The local agent has the simpler emergency detection algorithm based on tracking extreme values prescribed by the doctor. When a value exceeds the set extreme border the agent sends SMS to the doctor. The substitution raises energy consumption of the patient's smartphone, but it allows to continuously monitor the patient's state, which is very critical for the patient's life.

Another risk situation for the system is exhaustion of the power supply on mobile devices. The patient can easily detect low energy levels of wearable devices (ECG sensor and mobile phone), but it is harder to track for the stationary blood pressure measuring device. The blood pressure measuring device performs not only sensing, but data processing too, so it consumes more energy, than the device performing only sensing. In situations, when the device's energy level is low, data processing function can be moved to the corresponding substitute agent running on the mobile device. This will extend the life time of the blood pressure measuring device and therefore the life time of the whole system.

The case study demonstrates several situations that can disrupt the operation of the presented health monitoring system and shows benefits of using the agent substitution in these cases including reliable operation and possibility to extended the life time of the system.

4   Substitution Mechanism

This section summarizes the idea of the substitution mechanism proposed in [11], [12] and specifies details required for its implementation in SIB.

During operation a dataflow agent can lose connection with SIB. This situation may occur due to a failed communication, low battery level of a sensing or processing device and other circumstances.

In case of the agent disconnection SIB detects such a situation by the broken socket and starts substitution. In case of low battery level the agent can prevent disconnection by identifying this situation and requesting substitution explicitly. On such a request SIB proceeds substitution and notifies the agent that it can shut down with no negative effect on network operation.

For the sake of generality we consider agents as stereotype entities, each of which is able to interpret a program of a particular type. This program determines how to compute outputs and the new state of the agent from inputs and the current state.

We introduce two agent types: primary agent and substitute agent. In previous papers we used term "main" to reference primary ones. The primary agent participates in normal network operation. The substitute agent performs the work of a disconnected agent. The main difference between these types of agents is that they can execute different programs. For example, the substitution agent program can be simpler than the primary agent's one and can cover only partial functionality (like the local emergency detection agent in Sec. 3).

To substitute an agent with another one without negative impact on the network operation SIB uses the operation context, which characterizes the data processing by the agent at a particular moment in time. The context includes the current state, the program and specification of input and output triples of the agent.

The introduction of substitution mechanism requires addition of new operations including agent registration and unregistration, explicit substitution request and various substitution control messages. The way to introduce such operations is a subject for implementation decisions covered in the next section.

5   Implementation of Substitution Mechanism

In this section we present the original structure of RedSib, discuss notable aspects of the substitution mechanism implementation and give details related to structural and behavioural modifications of SIB.

5.1   RedSib Architecture

RedSib is the current version of SIB on Smart-M3 platform [9], which includes latest stability and performance improvements developed in University of Bologna. RedSib is executed as two processes (sib-tcp and redsibd) that exchange information via D-Bus interface. The sib-tcp daemon is responsible for communication with KPs using Smart Space Access Protocol (SSAP) over TCP/IP connections. The redsib daemon manages data storage and provides additional services including data access control and RDF++ reasoning.

Fig. 2 presents a simplified view on processing of commands coming from KP. The sib-tcp daemon receives SSAP message, decodes it and sends corresponding command to the redsib daemon. The received command is added to the end of the queue inside the communication module that executes commands one-by-one from the beginning of the queue. The module gets command and passes it to the corresponding module that executes the command on top the data storage. The result of the command is sent back to the KP upon operation completion if requested. When a command changes data, the subscription module sends a notification message to all KPs subscribed for tracking these particular triples.

5.2   Operation Level of Dataflow Network

There are at least two approaches to implement operations required by the substitution mechanism into the existing SIB architecture. The first approach implies modification of the SSAP protocol and implementation of corresponding operation handlers in SIB. The second approach proposes usage of specialized data structures inside data storage and their modification via existing data exchange methods.

The advantages of specialized commands addition include transactional integrity and clear API for creation of dataflow agents. However, the SSAP modification requires implementation of new commands in all KP libraries to give access to substitution mechanism functionality. The second approach does not require such modifications and does not limit application developers in the supported tools. The main downside of this approach is the need for KP to follow data modification protocol and execute several commands to perform the same action. It can be neglected by the use of agent templates or high-level libraries that provide convenient API. To make this mechanism available for every user of the platform, we implemented required operations on top of existing data exchange methods.

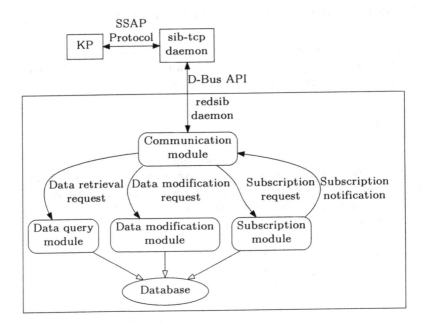

Fig. 2. Original RedSib architecture

5.3   Handler Implementation Alternatives

Following the existing data exchange mechanism implies the use of subscription for tracking the triples corresponding to the dataflow network operations (see Sec. 4). Subscription should be used by dataflow agents as well as by the substitution mechanism located is SIB. For the latter ones we use the term "internal subscriptions" as they are identical to the ones used by KPs but operated completely inside SIB.

There are two main approaches to organize these internal subscriptions. The first one implies creation of one internal subscription per event type; the second—one internal subscription per agent.

The main difference between these approaches is the amount of internal subscriptions. When having a small number of agents the first approach is preferable by the amount of internal subscriptions. But in case of the large agents number the situation is opposite. The second approach is scalable, therefor it was selected for the implementation.

5.4   SIB Architecture Modification

During the substitution mechanism implementation we added several new modules to RedSib. The modified SIB architecture is shown in Fig. 3.

1. Substitution Manager. This module provides support for dataflow network operations. Its functionality allows to

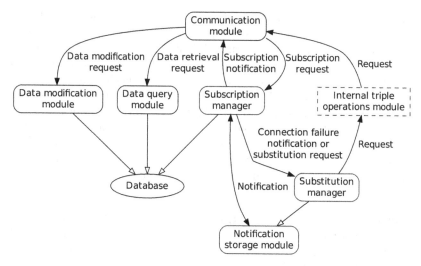

Fig. 3. Modified RedSib daemon architecture

- Register KP as a node of the dataflow network;
- Restrict access to input/output triples of the node;
- Substitute a failed agent with another one;
- Restore operation of the previously failed agent after its return to the network;
- Remove the agent from the network.

2. Notification Storage Module. The notification storage collects the data that would come by subscription in the situation when an agent has failed but the substitution process is not finished. To prevent data loss, SIB collects these data in the storage and pass them to the substitute agent when it becomes ready for processing.

3. Internal Triple Operations Module. The module provides operations (subscription, update and query) for usage inside the SIB code base. These operations work like the similar operations provided for KPs but directly interact with responsible modules.

5.5   Substitution Mechanism Operation

From the SIB side the substitution process is event-driven. Specifically, the following events activate certain actions in SIB:

1. Registration/unregistration of the agent;
2. Agent connection failure/Explicit substitution request;
3. Substitute agent activation;
4. Primary agent return;
5. Substitute agent deactivation.

1. Registration/Unregistration of the Agent. Dataflow network agents start their operation by adding descriptive entities to the data storage. A descriptive entity includes an unique identifier, a type, indication whether the agent is ready to process data and additional properties depending on the type. The primary agent should also specify the program for substitution agent along with its type. The substitution agent description includes the type of supported substitution program and indication, which primary agent it substitutes (initially—none). When the agent adds the descriptive entity to the data storage, the substitution mechanism adds this agent to the corresponding agent list. When the agent finishes its operation it removes the descriptive entity from the data storage, thereby unregistering itself from the substitution mechanism.

2. Agent Connection Failure/Explicit Substitution Request. Failure of the agent connection is detected by the fact of TCP connection breakage. The agent may also explicitly request substitution to prevent this situation (by changing the value of the "Active" property in the descriptive entity to "no"). By detecting any of these situations the substitution mechanism starts the substitution procedure.

Firstly, the substitution mechanism starts to collect subscription notifications with the use of the notification storage module. Then it sets "no" to the "Active" property of the primary agent descriptive entity unless it was set before or removes descriptive entity of the substitute agent. Then it looks for an appropriate free substitute agent capable of executing programs of a particular type. Upon finding the required agent, the mechanism grants write permissions on output and state triples of the primary agent to the found agent. At the next step the substitution mechanism requests the latter agent to substitute the primary agent by setting the "Substitutes" property in substitute agent descriptive entity to the unique identifier of the primary agent. If the substitution mechanism did not find the appropriate substitute agent, it continues to collect subscription notifications.

3. Substitute Agent Activation. After receiving the substitution request from the SIB the substitute agent retrieves the substitution program from the descriptive entity of the primary agent, parses it, setups subscriptions and notifies the substitution mechanism that it is ready to process data by setting "yes" value to "Active" property of the own descriptive entity. When the mechanism detects this change it sends collected subscription notifications one-by-one and deactivates the notification store module.

4. Primary Agent Return. When primary agent restores connection with the SIB it searches for the descriptive entity. Upon finding it the agent assumes that it should follow the reconnection procedure, setups subscriptions and sets "yes" value to the "Active" property of the entity. The substitution mechanism detects this change and starts to collect subscription notifications of the active substitution agent. Then it requests the substitution agent to stop operation by setting "none" value to the "Substitutes" property in its descriptive entity. If there is no substitute agent present and the notification store module is active at the moment of property change, then the mechanism sends all collected notification to the returned primary agent and deactivates the module.

5. Substitute Agent Deactivation. After receiving the shutdown request from SIB the substitution agent removes all subscriptions from SIB and waits for data processing to finish if required. Then it notifies the substitution mechanism by assigning "no" value to the "Active" property in the own descriptive entity and starts to wait for the next substitution request. The substitution mechanism detects this change and revokes write privileges on output and state triples of the primary agent from the substitute agent. Then the mechanism sends collected subscription notifications to the primary agent one-by-one and deactivates the notification store module.

6 Related Work

There are several case studies of ubiquitous healthcare applications that use the Smart-M3 platform. The intelligent environment including the workout tracking application is described in [5]. The case study also incorporates services from other domains, i.e., the SuperTux game represents the game domain, the mood renderer provides audio playback capabilities and the telephony status observer tracks the state of the mobile phone. The main goal of the study is to show how to achieve interoperability of applications from different domains. For example, if the user conforms to a training schedule, the game application may grant additional lives therefore making the game easier. The paper does not provide any details on reliability of the system.

Description of the personalized health tracking system is given in [13]. The system consists of wearable and external medical devices that provide data on the health state of the patient; environment and location sensors; aggregation services providing refined data; preliminary data analysis system that can initiate alarm if health measurements exceed the limits predefined by the doctor. The main goal of the study is to show how to handle the heterogeneity of incoming information and propose a way to overcome interoperability issues of medical devices from different vendors. The authors propose the ontology that describes all components and define how components should manage common knowledge to achieve the common goal. The reliability of the system is not discussed, on the contrary it is presumed that all KPs connect to the SIB at configuration time and stay connected all the time.

References to another case studies related to the Smart-M3 platform itself can be found in [4].

As seen from the mentioned case studies, the designers and developers of prototypes using the Smart-M3 platform mostly concern with how their systems provide conceived services ignoring the reliability of the solutions, although some services are able to manage temporary agent disconnection due to fact that required information is stored inside the SIB. When the agent returns to the network, it can restore its context using accessible data and continue operation. For example, the intelligent museum service [10] and personalized blogging service [7] follow such an approach. The drawback of such technique is that during the absence of the agent, service does not provide required functionality to the end-users and some data can also be missed and leave unprocessed.

The only work that focuses on reliability and availability aspects of Smart-M3 based services is [2]. Authors present the architectural approach for the problem dividing agent in the smart space into producers and consumers of data. The smart space may contain several agents producing the similar data. When an agent requires a data resource, it makes a request to the management KP asking to provide a reference to the corresponding producing agent. When one of active data providers disconnects from the smart space, the management KP notifies concerned data consumers about the data source change if such change is available or when corresponding data provider connects to the smart space. By applying such a scheme the authors implicitly assume that KP operation depends only on resources provided by other KPs, because no one can guarantee that other data would be available in SIB. This is different comparing with our approach. Another difference is in agent disconnection scheme: The authors use the challenge-response scheme based on data modification in the smart space. If the testable agent does not modify the corresponding triple in a minute interval, it is considered disconnected from network. This scheme does not preserve data coming from producers to consumers in case of disconnection.

7   Conclusion and Future Proposals

In the paper we described our implementation of the substitution mechanism for mobile agents functioning as nodes of dataflow network on Smart-M3 platform. We discussed modifications in RedSib architecture made to support agent substitution and summarized the most notable implementation details crucial for Smart-M3 developers and users. Additionally, we provided the case study that shows how the mobile services can benefit from using the substitution mechanism.

The substitution mechanism has been implemented in the last released version of RedSib and available in the public Smart-M3 repository: `https://github.com/smart-m3/redsib`.

The future proposals would include implementation of prototype services based on dataflow networks and analysis of their reliability. We also plan to develop a library to provide a consistent high-level API for dataflow agent development.

Acknowledgements. The study was supported by The Ministry of education and science of Russia, project 14.B37.21.0876. The authors would like to thank Sergey Balandin for his valuable guidance and "Internet of Things and Smart Spaces" working group of Open Innovation Association FRUCT for the provided feedback.

References

1. Balandin, S., Waris, H.: Key properties in the development of smart spaces. In: Stephanidis, C. (ed.) UAHCI 2009, Part II. LNCS, vol. 5615, pp. 3–12. Springer, Heidelberg (2009)
2. Bhardwaj, S., Ozcelebi, T., Syed, A., Lukkien, J., Ozunlu, O.: Increasing reliability and availability in smart spaces: A novel architecture for resource and service management. IEEE Transactions on Consumer Electronics 58(3), 787–793 (2012)
3. Fernández-López, H., Afonso, J.A., Correia, J., Simões, R.: Hm4all: a vital signs monitoring system based in spatially distributed zigbee networks. In: Proceedings of 4th International Conference or Pervasive Computing Technologies for Healthcare (PervasiveHealth), pp. 1–4. IEEE (2010)
4. Honkola, J., Laine, H., Brown, R., Tyrkkö, O.: Smart-M3 information sharing platform. In: 2010 IEEE Symposium on Computers and Communications (ISCC), pp. 1041–1046. IEEE (2010)
5. Honkola, J., Laine, H., Brown, R., Oliver, I.: Cross-domain interoperability: A case study. In: Balandin, S., Moltchanov, D., Koucheryavy, Y. (eds.) ruSMART 2009. LNCS, vol. 5764, pp. 22–31. Springer, Heidelberg (2009)
6. Janneck, J.W., Miller, I.D., Parlour, D.B., Roquier, G., Wipliez, M., Raulet, M.: Synthesizing hardware from dataflow programs. Journal of Signal Processing Systems 63(2), 241–249 (2011)
7. Korzun, D.G., Galov, I.V., Balandin, S.I.: Proactive personalized mobile multiblogging service on Smart-M3. Journal of Computing and Information Technology 20(3), 175–182 (2012)
8. Lo, B., Thiemjarus, S., King, R., Yang, G.Z.: Body sensor network—a wireless sensor platform for pervasive healthcare monitoring. In: The 3rd International Conference on Pervasive Computing, vol. 13, pp. 77–80 (2005)
9. Morandi, F., Roffia, L., DElia, A., Vergari, F., Cinotti, T.S.: RedSib: a Smart-M3 semantic information broker implementation. In: Proceedings of the 12th Conference of Open Innovations Association FRUCT and Seminar on e-Tourism. Oulu, Finland, pp. 86–98. SUAI, St.-Petersburg (2012)
10. Smirnov, A., Shilov, N., Kashevnik, A.: Ontology-based mobile smart museums service: Approach for small & medium museums. In: AFIN 2012, The Fourth International Conference on Advances in Future Internet, pp. 48–54 (August 2012)
11. Vasilev, A., Paramonov, I., Balandin, S.: Mechanism for robust dataflow operation on smart spaces. In: Proceedings of the 12th Conference of Open Innovations Association FRUCT and Seminar on e-Travel, Oulu, Finland, November 5-9, pp. 154–164. SUAI, St.-Petersburg (2012)
12. Vasilev, A., Paramonov, I., Dashkova, E., Koucheryavy, Y., Balandin, S.I.: Mechanism for context-aware substitution of smart-m3 agents based on dataflow network model. In: Proceedings of the 3nd IEEE International Workshop on Smart Communication Protocols and Algorithms (ICC 2013 WS - SCPA), Budapest, Hungary (2013)
13. Vergari, F., Cinotti, T.S., D'Elia, A., Roffia, L., Zamagni, G., Lamberti, C.: An integrated framework to achieve interoperability in person-centric health management. International Journal of Telemedicine and Applications 2011 (2011)

A Framework for Interacting Smart Objects

Arnab Sinha and Paul Couderc

INRIA, Rennes-Bretagne Atlantique,
Campus Universitaire de Beaulieu,
35042, Rennes Cedex, France
{arnab.sinha,paul.couderc}@inria.fr
http://www.inria.fr/en/en/teams/aces

Abstract. In this paper, we propose a framework enabling physical objects used everyday to participate in smart interactions. These objects are RFID tagged containing self description of their properties. The paper describes how smart context aware services can be supported directly by a collection of smart objects. One of the main advantage of our approach is the smart objects being piggybacked with necessary information, optimized enough to make inferences locally, without dependence on external system support.

Keywords: internet of smart objects, collection, object property, RFID tag, NFC, context representation, pervasive, ubiquitous computing.

1 Introduction

Smart objects and smart environments are core concepts in pervasive computing. A common understanding of the internet of things is the ability of more and more usual objects to be connected or referenced in the internet (or in cloud services). Since their services are dependent on the network infrastructure, occurances of failures are probable due to unavailability. Protecting privacy is an increasing concern. This motivated us to think of ways that would enable pervasive applications to take these collective decisions without external support. The objects must be piggybacked with the necessary information enabling them to take part in spontaneous decision making processes locally and avoiding remote communication. We call them as "self-described" objects due to semantic properties being attached to them. The intention is to primarily support limited but focussed decisions for pervasive applications. This information can be put into the RFID tag of the objects. RFID tagging is a developing trend [15,9,5], which mostly consists of storing a reference to remote information. Our approach proposes a step forward, to put some extra bits of information that could help in taking basic spontaneous decisions. This also provides a utilization of its limited memory space, a constraint that would decrease in the forthcoming years. The availability of information locally, would make the system more scalable while reducing on the network usage. Lastly, there are concerns of security with centralized databases containing huge amount of personal information. The access

S. Balandin et al. (Eds.): NEW2AN/ruSMART 2013, LNCS 8121, pp. 72–83, 2013.

trends and patterns could have possibilities of being stored, profiled and misused resulting in breach of privacy.

Let's consider the ready to assemble furnitures like IKEA^{TM}. The stores sell the disassembled pieces of a furniture packaged together. It needs to be assembled by the users taking the help of an instruction manual provided along. We propose that the important pieces be self described with NFC tags. Using a reader, an user can scan tags individually for information about them. Additionally, they are notified with information about their adjoining pieces from the remaining unscanned set. A NFC-enabled mobile could serve the purpose as a portable reader with an application that can be reused for providing this service independently of the puchased set of furniture.

Waste management is another domain for the application of collective decisions. Self describing the waste items with their properties could make the management simpler. For example, information like its composition would help to perform better sorting. A plastic bottle dropped in a glass bin can contaminate its entire contents and should be detected.

Analysing the situations closely, we can make few observations. Every item contains the necessary information for its self description. Based on the collective information about all the items present locally, the system suggests or makes some decision depending on the domain in consideration. This would be referred as **inference** in the rest of the paper. Inferences made could be pro or against by nature. In the first example, the application suggested to look for adjoining pieces while a furniture is in the process of assembling. The domain requires inferences, pro in nature recommending for adjoining pieces. While for the waste domain, inferences are made for incompatibilities as the application alerts when a problematic item is being added to a bin.

2   Approach Principles

Centralized Approach

If this classical approach is used, the domain knowledge would be stored in a server or centralized in an external database. The objects of the ubiquitous application would require to refer this central knowledgebase each time for making inferences. Hence, there should a communication infrastructure in place to transmit the data back and forth. Present RFID tags containing reference is an example of such approach.

Distributed Approach

This approach stores the domain knowledge in pieces spread across devices in physical space. One of such devices to hold these pieces are passive proximity communication tags like RFID/NFC which have limited memory space. Their handheld readers might also have limited computation capabilities. So, efficient context representation is required for optimum performance to make collective inferences.

Our Approach

In this paper, we have proposed a framework to encode the knowledge for making inferences in such a resource-limited distributed environment. Sometimes it may need to have a trade-off between the partial knowledge possessed by each item and keep the rest available locally. Its most important aspect is that the inferences can be done locally without references to any centralized knowledge-base. In the next section, we would explain our framework to encode information for item interactions.

Our framework proposes an encoding method to self describe items of a domain in an efficient way. Inferences are made among a collection of such items present locally. If the information in the tags are stored in the proposed format, our generic algorithm (described later) can be reused without any modifications and irrespective of the domain.

The functioning of our framework is also comparable to the Internet of Smart Objects [4]. The physical objects are RFID tagged containing information to transform into smart objects. These smart objects collectively form an Internet of Things. The aggregation of the information contained in the local IoT provides interesting inferences and services. Our situation can be better called as Intranet of Things (InoT) as inferences could be made with the smart objects located locally without using any network for communication [12].

3   Inference Model

The examples described in section 1 are not exhaustive. There are many other real world domains which could work on similar principles have been discussed later. This section, discusses the general model about how a domain knowledge be interpreted in terms of its properties. This knowledge is also encoded into the smart objects so that inferences could be made locally.

3.1   Defining Domain Properties for Inferring

Let's consider a sample domain D to explain our general model to infer the interactions. We will refer to this example throughout the paper. Suppose D consists of 10 marked important properties $\{p_1, p_2, ... p_9, p_{10}\}$ and some of these properties are incompatible to each other. Graph G represents the knowledge graphically as in figure 1. Each of the properties in G is represented as a vertex and an edge is drawn between two properties if there exists an incompatibility between them. G represents the global knowledge of all the properties of our sample domain. Our objective would be to distribute it in a way such that the entire domain knowledge is not required for making inferences. Subsequently, we have proposed a format for encoding the distributed knowledge for items having flexibility inferring simple to complex interactions. Making such inferences would be the objective for a collection of items.

From figure 1, we tabulate each property separately along with its incompatible ones as in table 1. The idea is similar to creating an adjacency matrix

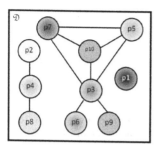

Fig. 1. Incompatible properties of the sample Domain

Table 1. Incompatibility of properties

Property	Incompatible with
p_1	-
p_2	p_4
p_3	$p_7, p_5, p_6, p_9, p_{10}$
p_4	p_2, p_8
p_5	p_7, p_3, p_{10}
p_6	p_3
p_7	p_5, p_3, p_{10}
p_8	p_4
p_9	p_3
p_{10}	p_3, p_5, p_7

from a graph. But their finer difference lies in the format for representing the information.

Initially we require to have a knowledgebase containing the interaction of the properties of the domain. Using it, the properties of the items are self described using RFID tags. A tag would contain the description string in the format {*properties : incompatible_properties*}. It consists of two fields with a colon as separator between them. For example, if an item I of D possesses property p_3, its tag would contain the information in the format:

$$p_3 : p_5, p_6, p_7, p_9, p_{10}$$

It is basically the third row of table 1. An item can have multiple properties. For such a case we perform an union of the individual fields. An item having the properties p_4 and p_6 would be self described as:

$$p_4, p_6 : p_2, p_8, p_3$$

An alternative solution for the above would be affixing multiple tags i.e. one tag each for the properties p_4 and p_6 written in the same format. If an item has too many properties to hold in the memory of one RFID tag, this is the easiest and cost effective method.

Until now we have suggested an encoding method to describe the items with their properties in the tag. Given a set of such items locally, our objective is making inferences, pro or against depending on the domain. Algorithm 1 provides an outline to perform this task. It is executed once each time an item I_i is detected by RFID reader. It is assumed they are self described in the same format as above. It could be used for inferring incompatibilities between the newly introduced item and the items present locally on pairwise basis. The reader reads the two fields of the tag in two different arrays as $Prop_i[]$ and $Iprop_i[]$. Additionally, it contains a set $S\{\}$ which is initialized to $\varnothing$ when the algorithm is executed for the first time. It caches the local context by incrementally storing the list of compatible item's properties, when added.

Referring back to the example in the beginning of this section, suppose we have four items I_1, I_2, I_3 and I_4 having properties p_2, p_1, p_3 and p_9 respectively.

Algorithm 1: Inferring incompatibility of items

Input: item I_i detected by the RFID reader
Output: infer if item I_i is compatible with existing set of items
Initialize: Set S ← ∅
while *item I_i detected by the RFID reader* **do**
    set flag to TRUE i.e. no incompatibilities inferred for item I_i
    Read the tag and store it's two fields in $Prop_i[]$ and $Iprop_i[]$
    for *each element Ip in $Iprop_i[]$* **do**
        if *$Ip \in S$* **then**
            set flag to FALSE i.e. Ip an incompatible property to item I_i, is
            incompatible with one or more existing set of items
    if *flag is TRUE* **then**
        for *each element p in $Prop_i[]$* **do**
            $S \leftarrow S \bigcup p$ i.e. adding the properties of item I_i to set S
    return flag

The interesting observation is that inferences would depend on the order of items added and the existing collection present locally. From the graph theoretic point of view, it can be stated that the algorithm prevents forming any subgraph of the graph G in figure 1.

3.2 Inferring Incompatibility in Groups

Until now, we have discussed inferences among pairs of properties. There may be some application domains where inferences involve groups of properties that are present together locally. For example in figure 1, properties p_3, p_5, p_7 and p_{10} can be considered incompatible as a group. The important aspect of being in group is that incompatibility is only ensured with the presence of every property of the group. Such groups are represented as cliques in a graph. A clique of a graph G is a complete subgraph of G. In the graph G of our example, the group of properties p_3, p_5, p_7 and p_{10} forms a clique. This information has to be distributed in a way such that it could be efficiently encoded in the RFID tag to self describe the item. It would be very inefficient to represent such a thing, pairwise. Hence, for the purpose we construct a string in the following format:

$$f_0{:}f_1{:}f_2{:}f_3{:} \cdots {:}f_n$$

The string consists of n fields with colon as separators with each containing a set of properties. The first field f_0 describes the properties of the item. This is similar to the previous example. The second field f_1 contains all individual pairs of incompatible properties. The groups of incompatible properties are encoded in the fields from f_2 to f_n. Provision for multiple fields would add the flexibility of having a property associated with more than one disjoint groups. From the example, an item can self describe itself having property p_3 as:

$$p_3 : p_6, p_9 : p_5, p_7, p_{10}$$

Summarizing the information for all the properties as it would be encoded is presented in table 2.

Table 2. Incompatibility of properties

Property f_0	Incompatible with $f_1:f_2:f_3: \dots :f_n$
p_1	-
p_2	p_4
p_3	$p_6, p_9 : p_5, p_7, p_{10}$
p_4	p_2, p_8
p_5	$-: p_3,\ p_7, p_{10}$
p_6	p_3
p_7	$-: p_3, p_5, p_{10}$
p_8	$p_4 : -$
p_9	$p_3 : -$
p_{10}	$-: p_3, p_5, p_7$

To accomodate this additional feature about groups of incompatible properties, algorithm 2 outlines a procedure to make such inferences. It has some additions to algorithm 1. It ensures that none of the incompatible pairs of properties as well as groups doesn't occur locally. Taking example from the graph in figure 1, the algorithm infers incompatibility when some of the properties among p_2, p_3, p_4, p_6, p_8 and p_9 are added locally and tries to form a subgraph. These are the properties that form incompatible pairs in the graph. Additionally, we have the properties p_3, p_5, p_7 and p_{10} forming a clique as described earlier. So collectively the situation would inferred safe until all but one of these properties exist locally. The algorithm combines both of the above to make inferences. An exception would be property p_1 which is disconnected from the rest of the graph. So it is always inferred safe regardless of other properties present locally.

4   A Smart Tool Box Application

The **Smart Tool Box** is an example of ubiquitous application which helps to maintain safety standards at sensitive sites like aircrafts [8]. Let's consider a workshop has a depot for tools which issues them to its workers in a toolbox. Each type of tool in the depot is assigned an unique identification. Every tool is NFC tagged which stores its type. When a worker requests for some tools, they are lent out in a tool box. Henceforth the box should always contain all the tools grouped together and warn the user if one or more tools are missing until returned back to the depot. This group information is written onto every tag before handing out the kit to the worker.

Algorithm 2: Inferring incompatibility for group of items

Input: item I_i detected by the RFID reader
Output: infer if item I_i is compatible with existing set of items locally
Initialize: Set S ← ∅
while *item I_i detected by the RFID reader* **do**

    set Flag to TRUE i.e. no incompatibilities inferred for item I_i
    Read field f_1 from the tag
    for *each property p in field f_1* **do**
        if $p \in S$ **then**
            set Flag to FALSE i.e. the property p is present locally hence I_i is incompatible

    for *each field f_i from f_2 to f_n* **do**
        set GroupFlag to FALSE i.e. assuming the all the incompatible properties in f_i are present locally
        for *each property Ip in f_i* **do**
            if $Ip \notin S$ **then**
                set GroupFlag to TRUE i.e. Ip an incompatible property in field f_i is not present locally

        if *GroupFlag is FALSE* **then**
            set Flag to FALSE i.e. all the properties in field f_i are present locally hence I_i is incompatible

    if *Flag is TRUE* **then**
        for *each element p in $Prop_i[]$* **do**
            $S \leftarrow S \bigcup p$ i.e. adding the properties of item I_i to set S

    return Flag

In our framework in section 3, an approach is proposed for self describing items with properties to make inferences. In the present context, the domain consists of tools used for maintainance. In the following subsections, we have illustrated how the domain properties could be represented efficiently and the framework used to perform grouping of tools at the depot, on request. In 4.2, we describe how our objective is achieved using a simple android application. Its effectiveness is discussed in 4.3.

4.1   Numeric Representation of Tools Domain

In our implementation, we have used numeric representation of properties written on their NFC tag in the format described in section 3.2. So each type of tool performing distinct function in the depot is assigned a unique natural number. Suppose the depot lends out specific type of toolbox is assigned number 2 along with tool types numbered 14, 5, 6, 26 and 9 as in figure 4.2. The self description string for the group written onto the toolbox in our proposed format would be

12:0:21405062609

Every field of the string description begins with a number specifying the fixed length of the properties. Numbers are padded with '0' on the left to make them all of equal length. According to the proposed format, field f_1 should contain the types having pairwise association with the toolbox. Hence the field f_1 contains 0 which represents properties of 0 length i.e. another way of saying that no property exists for the field.

4.2   The Application

We have developed a small android application to validate and verify the working of our current example. There is a subtle difference between the examples discussed presently with that in section 3. The group of tools in the toolbox strive to remain together throughout and our application tries to verify this situation which we referred to as an inference pro by nature in section 1. Our framework could be utilized for the current scenario but with a minor change. In our framework described in section 3, items are self described with it's own properties and the other's to which it would be incompatible. In this context, the idea is reversed for groups and algorithm 2 is modified accordingly. Figure 6 demonstrates the various phases of the application.

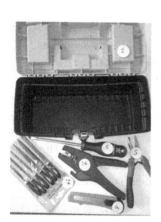

Fig. 3. Integrity Verification of the Tool Box

Fig. 2. Initial grouping of Tool Box and contents

Fig. 4. Acknowledgement on adding tool numbered 26

Fig. 5. Notification for tool(s) yet missing

Fig. 6. Smart Tool Box Application

4.3   Validation

Given the nature of our application, we have a validation approach with an use case implementation of working prototype. In this section, we have discussed the performance of the approach.

Coding Efficiency. The length of each field in the self description would depend on two factors. The maximum length among the group of numbers in the particular field are chosen as fixed length. So each number having smaller lengths are padded on their left adding to the overall length of the field. However if the maximum length among the numbers is less, then the encoded string automatically gets reduced.

Suppose we have the following sets:

A domain $D = \{x \epsilon N\}$ *and* $|D| = n$, where n is the number of properties represented.

Any field in the self-description of an item could be represented as set $F \subset D$ ...(i)

Let's define the following functions for some operations:

$|X|$ gives the cardinality or size of *set X*.

$max(X)$ gives the largest element of *set X*.

$len(element\ e)$ gives the character count of element e.

We can say that,

$max(F) \leq max(D)$

or, $len(max(F)) \leq len(max(D))$ [This gives the character count of highest elements of both sets] ...(ii)

Now, if we represent the field in (i) using the fixed length, we can write

$|F| * len(max(F)) \leq |F| * len(max(D))$

Hence we can make three observations for the encoded length of the field which is the L.H.S of the equation:

1. It depends on the number of properties grouped together represented as $|F|$.
2. The length of the largest number also contributes to the total length.
3. The maximum length is bounded upto the R.H.S of the equation.

Figure 7, provides an estimate on the upper bound of the characters required for encoding groups of properties in a field depending on the domain size. The maximum length of the properties for the domain are considered in the graph and hence the lines outline the upper bound on the total length of the field which can never be exceeded. However there may be circumstances when the total length is reduced if the group of numbers in the field comprises of smaller numbers.

For domains having average number of properties, they can be represented by the natural numbers until 99 having 2 characters as fixed length in the fields. For bigger domains with more properties, larger numbers can be used. But using natural numbers upto 999 would probably be sufficient for most domains as it has large interval of numbers. Also the encoding length is guaranteed to remain within the upper bound for the entire interval given the number of properties for a group.

The entire encoded string data to self describe an item contains numbers and the special character ':'. When the data is written onto RFID tags, we replace the colon with character 'a' and use 'b' as the data delimiter. Hence the string is transformed to hexadecimal and each of it's character consumes four bits of

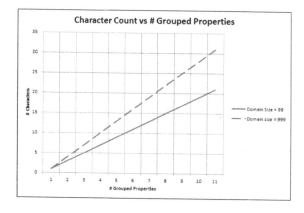

Fig. 7. Encoding performance

memory. Hence using numeric properties gives us an advantage whereas in some cases it could even take upto two bytes for a character [3].

Advantages : Deployment, Scalability and Privacy. The entire setup is very easy to build. It is effective in terms of cost and time. Initially, it would require identification of the properties of the domain i.e. different type of tools and assigning them unique identifier using cheap NFC tags. The interoperability of the tools among multiple deployments, spatially, is possible as long as their properties have the same significance. The grouping could be done later on-demand, based on the requirement of a worker. Reading and writing of the NFC tags are performed with applications using an android phone. When a large number of tools are grouped together, scalability factor comes into play in two different ways. Firstly, the memory space of a tag could be too limited for its self-description. This is resolved by sticking multiple tags instead of one. Secondly, the information is available locally for the entire setup. So, the absence of a network reduces the communication time and the only time taken is for the read/write of tags, performed one at a time. Also, this addresses the concern for privacy as the entire information is available locally instead of a server. When information are fetched from servers, the access patterns are sometimes recorded for profiling which is not possible in our framework. Last but not the least, this approach is simple, energy efficient due to use of passive tags and have moderate user interaction compared to the approach in [2].

5   Related Work

We have proposed a framework in the context of ubiquitous computing research supporting different application domains where smart physical objects are adaptive to decisions based on their collection of objects present locally. There are some similar work that make inferences dependent on collectivity [8,4,1,13,14].

[11] demonstrates sorting by a smart bin that accepts or rejects self described waste items containing its percentage composition. Our framework can be used to provide an alternative approach or additional feature to accumulate compatible waste to be recycled together using their self-description.

Antifakos et al. describes in [2], how user manual is avoided by most people and proposed an proactive approach using sensors for perceiving the actions and then provide guidance while assembling the pieces. In contrast, our infrastructure is lightweight and economical. It can be used with a simple NFC-enabled smartphone having the application installed.

Context representation is the language to express collective situation and knowledge. The authors in [6] mentions semantic based ontologies among the many in their extensive survey whereas [7] have demonstrated efficient encoding of semantic data using prime numbers for resource constraint devices which are commonly used in ubicomp solutions.

6   Conclusion

In this paper, we have suggested a framework to represent context for a collection of physical objects. The required knowledge is optimally distributed to contain with the objects which make them smart. We have also discussed how it could be utilized for various ubiquitous applications. Finally, we would like to highlight some of the advantages of our proposed framework:

Privacy and security are important concerns in ubiquitous computing especially in relation to cloud services. The self describing objects do not involve any sensitive information about the items or personal information about the users. Additionally, using numeric properties appear as obscure data to an intruder reading the tag. So, encryption is not necessary.

As stated earlier, our objective for the framework is context representation for inferring activities involving a collection of items. This has been clearly achieved in a decentralized manner by self describing the domain items. In this way we achieve to make inferences locally without any references to external knowledgebase or requiring communication infrastructure. An alternative could be storing the domain knowledge as a key in the local embedded system. But in case of changes, updating the tag contents at source is easier and it would be effected automatically.

Perspectives to this work would include supporting more applications such as smart drugs [8,10]. We also intend to seek for alternative to using numeric properties and better encoding approaches for compact representation.

References

1. Cavallaro, L., Di Nitto, E., Furia, C., Pradella, M.: A tile-based approach for self-assembling service compositions. In: Proceedings of the 15th IEEE International Conference on Engineering of Complex Computer Systems. Citeseer (2010)

2. Antifakos, S., Michahelles, F., Schiele, B.: Proactive instructions for furniture assembly. In: Borriello, G., Holmquist, L.E. (eds.) UbiComp 2002. LNCS, vol. 2498, pp. 351–360. Springer, Heidelberg (2002)

3. Glouche, Y., Couderc, P.: An autonomous traceability mechanism for a group of rfid tags. In: Proceedings of the 6th International Conference on Mobile Ubiquitous Computing, Systems, Services and Technologies (UBICOMM 2012) (2012)

4. Kortuem, G., Kawsar, F., Sundramoorthy, V., Fitton, D.: Smart objects as building blocks for the internet of things. IEEE Internet Computing 14(1), 44–51 (2010)

5. Ostman, H.: Item-level tagging in the grocery industry - are we there yet (April 2013), http://www.rfidarena.com/2013/4/2/item-level-tagging-in-the-grocery-industry-are-we-there-yet.aspx (accessed June 23, 2013)

6. Perttunen, M., Riekki, J., Lassila, O.: Context representation and reasoning in pervasive computing: a review. International Journal of Multimedia and Ubiquitous Engineering, 1–28 (October 2009)

7. Preuveneers, D., Berbers, Y.: Encoding semantic awareness in resource-constrained devices. IEEE Intelligent Systems 23(2), 26–33 (2008)

8. Römer, K., Schoch, T., Mattern, F., Dübendorfer, T.: Smart identification frameworks for ubiquitous computing applications. Wireless Networks, 689–700 (2003)

9. Saarijarvi, M.: Rfid: A world of opportunity - the ingenious use of rfid worldwide (June 2012), http://www.rfidarena.com/2012/6/12/rfid-a-world-of-opportunity-the-ingenious-use-of-rfid-worldwide.aspx (accessed June 23, 2013)

10. Siegemund, F., Flörkemeier, C.: Interaction in pervasive computing settings using bluetooth-enabled active tags and passive rfid technology together with mobile phones. In: Proceedings of the First IEEE International Conference on Pervasive Computing and Communications, PERCOM 2003, pp. 378–387. IEEE Computer Society, Washington, DC (2003)

11. Sinha, A., Couderc, P.: Using owl ontologies for selective waste sorting and recycling. In: OWLED (2012)

12. Sinha, A., Couderc, P.: Smart bin for incompatible waste items. In: Proceedings of the the 9th International Conference on Autonomic and Autonomous Systems (ICAS 2013) (2013)

13. Strohbach, M., Kortuem, G., Gellersen, H.: Cooperative artefacts - a framework for embedding knowledge in real world objects. In: Smart Object Systems Workshop at UbiComp (2005)

14. Strohbach, M., Gellersen, H., Kortuem, G., Kray, C.: Cooperative artefacts: Assessing real world situations with embedded technology. In: Mynatt, E.D., Siio, I. (eds.) UbiComp 2004. LNCS, vol. 3205, pp. 250–267. Springer, Heidelberg (2004)

15. Swedberg, C.: Survey shows half of u.s. retailers have already adopted item-level rfid (January 2012), http://www.rfidjournal.com/articles/view?9168 (accessed June 23, 2013)

A Subgraph Isomorphism Based Approach to Enable Discovery and Composition of Smart Space Elements

Oscar Rodríguez Rocha[1], Cristhian Figueroa[1,2], and Boris Moltchanov[3]

[1] Politecnico di Torino, Corso Duca Degli Abruzzi 24, 10129, Turin, Italy
{oscar.rodriguezrocha,cristhian.figueroa}@polito.it
[2] Universidad del Cauca, Calle 5 No. 4 - 70, Popayán, Colombia
[3] Telecom Italia, via G. Reiss Romoli, 274, 10148. Turin, Italy
boris.moltchanov@telecomitalia.it

Abstract. Nowadays, the variety of mobile services is growing together with the number of customers and the heterogeneity of their mobile devices, thus, monitoring the users of a mobile network has become a challenge. Considering *Smart Space Governing* to address this issue, in which the mobile network is considered an Smart Space (due to its size and complexity) and their elements (mobile devices and network monitoring services), it is possible to create rules to monitor the output of these elements. We call those rules as *Smart Space Compositions* and can be created through the platform's *Visual Editor*, that during the graphical creation process, provides the user with a list of similar existing compositions that can be reused at any time to improve the composition process. This paper describes the implementation by a Telecommunications Operator of this composition module supported by subgraph isomorphism techniques.

Keywords: Smart Spaces, Service Creation, Network Management.

1 Introduction

With the increasing variety of mobile services, the growing number of customers, and the heterogeneity of mobile devices, Telecommunications Operators constantly face a big challenge[1]: to monitor the behaviors of their mobile users in order to provide them with better mobile services. Taking as a basis, the concept of *Smart Space Governing* [2], the operator's mobile network is considered as an Smart Space (due to its size and complexity) and their elements (mobile devices and network monitoring services) as what we will refer to *Services* in this paper. The system allows the creation of Smart Space Compositions (through the platform's *Visual Editor*) that represent monitoring rules driven by the outputs of the *Services*. Finally, the platform can execute each scheduled Smart Space Composition on the *Execution Engine*. In this way, specific information from network mobile users can be constantly monitored during defined intervals. The platform has two repositories, one for storing the *Services* representing Smart

S. Balandin et al. (Eds.): NEW2AN/ruSMART 2013, LNCS 8121, pp. 84–93, 2013.
© Springer-Verlag Berlin Heidelberg 2013

Space Elements and the other one for storing Smart Space Compositions. Both enable the operation of the *Visual Editor* which proposes similar existing compositions to the user in real-time during the creation process. This paper focuses on describing the practical implementation of the composition process (Composer module), made by a Telecommunications Operator, supported by the use of subgraph isomorphism techniques.

The rest of the paper is organized as follows: Section 2 introduces the concept of Smart Space Governing. Section 3 presents the Composer of smart spaces elements and describes its layers. Section 4 shows the experimental results. Finally, Section 5 presents our conclusions and future work.

This work is an extension of the European project 4CaaSt [3]. Some variations have been introduced to the discovery mechanism, the main *Repository* and the *Visual Editor* module.

2 Smart Space Governing

The dynamism of todays mobile systems, plus the advance of hardware and software systems has introduced new advantages to Telecom Operators to create mobile services to target a wide variety and diversity of mobile users. However, this dynamic and heterogeneous environment that evolves the mobile systems has also increased the complexity and costs to properly managing and monitoring the process of creation and delivery of mobile services. To deal with this problem: first, software systems can be modeled as Smart Spaces by creating a virtual environment with features of multi-user, multi-device and dynamic integration that enhances a physical space by virtual services; and second those Smart Spaces can be properly managed and monitored by allowing the creation of compositions that represent rules driven by the outputs of the Smart Space Elements. This process of monitoring and manage a mobile system is called Smart Space Governing.

2.1 Smart Space Element as a Service

In this paper, we refer (for ease of study) to the Smart Space Elements (mobile devices and monitoring services of the network) as "Services". In this way, a *Service* contains a set of inputs and outputs and a label with the description of its functionality, thus, they can be compared not only by keywords relating to their labels, but also by taking into account its interfaces.

3 Composer

The composer module of the platform has been designed to assist users to create compositions of services in an Smart Space. It is divided in different layers:

3.1 Visual Layer

This layer contains only the *Visual Editor*, which is responsible for providing the interaction with the end-user.

Visual Editor: It allows to create service compositions by dragging and dropping components to the provided canvas (work area). Those components can be services (as stated before, Smart Space elements) or connectors (XOR-Split, AND-Split, AND-Join, OR-Join) also known as control-flow operators. In particular, services are stored in a service repository and can be retrieved trough keywords search (matching the keywords with the name of the services), or by specifying a set of interfaces (inputs/outputs) needed; and connectors are a fixed list available to the user.

After a service is dropped into the canvas, the system automatically generates and shows graphically to the user, a ranking list with the existing compositions stored in the repository containing such service in their structure; so the user can either click on a list item to import the existing composition into the canvas or continue with the manual composition. If user inserts a second service, the system refines again the ranking, reducing the list only to those compositions that accomplish the new rule of having the two components, and so on. We call this *Automatic Composition Discovery System*. In the case the user adds a connector, the discovery system executes a sub-graph isomorphism algorithm which helps to refine the ranking by structurally comparing the new user-generated composition with each of the compositions stored in the repository. Therefore, the ranking is refined again obtaining only a list of stored compositions with a higher level of similarity with the user-generated composition, so the user can select one existing composition before to create a complete new one.

3.2   Discovery Layer

Service Similarity Analyzer: This module receives as input query a service (s_q) and compares it with each of the services (s_{ri}) stored in the Services Repository SR. As result, this module returns a service distance (SD) which is computed by comparing the labels and the interfaces (inputs and outputs) of the services. The service distance is later used to generate the services ranking, and also to help in the compositions ranking. This module has two submodules the labels comparator and the interfaces comparator:

- **Labels Comparator:** It evaluates the distance between the labels of the services by comparing combinations of words and abbreviations. In this proposal, we have used the *NGram, check synonym,* and *check abbreviation* algorithms. *NGram* estimates the distance according to a number of common sequences of defined-length characters $(q - grams)$ between the tasks names; *check abbreviation* uses a custom abbreviation dictionary, and *check synonym* finds synonyms through a lexical database called *WordNet* [4]. In this way a "labels distance" (LD) [5] is computed taking into account the results of *NGram* (m_1), *check synonym* (m_2) and *check abbreviation* (m_3) algorithms.

$$LD = \begin{cases} 1 & \text{if } m_1 = 1 \vee m_2 = 1 \vee m_3 = 1 \\ m_2 & \text{if } 0 < m_2 < 1 \wedge m_1 = m_3 = 0 \\ 0 & \text{if } m_1 = m_2 = m_3 = 0 \\ \frac{m_1 + m_2 + m_3}{3} & \text{if } m_1, m_2, m_3 \in (0, 1) \end{cases} \tag{1}$$

– **Interfaces Comparator:** it calculates the similarity of two services according the correspondences between their interfaces (inputs and outputs). In this case, given a target service S_t it finds combinations of interfaces which "best cover" the interfaces of a query service S_q. In other words, the aim of this module is to determine S_t from inputs and outputs of a query service S_q, such that S_t shares the larger possible number of outputs with S_q and without exceeding its inputs [6].

Initially, the labels of the inputs are compared and organized in a similarity matrix where rows are the inputs of the S_q and columns are the inputs of the S_q. This matrix is then analyzed to determine the maximum similarities between pairs of inputs of the two compared services, which scored the greater similarity values among all the possible combinations of pairs of inputs guarantying that each input of S_q is matched with only one input in S_t and vice versa. In this way if $M_{m \times n}$ is defined as the similarity matrix of the inputs with a dimension $m \times n$ (m is the number of inputs of S_q, and n the inputs of S_t) and the set of maximum similarities determined as the grouping of similarities Sim_{ij} that meet the following rule:

$$Sim_{ij} : (\forall Sim_{kj}; k = (0, 1, 2, ..., m) \Rightarrow Sim_{ij} > Sim_{kj})$$
$$\wedge (\forall Sim_{il}; l = (0, 1, 2, ..., n) \Rightarrow Sim_{ij} > S_{il}) \tag{2}$$

Once the set of maximum similarities is defined, the isolated inputs (missed inputs) between S_q and S_t are established when the total number m of inputs from S_q exceed the number m of inputs from S_t. Taking into account those parameters and also the rule previously named, the similarity between interfaces (Sim_{IO}) can be computed by next equation:

$$Sim_{IO} = \frac{\sum Sim_{ij}}{min(m, n) + \beta * |m - n|} \tag{3}$$

Where Sim_{ij} is each value of similarity; $min(m, n)$ is the maximum number of input pairs established between the two compared tasks according the properties perviously named; and $\beta * |m - n|$ is a parameter that can determine the value of the similarity when there are missed inputs, and takes the value 0.8 when $m > n$ and 1 otherwise. The algorithm to compare interfaces is described next:

Algorithm 1. Algorithm for computing the similarity of interfaces

---

Require: S_q, S_t
Ensure: Sim_{IO}
1: $I_q[m] = getInputs(S_q)$
2: $m = cardinality(I_q)$
3: $I_t[n] = getInputs(S_t)$
4: $n = cardinality(I_t)$
5: $L_q[m] = getInputLabels(S_q)$
6: $L_t[m] = getInputLabels(S_t)$
7: **if** $L_q[m] = \emptyset \wedge L_t[n] = \emptyset$ **then**
8:     **return** $Sim_{IO} = 1$
9: **else if** $L_q[m] = \emptyset \vee L_t[n] = \emptyset$ **then**
10:     **return** $Sim_{IO} = 0$
11: **else**
12:     $ISM[m][n] = getInputSimilarityMatrix(IE_q[m], IE_t[n])$
13:     $MSA[] = getOutputSimilarityMatrix(ISM[m][n])$
14:     $MSS = addMaxSimArray(MSA[l])$
15:     **if** $m > n$ **then**
16:         **return** $Sim_{IO} = \frac{MSS}{n+(0.8*|m-n|)}$
17:     **else**
18:         **return** $Sim_{IO} = \frac{MSS}{m+|m-n|}$
19:     **end if**
20: **end if**

This algorithm begins by extracting the inputs of S_q, S_t and adding them to the array $I_q[]$, $I_t[]$ respectively. Then it calculates the number m of inputs from S_q, and n of inputs from S_t. Then, the algorithm obtain the labels of the interfaces for S_q and S_t and adds them to the arrays $L_q[m]$ and $L_t[n]$ respectively. Next, the similarity matrix of inputs $ISM[m][n]$ is obtained based on the comparison of the arrays of labels ($L_q[m]$ and $L_t[n]$) and the maximum similarities ($MSA[]$) are calculated. Finally, semantic similarity is calculated by the equation [7]. The similarity for outputs follows the same steps as the inputs.

Finally, the total service distance SD is calculated as:

$$SD = \alpha LD + \gamma(1 - Sim_{IO}) \qquad (4)$$

Where α and γ are values that can be set to give more relevance to the interfaces or labels of the compared services.

Composition Similarity Analyzer. This module is intended for comparing the similarity between each graph of the repository ($G_r \in R$) and a query graph (G_q), this problem is commonly known as graph isomorphism. Nowadays the faster known algorithm to address the isomorphism problem is the $VF2$ [7] which is based in matching two graphs G_1 and G_2 by determining a mapping function M to associate the nodes of the two graphs taking into account some restrictions.

M is expressed as the set of pairs (v, w) that represents the mapping of a node $v \in G_1$ with a node $w \in G_2$. Therefore a mapping M is a graph isomorphism iff M is a bijective function that preserves the structure of the branches of the two graphs, and it is a sub-graph isomorphism if M is a isomorphism between G_2 and a sub-graph of G_1. Next the original VF2 algorithm is presented:

Algorithm 2. VF2 Algorithm

Require: An intermediate state s, an initial state s_0 such that $M(s_0) = 0$
1: **if** $M(s)$ covers all the nodes of G_2 **then**
2:     **return** $M(s)$
3: **else**
4:     Compute the set $P(s)$ of the pair candidate for inclusion in $M(s)$
5:     **for all** $(v, w) \in P(s)$ **do**
6:         **if** factibility rules are correct in order to (v, w) be included in $M(s)$ **then**
7:             Compute the state s' obtained by adding (v, w) to $M(s)$
8:             Match (s') (execute this algorithm for s' as input intermediate state)
9:         **end if**
10:     **end for**
11:     Restore data structures
12: **end if**

The composition similarity analyzer module uses the VF2 algorithm in order to establish if a G_q is a sub-graph of one of the $G_r \in R$. In this way at the end of the execution of the VF2 we have a set of r graphs form the repository where the G_q is contained, i.e. a set of G_r such $G_q \subseteq G_r$. However, VF2 algorithm can only ensure that G_q is an exact subgraph isomorphism of each of the r graphs of the R without taking into account that each node of the graphs has a label and a set of interfaces (inputs and outputs), therefore we have adapted the VF2 sub-graph isomorphism in order to compare this kind of graphs by analyzing the similarity of labels and interfaces for each node (service) included in the graph of the service composition (algorithm 3):

Algorithm 3 receives a query graph G_q and compares it with each of the graphs G_{ri} stored in the repository R in order to obtain a ranked list ($Ranking$) of G_{ri} where G_q is a sub-graph, not only based on the structural sub-graph isomorphism, detected by VF2 algorithm, but also taking into account the similarity between each service that belongs to the compositions.

The algorithm begins by determining a set R' of the graphs G_{ri} stored in the repository R where G_q is a sub-graph by using the VF2 algorithm. Next each service s_q of G_q is compared with the correspondent service s_r of the sub-graph of G_{ri} as indicated by the mapping function $M(G_q, Gri)$. To do that, we have used the service comparison algorithm described before, which returns a service distance SD calculated as function of the labels and interfaces (inputs/outputs) comparison. Then, each service distance is added to the total composition distance CD, and additionally the connectors are compared according their type

---

Algorithm 3. Composition Ranking Algorithm

Require: $G_q, G_{ri} \in R$
Ensure: *Ranking* of G_{ri} where G_q is a sub-graph
1: **for all** G_{ri} **do**
2:    **if** G_q is a sub-graph of G_ri **then**
3:       Add G_{ri} to R' (R' is the set of graphs where G_q is a sub-graph)
4:       Calculate $M(G_q, G_{ri})$ (M is the mapping between each node of G_q and a sub-graph of G_{ri})
5:    **end if**
6: **end for**
7: **for all** $G_{ri'} \in R'$ **do**
8:    **for all** pair of services $(s_q, s_{ri}) \in M(G_q, G_{ri'})$ **do**
9:       $SD = compareServices(s_q, s_{ri})$ (SD is the service distance based on labels and interfaces comparisons. Algorithm 2)
10:       Add SD to CD (CD is the total composition distance)
11:    **end for**
12:    **for all** pair of connectors $(c_q, c_{ri}) \in M(G_q, G_{ri'})$ **do**
13:       get the type of the compared connectors $type(c)$ (AND, OR, XOR (Split, Join))
14:       **if** $type(c_q) \neq type(c_{ri})$ **then**
15:          $ConD = 1$
16:          Add $ConD$ to CD ($ConD$ is the connector distance when the type of connectors is different)
17:       **end if**
18:    **end for**
19:    Order $G_{ri'}$ according CD and add it to *Ranking*
20: **end for**
21: **return** *Ranking*

(i.e. AND, OR, XOR (Join, Split)). Finally the graphs $G_{ri'}$ are ranked according their CD with respect to the G_q.

3.3   Storage Layer

Services and compositions are stored in this layer trough two repositories: one for services (inside the Service Similarity Analyzer) and another one for compositions (in the Composition Similarity Analyzer). It also provides the means for automatically generate similarity rankings of existing compositions according the elements selected by the user on the graphical interface. This module is a reduced version of the original one (4Caast marketplace [8] as it can achieve faster speeds when responding to the component suggestion queries made by *Visual Editor*.

This contains services and compositions. A service $S = (L, O, l)$ in our repository is represented by a labeled node having a set I of inputs, a set O of outputs, and a label L which is a string to name the service. A composition is represented by a graph $G = (S, C, E)$ where E is the set of edges, S the set of nodes, and C a set of connectors. The nodes $s \in S$ embody the services of the composition, the

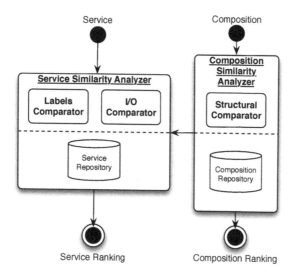

Fig. 1. Storage Layer

edges $e \in E$ are links between the services of the composition, and the connectors $c \in C$ are logical gates (AND (split -join), OR (split - join)) that implement the behavior of the composition. Figure 1 shows the basic architecture of the storage layer which is composed by two main modules, the services comparator and the composition comparator in order to help the user in the service composition.

4   Experimental Results

The experimentation is based on the performance evaluation of the algorithm for ranking compositions used in the automatic composition system. To do that, a repository with 100 graphs of compositions was created and a set of 6 queries from that repository were selected in order to evaluate the time required by the algorithm to rank the most similar compositions to each query. Graphs evaluated in the repository contain 5 to 100 nodes, and queries contain 5, 10, 20, 30, 50 nodes respectively.

Figure 2 shows the performance results of the median time required for the algorithm to rank the compositions stored in the repository according each one of the queries. The results show that the algorithm provides good performance because it allows to compare a graph in the repository with a query in times ranging between 61 and 163 ms, and ranking the full set of 100 compositions according each query in times ranging from 150 to 5000 ms (i.e. the comparison between each query and all the compositions in the repository). Additionally, also it should be noted that figure 2, shows a tendency line (dotted line) with a slight exponential behavior specially when the number of nodes of the graphs

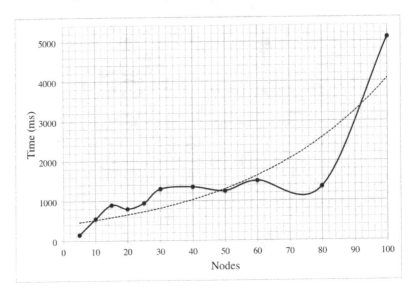

Fig. 2. Performance evaluation of the composition similarity algorithm

exceeds 80. Nevertheless, it is not a problem because in our assessment practices, the compositions created by users, showed barely reach 50 nodes.

5   Conclusion

An experimental platform implemented by a major mobile network operator in Italy to monitor its mobile users through the creation of rules, was presented. Specifically, this work provides a description of how the composer module has been designed and implemented: by taking advantage of subgraph isomorphism techniques to provide suggestions to the users. Such a mechanism can help users to create compositions by reusing other compositions stored in the repository while avoiding redundancies. Additionally, as it can be seen from the evaluation of the algorithm, it demonstrated to be fast enough to be executed each time the user creates a composition, i.e. users should not wait long time to receive feedback from the system. As future work, we plan to evaluate the effectiveness of our approach by measuring the customer satisfaction in order to improve the suggestions presented to the user. Additionally, we are working in an indexing mechanism in order to improve the performance, avoiding to compare the compositions created by the user to the full set stored in the repository; in this way the system can considerably reduce the total execution time.

References

1. Wallin, S., Leijon, V.: Telecom network and service management: An operator survey. In: Pfeifer, T., Bellavista, P. (eds.) MMNS 2009. LNCS, vol. 5842, pp. 15–26. Springer, Heidelberg (2009)

2. Rodríguez Rocha, O., Suarez-Meza, L.J., Moltchanov, B.: Smart space governing through service mashups. In: Andreev, S., Balandin, S., Koucheryavy, Y. (eds.) NEW2AN/ruSMART 2012. LNCS, vol. 7469, pp. 119–127. Springer, Heidelberg (2012)

3. 4CaaSt: 4caast european project (2012), `http://4caast.morfeo-project.org/`

4. Miller, G.: Wordnet: A lexical database for english. Commun. ACM 38(11), 39–41 (1995)

5. Patil, A.A., Oundhakar, S.A., Sheth, A.P., Verma, K.: Meteor-s web service annotation framework. In: 13th International Conference on World Wide Web. WWW 2004, pp. 553–562. ACM, New York (2004)

6. Benatallah, B., Hacid, M., Rey, C., Toumani, F.: Semantic reasoning for web services discovery. In: 2nd International Semantic Web Conference. ISWC 2013. ACM, Florida (2003)

7. Patil, A., Oundhakar, S., Sheth, A., Verma, K.: A (sub)graph isomorphism algorithm for matching large graphs. IEEE Tran. Pattern Anal. Mach. Intell. 26, 1367–1372 (2004)

8. 4CaaSt: 4caast project's marketplace (2011)

Intelligent Mobile Tourist Guide

Context-Based Approach and Implementation

Alexander Smirnov[1], Alexey Kashevnik[1], Sergey I. Balandin[2], and Santa Laizane[3]

[1] St.Petersburg Institute for Informatics and Automation RAS (SPIIRAS), Russia
{smir,alexey}@iias.spb.su
[2] FRUCT Oy, Helsinki, Finland
Sergey.Balandin@fruct.org
[3] University of Oulu, Center for Internet Excellence, Finland
santa.laizane@cie.fi

Abstract. Nowadays there is a wide range of different mobile solutions that support travelers before, during and after the trip. However, majority of these solutions focus either on recommending tourist attractions or on providing of some tourist services, but there is a lack of studies for unified approach that combines both needs. This paper describes mobile tourist guide - a complex system that enables comprehensive up-to date information search along with personalized recommendations and services. The key principle of the developed solution is based on fact that to provide really relevant tourists information, it should be based on analysis of current situation. Prototype of the mobile tourist guide has been developed using Smart Space infrastructure to facilitate integration of services and internal processes in such complex system. This paper aims to describe context-based information implementation in the complex mobile tourist guide, developed using Smart Space infrastructure.

Keywords: Smart space, mobile tourist guide, context management, e-tourism, IoT, Smart-M3, Karelia ENPI.

1 Introduction

Pervasive adoption of information and communication technologies (ICTs) throughout the tourism industry has brought fundamental implications for the way travel is planned and the tourism products are created and consumed [1]. Increasingly, travel-related information has been searched online using a wild range of tools like mobile devices. As pointed out in [2], mobile phones are becoming a primary platform for information access.

Mobile tourism is an emerging trend in the field of tourism involves the use of mobile devices as electronic guides [3]. Mobile guides based on PDAs, smartphones and mobile phones play an increasingly important role in tourism, giving tourists ubiquitous access to relevant information especially during the trip [4].

Today there is a vast variation of different mobile solutions already developed to support travelers before, during and after the trip, for instance in a form of city

S. Balandin et al. (Eds.): NEW2AN/ruSMART 2013, LNCS 8121, pp. 94–106, 2013.

attractions, sightseeing, exhibition or museum guides. However, majority of these solutions focus either on recommending tourist attractions or on providing some tourist services, but as argued in [5] these service types should not be approached separately and recommendation systems offering a unified perspective are needed. Thus, a prototype of a complex mobile tourism guide enabling comprehensive up-to date information search, for instance, information about points of interests, attractions, as well as providing personalized recommendations and services, including ridesharing and taxi calling services was developed to enrich tourist experiences. To facilitate integration of services and internal processes with other systems, and program modules, Smart Space infrastructure has been used for the prototype development.

Up-to-date and personalized information available anytime and anywhere is substantial for demanding tourists seeking exceptional value for money and time [6]. Thus, personalized recommendations and services provided by the mobile guide should be based on current situation of tourists to provide tourists relevant information. Hence, context-based information is crucial for such system development.

In this paper, context-based information implementation in the complex mobile tourist guide, developed using smart space infrastructure has been described.

The rest of the paper is structured as follows. Section 2 reviews context-based applications developed using smart space infrastructure. Section 3 briefly describes the mobile tourist guide. In Section 4 ontological knowledge representation approach is presented. Section 5 summarizing mobile tourist guide implementation is followed by conclusion section that summarizes main results and findings of the study.

2   Related Work

As pointed out in [7] context in tourism is an information facilitator in the negotiation process between all available information and the information that tourist require at a given moment of time based on current situation.

Authors of [8] present a context based design for guiding visitors in museums based on mobile devices. A set of existing systems have been considered compared and classified according to the presented context-based approach.

P. Jeon proposes a context-aware mobile platform for use of various surrounding local services with minimal user intervention in the forthcoming ubiquitous environment [9]. Context management technology is used for modeling current situation and proposing the user appropriate services.

Authors of [10] describe implementation of a context-aware mobile guide for outdoor as well as indoor locations. It uses GPS to identify user's location in outdoor environments, communicates with other objects in the environment through Bluetooth. The information that is shown in the user interface can be obtained in two different ways: stored on the mobile guide, or queried from the artifacts that are in the direct surroundings of the mobile guide through wireless communication.

The travel guide Triposo [11] is a free mobile guide service available for Apple and Android devices. A user can download the application and appropriate database (which is updated ones each two months) to the mobile device beforehand and use it during the trip without Internet connection. The application supports logging of travelling. It includes databases from the following sources World66, Wikitravel, Wikipedia, Open Street Maps, TouristEye, Dmoz, Chefmoz and Flickr [12]. Each guide contains information on sightseeing, nightlife, restaurants and more.

Millions of traveler reviews, photos, and maps can be accessible in TripAdvisor [13]. Tourists can plan their trips taking into account over 100 million reviews and opinions by travelers. TripAdvisor makes it easy to find the lowest airfare, best hotels, great restaurants, and fun things to do, wherever you go. The mobile application is free, it supports all mobile platforms.

Smart Travelling [14] is an online travel guide, supports about 30 cities worldwide including the most interesting capitals of European countries and USA. The guide includes a database of restaurants, cafes; hotels, shopping-tips and other places of interests. The mobile application for iPhone is accessible through AppStore. Integration with Google maps allows user to see the current location in the map and helps to navigate to each and every tip in destination cities. Application allows the user to download the content and use guide without Internet connection.

ARTIZT [15], an innovative museum guide system, where a ZigBee[16] protocol is used for determine user's position information. Visitors use tablets to receive personalized information and interact with the rest of the elements in the environment. The system achieves a location precision of less than one meter. The context is used to provide needed at the moment personalized information to the user.

Decentralized smart space in the proposed approach allows mobile device of every tourist acquire up-to-date information from different services (e.g., attraction, region, or worldwide services) and based on it make own decision about the suggestions to the tourist taking into account his/her preferences and current situation in the region.

3   Mobile Tourist Guide Description

The mobile tourist guide aims to provide a comprehensive up-to-date information and assistance in the travel-related decision making, especially during the trip to enrich tourist experiences. Hence, variety of services has been integrated to the mobile tourist guide application through the smart space infrastructure. The tourist can receive information about points of interest, attractions, and routes based on his preferences and contextual information of visiting region. Besides, the tourist guide application facilitating mobility and providing information about available transportation means alternatives have been developed. General architecture of the mobile tourist guide is presented in the Fig. 1.

As actual, up-to-date information is the lifeblood of tourism [17], accessibility of travel-related information in travel decision making may be perceived as crucial. To provide tourists comprehensive information, a variety of sources included in a great extent of user-generated content has been used. So far following information sources containing textual, visual and audio information have been integrated in the system: WikiLocation, GeoNames, Wikivoyage, Wikipedia, Wikimapia, Geo2Tag, Flickr, Panoramio (Fig. 2).

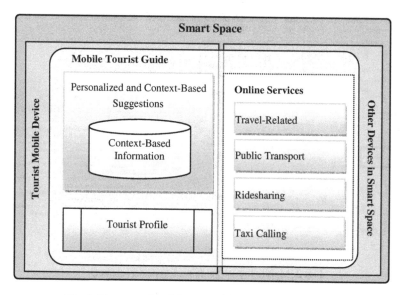

Fig. 1. The general architecture of the mobile tourist guide

Open content with little restrictions for further applications has been used also in other travel guides, for example Triposo where data from World66, Wikitravel, Wikipedia, Open Street Maps, TouristEye, Dmoz, Chefmoz and Flickrhave been integrated manually to the internal database and providing the tourist on demand [12].

In the Fig. 3 system prototype screenshots have been presented. Based on the tourist's current location ("Saint Petersburg", at the left screenshot), attractions within a certain radius have been presented (middle screenshot).

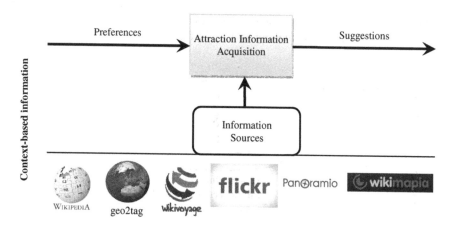

Fig. 2. General scheme of attraction information acquisition

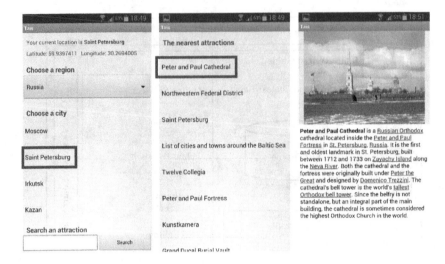

Fig. 3. Mobile tourist guide prototype screenshots

Tourist can click on interested attraction and see information about it (right screenshot). Besides, information search regardless location, for example freely choosing region and city is also supported by the system.

As a result of internal information processing processes in the system, extracted and personalized information has been presented to facilitate travel-decision making. Services facilitating tourist mobility have been integrated to the tourist guide application, they consist of: public transport searching (Fig. 4), taxi calling (Fig. 5) and ride-sharing (Fig. 6). The system enables tourists to find appropriate public transportation (plane, bus, train, ferry, and etc.), cab or fellow traveler with whom to share the ride and related expenses.

The taxi calling service finds the nearest taxi based on shared through the smart space by core service information about tourist location and his/her preferences about vehicle type, smoking or non-smoking driver (see Fig. 5).

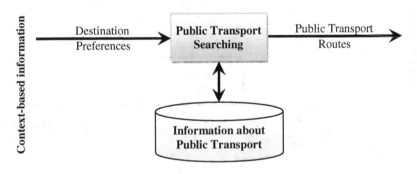

Fig. 4. The general scheme of the public transport searching

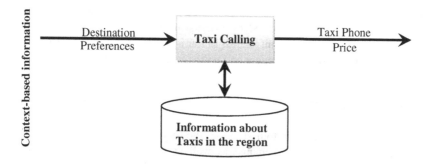

Fig. 5. The general scheme of taxi calling service

The ridesharing service finds fellow travelers for the tourist based on information about tourist location, attraction location and his/her preferences about maximum waiting time of the fellow traveler, walking distance to the meeting point, vehicle type, gender of the driver, smoking or non-smoking driver / passenger, and other specific to ridesharing information with the ridesharing service (see Fig. 6).

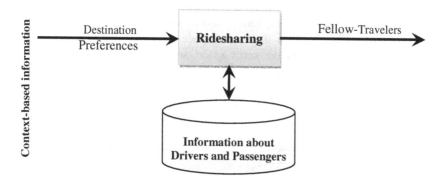

Fig. 6. The general scheme of ridesharing service

Fig. 7. Ridesharing service example

The ridesharing service example is presented in Fig. 7. To reach the destination tourist have to walk to a calculated by ridesharing service meeting point, then the driver pick him/her up to the point nearest to tourist destination and driver path. After that the driver drops off the passenger and he/she walk to the destination.

4   Context-Based Approach for Mobile Tourist Guide

In the development of mobile tourist guide, the ontology has been used to describe knowledge and information exchange as well as to facilitate service and device interoperability in the smart space. The conceptual model of the proposed ontological approach is based on the earlier developed ideas of knowledge logistics [18]. Ontology describes the main terms used for the smart space description and relationships between its different constituents. The tourist context is formed based on the interaction process between the tourist's mobile device and different services through the smart space. The context is the current situation description in terms of the ontology.

Mobile devices interact with services and get information from resources through the smart space (see Fig. 8).The tourist installs the guiding application (core service) to his/her mobile device. When the tourist registers in the application, core service creates a profile which contains long-term context information of the tourist. This profile allows specifying and complements tourist requirements and personifying the information and knowledge flow from smart space services to the tourist. The client shares the tourist profile with the smart space, so, that this information could be acquired by other smart space services.

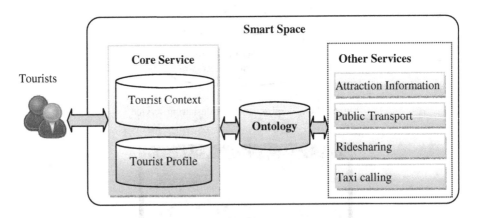

Fig. 8. Ontological approach for tourist guiding application

The context information for mobile tourist guide service consists of the following elements (see Fig. 9).

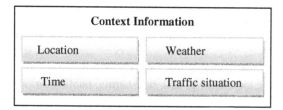

Fig. 9. Context Information for Mobile Tourist Guide Service

Tourist location and time acquired by core service from the tourist mobile device. This information is used for suggesting the tourist the nearest museums, beaches in the afternoon and cafes, restaurants, night bars in the evening. These information is shared by core through the smart space with the tourist attraction information service using the ontology.

Current weather and traffic situation is acquiring by the region specific services by core service. These information is used for propose the tourist attractions to see (is it applicable to propose outdoor attractions at the moment), and choose the transportation means (e.g. in case of traffic jams it is reasonable to use underground public transport).

The tourist profile (Fig. 10) accumulates and stores main tourist related information in the smart space. It contains tourist role and his/her preferences.

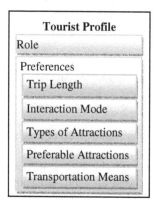

Fig. 10. Information Model of Tourist Profile

The role determines a template for suggesting the tourist attractions visiting plan (e.g., business, education, health, adventure, cultural, eco-tourists, leisure, visit friends and relatives, youth, religious, shopping, sport). Preferences include the following elements:

— trip length (to provide suggestions about attractions visiting);
— interaction mode (e.g. textual, audio, and video information presenting on tourist mobile device);

- types of attractions (determine attractions types which is interested for the tourist, e.g. Renaissance painting, sculpture of XIX century);
- preferable attractions (contains a list of interested and mandatory for visiting attractions);
- transportations means (contains the tourist preferences related to the types of vehicles for changing location, e.g. taxi, ridesharing, public transport).

5   Mobile Tourist Guide Implementation

Implementation of the proposed services has been developed based on Smart-M3 information platform [19, 20] which is open source and accessible at Sourceforge [21]. Service implementations has been developed using Java KPI library [22]. Mobile clients have been implemented using Android Java Development Kit [23].

The main scenario of tourist attraction information service is presented in Fig. 11. Tourist attraction information service acquires the tourist location from smart space and generates the list of attractions around this location. For searching attractions Wikipedia is used. It has a lot of articles which describe attractions with linked geographical coordinates which allows determining attractions names. For searching this article MediaWiki and GeoNames API can be used. For the attraction information service GeoNames API is used as they provide more powerful functionality in comparison with MediaWiki. Then textual and graphical information about attractions can be acquired as from Wikipedia as from another services (e.g. Wikivoyage, Wikipedia, Wikimapia, Geo2Tag, Flickr, Panoramio).

Fig. 11. Information Model of Tourist Profile

The transport services include ridesharing service, taxi service and service for calculation routes for the tourists based on public transport schedule. Development and implementation of ridesharing service is described in detail in [24]. Taxi service is under development now. Public transport service has been developed for Russian Karelia region.

For calculation of routes based on public transport schedule the database for Russian Karelia region provided by Yandex Schedule has been used. This database has the following structure:

```
<stations>List of stations</stations>
<transportation_facility_types>
<type><id>0</id><name>Bus</name></type>
<type><id>1</id><name>Train</name></type>
```

```
</transportation_facility_types>
<transportation_facilities>
<transportation_facility>
<id>b_1</id><type>0</type>
<route_number>10</route_number>
<route>
<station><id>0</id>
<arival_time>10:20</arival_time>
<departure_time>10:22</departure_time>
<next_station><id>1</id></next_station>
</station>
</transportation_facility>
</transportation_facilities>
```

The database has been imported to PostgreSQL database. For calculating of routes the Dejkstra algorithm [25] is used. A result of public transport schedule service has the following structure:

```
<facility arr_time="03:40" distance="1.046" type="Пешком">
  <departure_station lat="60.97" lon="32.93"/>
<arrival_station lat="60.97" lon="32.96" name="Олонец"/>
</facility>
<facility arr_time="07:25" dep_time="03:40" distance="5"
id="c2176_kvc_tula" time="03:45" type="Автобус">
<departure_station lat="60.97" lon="32.96"name="Олонец"/>
<arrival_station lat="61.69" lon="33.61" name="Пряжа"/>
</facility>
<facility arr_time="10:25" dep_time="07:30" distance="16"
id="c3336_kvc_tula"time="02:55" type="Автобус">
<departure_station lat="61.69" lon="33.61"name="Пряжа"/>
<arrival_station lat="61.68" lon="31.26"name="Леппясилта"/>
</facility>
<facility dep_time="10:25" distance="1.549" type="Пешком">
<departure_station lat="61.68" lon="31.26"name="Леппясилта"/>
<arrival_station latitude="61.67865" longitude="31.22759"/>
</facility>
```

The result is shared by public transport schedule through the smart and core service sort these transportation means in accordance with tourist preferences and provides it to the tourist.

6   Conclusion

Even though there is a vast amount of different mobile solutions developed already, new mobile solutions are still needed as majority of existing solution focuses either on recommending tourist attractions or tourism services. Hence, to enrich tourist experiences and facilitate travel-related decision making, in the paper described mobile tourist guide that has been developed as a complex system what enables comprehensive up-to date information search, as well as provides personalized recommendations and services.

Used approach seems to be promising from various perspectives. One of the concerns related to the use of mobile applications in the foreign country is roaming price. In this regard, there are some positive indications in terms of price reductions. For instance, at the moment there are several offers from telecommunication companies for Russian tourists travelling to Europe like one day unlimited Internet connection in Finland for a bit more than one euro. This may result in demand increase for such mobile solutions. Besides, in the mobile tourist guide in a great extent user-generated content has been used as the information source (e.g. Flickr, Panoramio, Wikipedia). Through the use of different sorts of social media generated user content, for instance travel-related information, personal opinions, recommendations and experiences plays an important role in travel decision making process as consumers increasingly trust better their peers, rather than marketing messages [6]. Furthermore, to provide tourists relevant information, travel-related recommendations and services provided by the mobile tourist guide are based on tourist current situation. Context-based information is crucial for such system development as demanding tourists are seeking exceptional value for money and time [6]. Smart Space infrastructure used for the system development enables seamless integration of different services, thus creating a great platform for further system development.

The main goal of this paper was to describe context-based information implementation in the mobile tourist guide what has been developed using Smart Space infrastructure to facilitate interoperability of services and supporting processes. Nevertheless, further evaluation and studies on described approach are still needed. Besides, current research has indicated very promising further system development directions like social networking (e.g. Facebook, vKontakte), user forums or other sorts of social media integration in the system. Also user studies with prospective system users are needed.

Acknowledgements. This research is a part of Karelia ENPI CBC programme grant KA322 «Development of cross-border e-tourism framework for the programme region (Smart e-Tourism)», co-funded by the European Union, the Russian Federation and the Republic of Finland. The presented results are also a part of the research carried out within the project funded by grant #13-07-12095 of the Russian Foundation for Basic Research. Authors are grateful for TiViT IoT SHOK program that provided required support of this research.

Authors would like to thank Yandex Schedule team and particularly Mr. Dmitry Kryukov (the head of Yandex Schedule Service) for providing public transport schedule for the Republic of Karelia, Russia.

References

1. Neuhofer, B., Buhalis, D., Ladkin, A.: Conceptualising technology enhanced destination experi-ences. Journal of Destination Marketing & Management 1, 36–46 (2012)
2. Ricci, F.: Mobile recommender systems. International Journal of Information Technology and Tour-ism 12, 205–231 (2010)
3. Kenteris, M., Gavalas, D., Economou, D.: An innovative mobile electronic tourist guide application. Personal and Ubiquitous Computing 13, 103–118 (2009)
4. Höpken, W., Fuchs, F., Zanker, M., Beer, T.: Context - based adaptation of mobile appli-cations in tourism. Information Technology & Tourism 12, 175–195 (2010)
5. Gavalas., D., Konstantopoulos, C., Mastakas, K.: Mobile recommender systems in tour-ism. Accepted for publishing in: Journal of Network and Computer Applications (2013)
6. Buhalis, D., Law, R.: Progress in information technology and tourism management: 20 years on and 10 years after the Internet – The state of eTourism research. Tourism Man-agement 29, 609–623 (2008)
7. Lamsfus, C., Grün, C., Alzua-Sorzabal, A., Werthner, H.: Context-based matching to enhance tourists´ experiences. The European Journal for the Informatics Professional 2, 14–21 (2010)
8. Raptis, D., Tselios, N., Avouris, N.: Context-based design of mobile applications for mu-seums: a survey of existing practices. In: Proceedings of the 7th International Conference on Human Computer Interaction with Mobile Devices & Services, Salzburg, Austria (Sep-tember 2005)
9. Jeon, P.B.: Context Aware Intelligent Mobile Platform for Local Service Utilization. In: The 2012 IEEE/WIC/ACM International Conference on Intelligent Agent Technology, Macau, pp. 635–638 (December 2012)
10. Luyten, K., Coninx, K.: ImogI: Take Control over a Context Aware ElectronicMobile Guide for Museums. HCI in Mobile Guides. University of Strathclyde, Glasgow (2004)
11. "New travel guide Triposo brings algorithms to apps," The Independent (September 2011), http://www.independent.co.uk/travel/news-and-advice/new-travel-guide-triposo-brings-algorithms-to-apps-2353352.html (last access date May 8, 2013)
12. Triposo, http://www.triposo.com (last access date May 8, 2013)
13. Tripadvisor, http://www.tripadvisor.ru (last access date May 8, 2013)
14. Smart Travelling web page, http://www.smart-travelling.net/en/ (last access date May 8, 2013)
15. García, O., Alonso, R.S., Guevara, F., Sancho, D., Sánchez, M., Bajo, J.: ARTIZT: Apply-ing ambient intelligence to a museum guide scenario. In: Novais, P., Preuveneers, D., Cor-chado, J.M. (eds.) ISAmI 2011. AISC, vol. 92, pp. 173–180. Springer, Heidelberg (2011)
16. ZigBee Aliance, http://www.zigbee.org/ (last access date May 8, 2013)
17. Buhalis, D.: Strategic use of information technologies in the tourism industry. Tourism Management 19, 409–421 (1998)
18. Smirnov, A., Pashkin, M., Chilov, N., Levashova, T., Krizhanovsky, A.: Knowledge Lo-gistics as an Intelligent Service for Healthcare. Methods of Information in Medicine 44(2), 262–264 (2005)
19. Honkola, J., Laine, H., Brown, R., Tyrkko, O.: Smart-M3 Information Sharing Platform. In: Proc. IEEE Symp. Computers and Communications (ISCC 2010), pp. 1041–1046. IEEE Comp. Soc. (June 2010)
20. Morandi, F., Roffia, L., D'Elia, A., Vergari, F., Cinotti, T.: RedSib: a Smart-M3 semantic information broker implementation. In: Proc. 12th Conf. of Open Innovations Association FRUCT and Seminar on e-Tourism, SUAI, pp. 86–98 (2012)

21. Smart-M3 at Sourceforge, `http://sourceforge.net/projects/smart-m3` (last access date May 8, 2013)
22. Smart-M3 Java KPI library at Sourceforge, `http://sourceforge.net/projects/smartm3-javakpi/` (last access date May 8, 2013)
23. Android Java Development Kit, `http://developer.android.com/sdk/index.html` (last access date May 8, 2013)
24. Smirnov, A., Shilov, N., Kashevnik, A., Teslya, N.: Smart Logistic Service for Dynamic Ridesharing. In: Andreev, S., Balandin, S., Koucheryavy, Y. (eds.) NEW2AN/ruSMART 2012. LNCS, vol. 7469, pp. 140–151. Springer, Heidelberg (2012)
25. Dijkstra, E.W.: A note on two problems in connexion with graphs. Numerische Mathematik 1, 269–271 (1959)

Geo-coding in Smart Environment: Integration Principles of Smart-M3 and Geo2Tag

Kirill Krinkin[1] and Kirill Yudenok[2]

[1] FRUCT LLC,
Saint-Petersburg, Russia
`kirill.krinkin@fruct.org`
[2] Saint-Petersburg Electrotechnical University,
Saint-Petersburg, Russia
`kirill.yudenok@gmail.com`

Abstract. Geo-tagging and smart spaces are two promising directions in modern mobile market. Geo-tagging allows to markup any kind of data by geographical coordinates and time. This is the basis for defining geographical context which can be used in different types of applications e.g. semantic information search, machine-to-machine (M2M) interactions. Smart spaces as the basis for seamless distributed communication field for software services provides semantic level for data processing. Most desired feature of coming software is pro-activeness and context awareness, i.e. services will be able to adapt to the user's needs and situations and be able to manage decisions and behaviors on behalf of the user. The paper is dedicated discussion of integration most popular open platforms for smart spaces and geo-tagging (Smart-M3 and Geo2Tag) as possible solution for creation context aware proactive location based (LBS) services.

Keywords: geo-tagging, geo-coding, smart-m3, Geo2Tag, LBS.

1 Introduction

Nowadays we have two most promising software trends – location based services and pervasive smart environments (smart spaces). Both of them will be a base for user- and machine- oriented proactive services. Smart spaces should provide continuous distributed semantic data and communication field for software services, which is being run on personal devices and autonomous computers and robots. Most desired feature of coming software is pro-activeness and context awareness, i.e. services will be able to adapt to the user's needs and situations and be able to manage decisions and behaviors on behalf of the user [1]. One of the important part of context is location based data. This data is being used for two purposes: for clarifying semantic meaning of queries (when service retrieves the data from smart environment) and for limitation of space of search (usually there is no point to make global search). Geo-coding (or geo-tagging) is the technique of markup real or virtual object by adding geographical coordinates and time. If we consider software, we have only virtual (or digital) objects

S. Balandin et al. (Eds.): NEW2AN/ruSMART 2013, LNCS 8121, pp. 107–116, 2013.

like media, events, documents etc. So far, smart spaces and geo-tagging systems are being developed mostly separately, there are only few works [2, 3, 4] where software design of smart spaces and geo-tagging integration is discussed. In this paper requirements and high level design for Integrated Geo-Coded Smart-Space (GCSS) middleware are discussed. Rest of paper is constructed next way: in second part we analyze related works, in third part requirements and architecture are proposed.

From practical point of view we use Smart-m3[1] and Geo2Tag Platform[2] as most developed open source middleware for smart spaces and geo-tagging. In the last part of paper we conclude proposed architecture and define directions for development.

2   Overview of Smart-M3 and Geo2Tag Platforms

2.1   Smart-M3 Platform

Smart-M3 is an open source software platform [5, 6] that aims to provide Semantic Web information sharing infrastructure between software entities and various types of devices. The platform combines ideas of distributed, networked systems and Semantic Web [7, 8]. The major application area for Smart-M3 is the development of smart spaces solutions, where a number of devices can use a shared view of resources and services [9]. Smart spaces can provide better user experience by allowing users to easily bring-in and take-out various electronic devices and seamlessly access all user information in the multi-device system from any of the devices [10].

The simplified version of the Smart-M3 smart spaces reference model is shown in Fig. 1. The Knowledge Processors (KPs) represent different applications that use the smart space. The smart space core is implemented by one or several Semantic

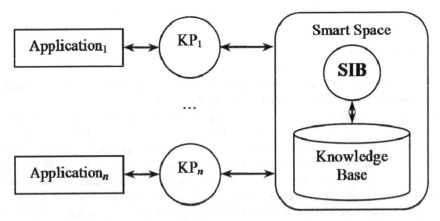

Fig. 1. Smart space based on Smart-M3: simplified reference model

---

[1] http://en.wikipedia.org/wiki/Smart-M3
[2] http://geo2tag.org

Information Brokers (SIBs) interconnected into the common space. The information exchange is organized through transfer of information units (represented by RDF triples) from KPs to the smart space and back. The information submitted to the smart space becomes available to all KPs participating in the smart space. The KPs can also transfer references to the appropriate files/services into the smart space, since not all information can be presented by RDF triples (e.g., a photo or a PowerPoint presentation). As a result the information is not really transferred but shared between KPs by using smart space as a common ground [11].

2.2   Geo2Tag Platform

Geo2Tag is centralized system with server, that storage all information and provide access to it by REST API. It has following main advantages – it is open source and it doesn't depend on concrete web-server or DB type. Server consists of two main parts – server application and database (DB). Server application is a web application written by FastCGI framework. It main task is processing clients requests to REST API. Database contains information about users and their geodata.

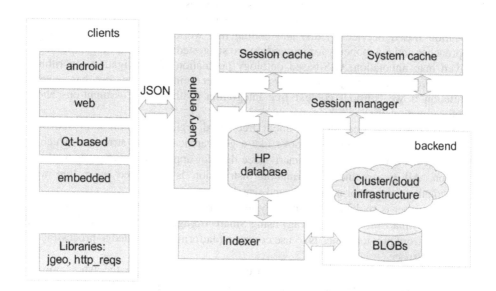

Fig. 2. Geo2Tag architecture

Platform use following basic data types:

- User – data that represent human user of Geo2Tag – login and password;
- Channel – object that contain name, description and set of tags. User subscription for the channel means that user can read tags from it;

- Tag – object that contain URL of multimedia object, name, description, time of creation and coordinates(latitude and longitude);

Main Features of Geo2Tag platform:

- user management: registration, login, log-off, session management;
- data retrieval about users and matching personal geographical spaces to the personal smart spaces;
- channel management: subscription/unsibscription;
- sending geographical data from smart-space to the geo-tagging system;
- getting data from geo-tagging system;
- spatial filtration;

More information about platform and design details can be found at related work – [12].

3   Related Work

The most overall point at the integration smart spaces facilities and geo-coding has been developed by Pervasive Computing Research Group. They focused on indoor Location Based Services (LBS) and coding real world objects. In [4] next basic approaches and components for integration are suggested: spacial ontology, ontology-driven map annotation, GIS-based ontology population and navigation algorithms. This approach is very justified from application point of view but has several limitation if we consider context free integration (without any assumptions about application domain).

In [13] we can find an idea about tree-based region distribution of semantic information in global space. It looks like a geographical fractal structure, which is providing the same structure of smart space data for application in any geographical position. All geographical information distribution is organized as a tree with orthogonal algorithm for navigation and search.

In this paper is spoken about smart system creation by combining the work of two platforms (Smart-M3 and Geo2Tag) using Smart-M3 agent (KP). Main difference of this work from previous ones is to use common platform for knowledge processing of Smart-M3 space.

The platform and its knowledge processors (agents) take on the whole job of collecting, analyzing and processing of space knowledge described by ontologies and deductions. When developing the platform agents we can use existing GIS-based or spatial ontology according to the Geo2Tag LBS-system or use your own, such as Geo2Tag system representation ontology.

The use and presentation of geo-data in the Smart-M3 platform adds a new property to the data as location in space and time that will allow to place objects in the real world as well will also give them the ability to search in a space such as a room or house space.

4   System Requirements and High-Level Design

There are several promising use-cases of GCSS:

- geographical markup of smart space data;
- search set reduction;
- search context rectification.

In next subsections high requirements and architecture of GCSS are discussed.

The main task of the agent - the Smart-M3 and Geo2Tag platforms union. One can say that the agent is an extension of the Smart-M3 platform, since the fully interoperable with Geo2Tag LBS platform, expanding the space with new data – geo-data.

The main user interaction with agent is to run it and specify the connecting settings for the Smart-M3 and Geo2Tag platforms and also monitoring and control of its operations. The agent then works independently checking receipt of new geo-data, producing a conversion in triplets and publish them into space.

4.1   High-Level Requirements

GCSS should implement effectively main features from both (smart space and geo-coding) type of systems, which are:

- providing interfaces for semantic data and access;
  Providing an interface for connecting to the platform and access to its semantic data. This feature is implemented in the Smart-M3 API (Qt, Python).
- distributed storage for semantic information;
  Smart-M3 platform has its own data storage, this storage may be replaced with a more stable and productive. Current stable Smart-M3 0.4.1 version uses Berkeley DB storage.
- interfaces for association semantic objects with geo-tags;
  Development of two-way geo-data conversion mechanism in the semantic objects (space triplets) and the creation of Geo2Tag platform ontology.
- spacial and temporal filtration.
  Development of the necessary algorithms for searching and filtering of smart space semantic data.

Also non-functional requirements should be taken in account:

- Performance – ability to work with big amount of semantic objects geo-tags; for some purposes we need to implement features like cloud based massive offline processing and local context indexing/caching.
- Compatibility – the GCSS should be accessible by legacy interfaces (i.e. SSAP or REST), which is required for seamless integration with existing systems.

At the moment, there are all functionality to work with the Smart-M3 platform and for development of its agents (KP), there is no mechanism for geo-data converting to the space triplets as well as algorithms for their search and filtering.

The latest versions of Smart-M3 and Geo2Tag platforms fairly stable, but have some defects such as those associated with security or performance of the Smart-M3 platform.

4.2   Layered Architecture

In our work we rely on existing middleware for smart-space environment and geo-tagging. As smart space infrastructure Smart-M3 is being used; for geo-coding we use Geo2Tag Platform. Selection of those platforms is caused the availability all source codes and solution maturity.

On Fig 1. High-level layered design for GCSS is presented. There are four levels of system.

Each level of the system is responsible for the functions and includes its own interface. The following are the layers of the system GCSS:

- **Interfaces** – it contains smart-spaces (Smart-M3) and geo-coding (Geo2Tag) front-ends (FE) and their functionality; This level is responsible for data representation and processing for applications and services;
- **Integration** – this level contains components for translating geographical data (geo-tags) from Geo2Tag format to Smart-Space format and vice versa;
- **Domain engines** – level contains particular implementations of smart-space geo-coding middleware (in our case it is Smart-M3 and Geo2tag);
- **Data cloud backend** – optional components, which is being used for providing advanced services like off-line data (pre-)processing, storage for BLOB objects, indexing, caching etc.

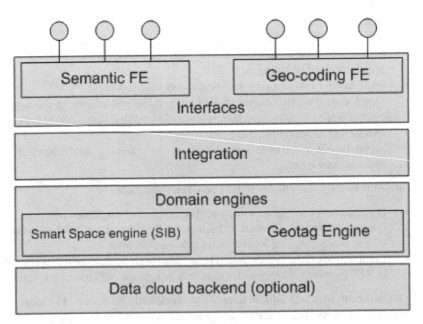

Fig. 3. Layered design of GCSS

The main suppliers of GCSS system functionality are a Smart-M3 platform and Geo2Tag LBS system. Details on this platforms and their main features can be found in [5, 6, 12].

Integration level is responsible for the platforms data conversion into one common format. For the system data storage responds GCSS Domain engines level, which is composed of the main data storage of the Geo2Tag and Smart-M3 platforms. To work with a large volumes of geo-data and increasing the overall performance of the system is planned to add the ability to work with Data cloud, as a repository for offline processing.

4.3   Location Based Engine

According to Smart-M3 architecture all knowledge of smart space is being presented as RDF triples. Set of RDF-tuples describing particular domain is an ontology. So, to present geo-coding data we need to use ontology also.

On the Fig 2. Geo2tag ontology is presented, it is defined according to Web Ontology Language Specification (OWL) as the tree of classes ans properties. The root of ontology tree is the class *User*. It contains all geographical data from personal geo-space. With this approach we can use personal and shared geo-tags, last can be implemented by introducing common user or group. Each instance of *User* class is connected with one or more *Channels* (Geo2tag terminology is used [6]). User has a relation *subsribesTo* so that follow different sets of geo-tags. The *Channel* could contain any amount of geo-tags by using property *containsA*.

Usially the *User* contains constant or rarely seldom changing properties about user credentials in geo-tagging system. This class can be presented according to FOAF.

The *Channel* includes "name" and "description" properties for identification of channel purpose and contains (through the property containsA) geo-tags itself, which are presented by time coordinates and data.

The main object of this Geo2tag ontology is a geo-tag (class *Tag*). Geo-tag is basic information, which manages and handles the Geo2Tag platform and it is the basic knowledge of the space, which describes the location of its subjects.

The size of a one geo-tag nearing 1K, Geo2Tag system allows you to store an unlimited number of data values, for it has developed special synchronization and optimization algorithms of system database [14]. After converting them into space triplets, the tag size will not increase, but now to work with them in the space will be used Smart-M3 platform storage. To improve the performance of data search is planned to develop the necessary algorithms for searching and filtering smart space data.

The property *data* plays significant role in integration mechanism. It contains set of identifiers and properties, which allow to specify objects or relations in smart-space.

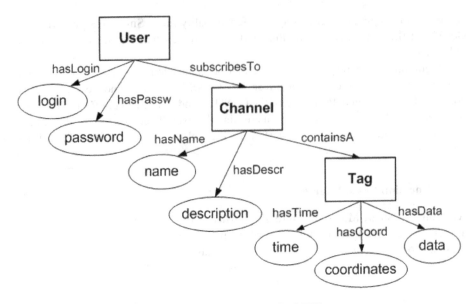

Fig. 4. Geo2Tag ontology in GCSS

On the integration level we introduce Geo2tag Agent. Geo2Tag agent is required for monitoring and synchronizing data between geo- and semantic spaces. Agent is Knowledge Processor (KP) in terms of Smart-M3.

Basic architecture of Geo2Tag KP is presented on Fig 3. There are next main components:

- Geo2Tag service handler;
- Smart-M3 handler;

Geo2Tag platform responsible for storing and processing of LBS-system geo-data, Geo2Tag-agent converts these data for use them in the space.

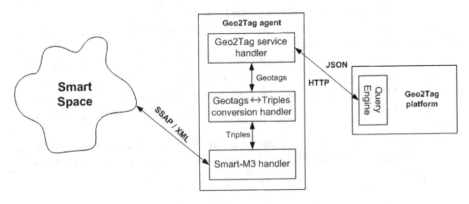

Fig. 5. Geo2Tag knowledge processor architecture

Let's discuss collaboration with Geo2Tag platform. Geo2Tag Platform is the centralized high performance geo-tagging database with dedicated server, which is provided REST API for access geo-tags. All communications could be implemented only over HTTP protocol, which could be a reason for performance issues. All geo-tagging data is presented as JSON objects.

As mentioned earlier, the basic data format of the Smart-M3 platform – RDF-triples <S, P, O>. To work with Geo2Tag platform geo-data it is necessary to convert them to the data format of the Smart-M3 platform, for these purposes serves a separate component of the Geo2Tag-agent – «Geotags <-> Triples conversion handler», whose main task is to convert the geo-data to the space RDF-triplets.

Geo-tag consists of a tuple <t, L, B, H, data>, where

- t – time;
- L, B, H – coordinates;
- data - text data (~ 1K).

From this we can easily generate triplet of <time, coordinates, data> type, which will describe the space geo-data. Geo-tags can also be combined into channels and users can subscribe to the tags channels. Which is easily described with triplets, using the agent ontology.

Smart-M3 handler retrieves JSON objects and translated them into Smart-M3 triples, and handles operations with smart space such as subscribing, publishing. There are several methods for ontological data processing here:

- ontological model, where all data is being kept in agent memory as RDF graph similar to RDF graph of smart space.
- object-oriented model, where all RDF data is transforming from RDF triples into objects with properties and methods. And reverse transformation is being used only for publishing data into smart space. We suggest to use this type of geo-tagging data processing.

One of the disadvantage of data representation into Smart-M3 is inability of presenting media data. It cannot be presented as triples. Geo2Tag architecture is designed for supporting such king of data. This features could be used for extending Smart-M3 functionality. Main problem should be discussed is a privacy and data protection.

5 Conclusion

It this paper we proposed high level design of smart-space and geo-coding middleware integration. This integration could be made by using special Knowledge Processor, which monitors both spaces and translates data from one to another and vice versa. There are still open question for future development: overall system performance, effective object monitoring, temporal and spatial filtration, integration with media objects.

Acknowledgments. The presented work is done as a part of the research and development work done in the project funded by grant #11555p/20929 of FASIE Foundation.

References

1. Floch, J., Angermann, M., Jennings, E., Roddy, M.: Exploring Cooperating Smart Spaces for Efficient Collaboration in Disaster Management. In: Proceedings of the 9th International ISCRAM Conference – Vancouver, Canada (2012)
2. Nabian, N., Ratti, C., Biderman, A., Grise, G.: MIT GEOblog: A platform for digital annotation of space for collective community based digital story telling. In: 3rd IEEE International Conference on Digital Ecosystems and Technologies, pp. 353–358. IEEE, Piscataway (2009)
3. Rishede, J., Man, T., Yiu, L.: Effective Caching of Shortest Paths for Location-Based Services, SIGMOD 2012, Scottsdale, Arizona, USA (2012)
4. Kolomvatsos, K., Papataxiarhis, V., Tsetsos, V.: Semantic Location Based Services for Smart Spaces. In: 2nd International Conference on Metadata and Semantics Research (MTSR), Corfu, Greece (2007)
5. SourceForge, Smart-M3, http://sourceforge.net/projects/smart-m3
6. Honkola, J., Laine, H., Brown, R., Oliver, I.: Cross-Domain Interoperability: A Case Study, Nokia Research Center, Helsinki, Finland (2009)
7. Berners-Lee, T., Hendler, J., Lassila, O.: The semantic web. Scientific American (2001)
8. Oliver, I., Honkola, J.: Personal Semantic Web Through a Space Based Computing Environment Middleware for Semantic Web 08 at ICSC 2008, Santa Clara, CA, USA (2008)
9. Oliver, I., Honkola, J., Ziegler, J.: Dynamic, Localized Space Based Semantic Webs. In: WWW/Internet Conference, Freiburg, Germany (2008)
10. Boldyrev, S., Oliver, I., Honkola, J.: A mechanism for managing and distributing information and queries in a smart space environment, MDMD 2009. Milan, Italy (2009)
11. Roffia, L., D'Elia, A., Vergari, F., Manzaroli, D., Bartolini, S., Zamagni, G., Cinotti, T.S., Honkola, J.: A Smart-M3 lab course: approach and design style to support student projects. In: Balandin, S., Ovchinnikov, A. (eds.) 8th FRUCT Conference of Open Innovations Framework Program FRUCT, pp. 142–153. Saint-Petersburg State University of Aerospace Instrumentation (SUAI), Lappeenranta (2010)
12. Bezyazychnyy, I., Krinkin, K., Zaslavskiy, M., Balandin, S., Koucheravy, Y.: Geo2Tag Implementation for MAEMO. In: 7th Conference of Open Innovations Framework Program FRUCT, Saint-Petersburg, Russia (2010)
13. Dearle, A., et al.: Architectural Support for Global Smart Spaces. In: Chen, M.-S., Chrysanthis, P.K., Sloman, M., Zaslavsky, A. (eds.) MDM 2003. LNCS, vol. 2574, pp. 153–164. Springer, Heidelberg (2003)
14. Zaslavsky, M., Krinkin, K.: Geo2tag Performance Evaluation. In: Proceedings of the 12th Conference of Open Innovations Association FRUCT and Seminar on e-Travel, Oulu, Finland (2012)

Geofence and Network Proximity

Dmitry Namiot[1] and Manfred Sneps-Sneppe[2]

[1] Lomonosov Moscow State University,
Faculty of Computational Mathematics and Cybernetics,
Moscow, Russia
{dnamiot@gmail.com}
[2] ZNIIS,
M2M Competence Center,
Moscow, Russia
{manfreds.sneps@gmail.com}

Abstract. Many of modern location-based services are often based on an area or place as opposed to an accurate determination of the precise location. Geo-fencing approach is based on the observation that users move from one place to another and then stay at that place for a while. These places can be, for example, commercial properties, homes, office centers and so on. As per geo-fencing approach they could be described (defined) as some geographic areas bounded by polygons. It assumes users simply move from fence to fence and stay inside fences for a while. In this article we replace geo-based boundaries with network proximity rules. This new approach let us effectively deploy indoor location based services and provide a significant energy saving for mobile devices comparing with the traditional methods.

Keywords: location, privacy, lbs, mobile, HTML5, geo coding, boundary geofence.

1 Introduction

Geo-fencing enables remote monitoring of geographic areas surrounded by a virtual fence (geo-fence), and automatic detections when tracked mobile objects enter or exit these areas [1]. A huge set of LBS (location based services) use geo-fence observation as a key feature. Location plays a basic role in context-aware applications. Geo-fences are user-defined areas defined around a Location. Locations here are cities, towns, other identifiable landmarks as well as vehicle parks of the user organization. Usually, the user is able to define the bounding of geo-fence area. For example, in simplest case it is just a radius defines some circular area. On practice, in the vehicle tracking system, a vehicle is determined to be at a particular Location if it is within this geo-fence (e.g., within the given radius for circular area).

Any geo-fence implementation requires obviously some form of location monitoring. Technically, this monitoring could be performed either right on the mobile device or via some centralized scheme (e.g., telecom operator observes the location for own subscribers).

S. Balandin et al. (Eds.): NEW2AN/ruSMART 2013, LNCS 8121, pp. 117–127, 2013.

The main sources for user's raw coordinates on mobile phones as Global Positioning System (GPS) and Wireless Positioning System (WPS) using cell tower and Wi-Fi access points (AP) [2].

One of the biggest and well known problems with the location monitoring is energy consumption. It is, probably, the biggest limitation factor. Typical battery capacity of smart phones today is barely above 1000 mAh (e.g., the lithium-ion battery of HTC Dream smart phones has the capacity of 1150 mAh). GPS, the core enabler of LBS, is power-intensive, and its aggressive usage can cause complete drain of the battery within a few hours [3]. A typical GPS invocation consists of a locking period and a sensing/reporting period. The lengths of these two periods are about 4-5 seconds and 10-12 seconds, respectively. More importantly, the average power draws for the two above-mentioned periods are about 400 mW and 600 mW, respectively. For a typical battery capacity of 1000 mAh such high power consumption is very expensive as continuous GPS sensing can deplete the battery in merely 6 hours [4].

Figure 1 illustrates battery depletion test with GPS mode on.

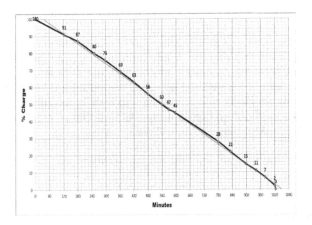

Fig. 1. Battery depletion [5]

Figure 2 compares GPS and non-GPS modes as well as illustrates the power spikes. The battery drain effect could be presented more dramatically in case of several applications that work in parallel. Note, that several independent location monitoring applications present the more realistic picture.

Many research papers declare the goal to develop frameworks that continuously provides location context with minimum energy consumption.

We can mention here, for example, deploying fingerprints to recognize semantic places with high level accuracy using radio beacons (e.g., cell towers, WiFi APs, and Bluetooth), counting the surrounding factors (e.g., light, texture, and sound patterns) – so called context. As per classical definition [6], context-related information can consist of a users profiles and preferences, their current location, the type of connection that to the mobile network, the type of wireless device being used, the objects that are currently in the user's proximity, and/or information about their behavioral history. Actually, most of the authors define context awareness as complementary element to location awareness, whereas location may serve as a determinant for resident processes. By this reason, all the context-aware applications are linked to location exchange.

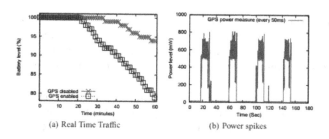

(a) Real Time Traffic   (b) Power spikes

Fig. 2. GPS vs. non-GPS mode

To optimize energy consumption for continuous sensing, various approaches have been proposed. These include sensor selection by movement detector using accelerometers [7, 8], minimizing energy consumption within accuracy requirements [9, 10], utilizing a prediction-based approach [11], etc.

We can select the following common directions (areas) for energy saving during the location monitoring:

1) Adaptive selection of location sensing mechanisms. Actually, it should be selected dynamically (e.g., GPS or network fingerprints). Location sensing mechanisms could have performance tradeoffs in terms of accuracy, power consumption, and availability.
2) Usage of context information. Modern LBS should be context-aware too.
3) Adding cooperation for multiple LBS on client's side. They should communicate by some way in order to avoid redundant location sensing invocations [12].

But in general, any client side monitoring is and always will be energy consuming operation. The more prospect area by our opinion is the centralized location monitoring [13]. It is one of the few areas where telecom operators can get advantages over Internet companies and effectively use own base of connected devices. One possible example: Sprint Geofence API [14]. Another example is Open API platform for Alcatel-Lucent [15]. Unfortunately, this prospect line in location monitoring is not elaborated yet from the practical point of view. The biggest problem by our opinion is the lack of common standards. One possible candidate for such standard in telecom was Parlay, but at this moment we cannot name one widely accepted candidate. That is why most of the scientific papers and practical implementations are devoted to the client side location monitoring.

The rest of the paper is organized as follows. Section 2 contains an analysis of existing projects devoted to network proximity. In Section 3, we consider our Spotique service and related applications.

2   Network Proximity

The main idea behind our Spotique service described below is the replacement of geo data with network proximity. We will try to describe geo fence (geographically restricted areas) with network proximity rules and replace geo locations monitoring with networks proximity monitoring. The reasons behind this movement are very transparent.

At the first hand, it will work indoor. At the second, it is based on the actions, performed by the most smart-phones anyway. Most of the mobile users keep Wi-Fi on all the time. And Wi-Fi scanning is a part of the network proximity. So, from the energy saving point of view, there are no extra operations. Network scanning is centralized. So, for any mobile phone all the installed LBS based on the network proximity will use the same data (share the same processes) automatically (see above-mentioned remark about cooperation for multiple LBS on the client side).

Network proximity based systems support dynamic LBS. It is described in [16], for example. If our "location" is linked by some way to Wi-Fi access point (network node), than not only this node could move. It could be opened (closed) dynamically right on the mobile phone:

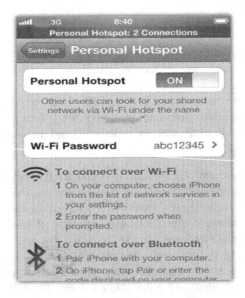

Fig. 3. Wi-Fi Access Point on the phone

One example for LBS based on the network proximity principles is SpotEx [17]. In this concept, any existing or even specially created wireless network node could be used as a presence sensor, which can open (discover) access to some dynamic or user-generated content. As a key service point, SpotEx introduces an external database with some rules (productions or if-then operators) related to the Wi-Fi access points. Typical examples of conditions in our rules are: AP (access point) with SSID Café is visible for mobile device; RSSI (signal strength) is within the given interval, etc. Based on such conclusions, we then deliver (make visible) user-defined messages to mobile terminals. In other words, the visibility of the content depends on the network context (Wi-Fi network environment).

Technically, SpotEx presents proximity information via some set of rules. Each rule is a logical production (if-then operator). The conditional part includes the following objects:

Wi-Fi network (SSID, mac-address)
RSSI (signal strength - optionally)
Time of the day (optionally)
ID for the client (mac-address)

It means that collection of all rules is a set of operators like:

IF IS_VISIBLE('mycafe') AND FIRST_VISIT() THEN
{present the coupon info }   [18]

LifeTag [19] uses collected database of so called Wi-Fi "fingerprints", including MAC addresses and the received signal strengths (RSSI) of nearby access points for discovering the user's behavioral patterns.

What could be used as fingerprint? One simplest approach could be based on the time any particular Wi-Fi access point is visible from the mobile phone [20]. The MAC addresses of visible access point let us logically estimate the location ("not far from that access point"). The mobile application on the phone can record periodically MAC addresses from received Wi-Fi beacons. A fingerprint is acquired by computing the fraction of times each unique MAC address was seen over all recordings. A tuple of fractions (each tuple element corresponding to a distinct MAC address) forms the Wi-Fi fingerprint of that place.

Fingerprint matching is performed by computing a metric of similarity between a test fingerprint and all candidate fingerprints. The comparison between two fingerprints, f_1 and f_2, is performed as follows. Denote M as the union of MAC addresses in f_1 and f_2. For a MAC address $m \in M$, let $f_1(m)$ and $f_2(m)$ be the fractions computed as above. Then the similarity S of f_1 and f_2 is computed as:

$$MinMax(m) = min(f_1(m), f_2(m))/max(f_1(m), f_2(m))$$

$$S = \sum_{m \in M} (f_1(m) + f_2(m)) * MinMax(m)$$

The intuition behind this metric is to add a large value to S when a MAC address occurs frequently in both f_1 and f_2.

The purpose of the fraction is to prevent adding a large value if a MAC address occurs frequently in one fingerprint, but not in the other. Note, that this calculation does not use signal strength measurement at all.

For geo-fence analogue we can compare current fingerprint and pre-recorded fingerprints for boarding points. Any given metric for similarity let us describe proximity (e.g., "close enough").

A classical approach to Wi-Fi fingerprinting [21] involves RSSI (signal strength). The basic principles are transparent. At a given point, a mobile application may hear ("see") different access points with certain signal strengths. This set of access points and their associated signal strengths represents a label ("fingerprint") that is unique to that position. The metric that could be used for comparing various fingerprints is k-nearest-neighbor(s) in signal space. It means that two compared fingerprints should have the same set of visible access points and they could be compared by calculating the Euclidian distance for signal strengths.

Fingerprinting is based on the assumption that the Wi-Fi devices used for training and positioning measure signal strengths in the same way. Actually, it is not so (due to differences caused by manufacturing variations, antennas, orientation, batteries, etc.). To account for this, we can use a variation of fingerprinting called ranking. Instead of comparing absolute signal strengths, this method compares lists of access points sorted by signal strength. For example, if the positioning scan discovered $(SS_A; SS_B; SS_C) = (-20; -90; -40)$, then we replace this set of signal strengths by their relative ranking, that is, $(R_A; R_B; R_C) = (1; 3; 2)$ [21]. As the next step, we can compare the relative rankings by using the Spearman rank-order correlation coefficient [22].

We can use signal strength features for distance estimation in terms of the Euclidean distance in signal strength space and the Tanimoto coefficient [23].

As a prerequisite we compute the vector of average signal strength per access point S'_x from the list of signal strength vectors S_x. In the Euclidean version the distances are defined as follows for each pair of average signal strength vectors S'_a, S'_b, with entries for non-measurable access points in either vector set to -100dBm:

$$d_{a,b} = \|S'_a - S'_b\|$$

For the Tanimoto coefficient version, the distance is computed as follows so the value increases as the vectors becomes more dissimilar:

$$d_{a,b} = 1 - (S'_a \cdot S'_b)/(\|S'_a\|^2 + \|S'_b\|^2 - S'_a \cdot S'_b)$$

Technically, it means that we can describe our geo-area as a set of basic point with statically calculated fingerprints. And later we can compare current fingerprint for mobile device with our basic fingerprints. By the similar schemes work almost all Wi-Fi based positioning systems. But there are two main problems. At the first hand, the task for creating basic "Wi-Fi mail stones" could be expensive. Also we will need to recalculate them every time our network environment is changed. At the second, we will face the same problems with energy consumption during the client side calculations. To overcome this we can use a fact that for the proximity calculation we do not need the distance. For the simple proximity calculation we can use some form of graph for signal strength versus distance for one Wi-Fi access point (Figure 4).

And for several Wi-Fi access points we can combine individual metrics.

As per energy consuming, we think that the most proper direction is to remove measurements processing from the mobile phone completely. It is what our Spotique service is about.

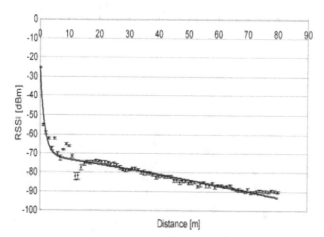

Fig. 4. RSSI vs. distance [24]

3   Spotique Service

Our Spotique service let broadcast hyper-local message to mobile clients. A typical usage scenario is: a user with a Wi-Fi device walking near a shop sees an ad for the hot offer on his Wi-Fi device, and also captures a coupon from the shop. The user then enters the shop and redeems his coupon by displaying it on the screen. By the same principles we can distribute information is Smart City projects, etc.

Spotique disconnects location-related calculations from mobile phone and uses server-side proximity detection based on Wi-Fi beacons. Wi-Fi client (mobile phone in our case) can periodically send so called probe request frame [25]. As per Wi-Fi spec, a station sends a probe request frame when it needs to obtain information from another station. For example, a client would send a probe request to determine which access points are within range. It is so called passive Wi-Fi tracking.

One benefit of the beacon-based approach is that it is implicitly location-aware. We are dealing with devices whose locations are known and whose accessibility is limited by the propagation of 802.11 signals. This approach works regardless of whether the client Wi-Fi devices are already connected to an existing Wi-Fi network, or the client is completely disconnected from all Wi-Fi networks. For mobile phones it is completely enough just to keep Wi-Fi networks mode switched on.

An additional benefit of this approach is that we eliminate the need to explicitly locate the client, which as a side-effect improves the privacy model. This scheme does not require from clients to send messages that explicitly reveal their location.

Technically, such external beacon-based monitoring can provide the following information about Wi-Fi based devices in proximity:

MAC – address
RSSI (signal strength)

There are several out-of-the-shelf components that can provide probe request detection [26, 27]:

Fig. 5. Wi-Fi beacon detection [27]

In our scheme we will use RSSI info for getting in-proximity devices and MAC-address for the identification. From the database on detected Wi-Fi devices we can get MAC-addresses for units in the proximity. Note that for the security reasons we can replace the real MAC-address with some hash.

But how can we detect our subscribers among them? For doing this we will deploy Cloud Messaging [28]. Google Cloud Messaging for Android (GCM) is a service that lets developers send data from servers to their applications on Android devices. This could be a lightweight message telling the Android application that there is new data to be fetched from the server (for instance, a movie uploaded by a friend), or it could be a message containing up to 4kb of payload data (so apps like instant messaging can consume the message directly). The GCM service handles all aspects of queuing of messages and delivery to the target Android application running on the target device. There is the similar service in Apple's world too (Apple Push Notification).

GCM allows 3rd-party application servers to send messages to their Android applications. GCM deployment depends on two things: Application ID and Registration ID. Application ID assigned to the Android application that is registering to receive messages. The Android application is identified by the package name from the manifest. This ensures that the messages are targeted to the correct Android application.

Registration ID is unique ID issued by the GCM servers to the Android application that allows it to receive messages. Once the Android application has the registration ID, it sends it to the 3rd-party application server, which uses it to identify each device that has registered to receive messages for a given Android application. In other words, a registration ID is tied to a particular Android application running on a particular device.

An Android application on an Android device doesn't need to be running to receive messages. The system will wake up the Android application via Intent broadcast when the message arrives, as long as the application is set up with the proper broadcast receiver and permissions.

And here is the main idea for Spotique: during the registration for GCM collect MAC-address for Wi-Fi tracking. Spotique presents mobile application that lets users subscribe to the topics (read – local businesses) they are interested. Local businesses define our topics. Each subscription tights in the server side database the following things: topic, MAC-address for the subscriber (for his mobile phone) and Registration ID for GCM.

Each time when passive tracking system at the local business location obtained MAC-address for in-proximity mobile device, we can check that MAC-address against subscription database. And as soon as the user (mobile phone, actually) is subscribed, we can push to his phone our custom message, using his Registration ID from the same database [29].

In this schema we eliminate the need to explicitly locate the client. This scheme does not require from clients to send (or publish in some social network) messages that explicitly reveal their location.

Note again, that for improving the privacy we do not need even to save in our database original MAC-addresses. It is enough to keep some hash-code instead of the real address.

For Apple Push Notification (APN) service each device establishes an accredited and encrypted IP connection with the service and receives notifications over this persistent connection. If a notification for an application arrives when that application is not running, the device alerts the user that the application has data waiting for it. Developers ("providers") originate the notifications in their server software. The provider connects with APNs through a persistent and secure channel while monitoring incoming data intended for their client applications. When new data for an application arrives, the provider prepares and sends a notification through the channel to APNs, which pushes the notification to the target device. For the future development with APN we will keep the same principles for subscription as the above described GCM model.

4   Conclusion

This paper discusses geo-fence limitations for mobile applications and offers as a replacement network proximity model. We discuss several models related to Wi-Fi proximity. Namely, they are fingerprints and rule based proximity. We describe a new location based service Spotique. It is based on passive Wi-Fi tracking and Cloud Messaging. This new approach let us effectively deploy indoor location based services and provide a significant energy saving for mobile devices comparing with the traditional methods. In Spotique we eliminate the need to explicitly locate the client, which as a side-effect improves the privacy model. This scheme does not require from clients to send messages that explicitly reveal their location. The proposed approach eliminates one of the biggest concerns for location based systems adoption – privacy.

References

1. Reclus, F., Drouard, K.: Geofencing for fleet & freight management. In: 2009 9th International Conference on Intelligent Transport Systems Telecommunications (ITST), pp. 353–356 (2009)
2. LaMarca, A., et al.: Place lab: Device positioning using radio beacons in the wild. In: Gellersen, H.-W., Want, R., Schmidt, A. (eds.) PERVASIVE 2005. LNCS, vol. 3468, pp. 116–133. Springer, Heidelberg (2005)
3. Constandache, I., Gaonkar, S., Sayler, M., Choudhury, R.R., Cox, L.: Enloc: Energy-efficient localization for mobile phones. In: Proceedings of IEEE INFOCOM Mini Conference 2009, Rio de Janeiro, Brazil (2009)
4. Zhuang, Z., Kim, K.-H., Singh, J.: Singh Improving Energy Efficiency of Location Sensing on Smartphones. In: MobiSys 2010, San Francisco, California, USA, June 15-18 (2010), Copyright 2010 ACM 978-1-60558-985-5/10/06
5. http://forums.watchuseek.com/f296/garmin-fenix-ongoing-review-several-parts-746366-11.html (retrieved: March 2013)
6. Hristova, N., O'Hare, G.M.P.: Ad-me: Wireless Advertising Adapted to the User Location, Device and Emotions. In: Thirty-Seventh Hawaii International Conference on System Sciences (HICSS-37) (2004)
7. Lu, H., et al.: The jigsaw continuous sensing engine for mobile phone applications. In: Proc. 8th ACM Conf. Embedded Netw. Sens. Syst., SenSys 2010, pp. 71–84. ACM (2010)
8. Paek, J., Kim, J., Govindan, R.: Energy-efficient rate-adaptive gpsbased positioning for smartphones. In: Proc. 8th MobiSys 2010, pp. 299–314. ACM (2010)
9. Namiot, D.: Geo messages. In: 2010 International Congress on Ultra Modern Telecommunications and Control Systems and Workshops (ICUMT), pp. 14–19 (2010), doi:10.1109/ICUMT.2010.5676665
10. Priyantha, B., Lymberopoulos, D., Liu, J.: LittleRock: Enabling Energy-Efficient Continuous Sensing on Mobile Phones. IEEE Pervas. Comput. 10(2), 12–15 (2011)
11. Song, L., Kotz, D., Jain, R., He, X.: Evaluating next-cell predictors with extensive wi-fi mobility data. IEEE Trans. Mobile Comput. 5(12), 1633–1649 (2006)
12. Chon, Y., Talipov, E., Shin, H., Cha, H.: Mobility Prediction-based Smartphone Energy Optimization for Everyday Location Monitoring. In: SenSys 2011, Seattle, WA, USA, November 1-4 (2011), Copyright 2011 ACM 978-1-4503-0718-5/11/11
13. Namiot, D., Sneps-Sneppe, M.: Where Are They Now – Safe Location Sharing. In: Andreev, S., Balandin, S., Koucheryavy, Y. (eds.) NEW2AN/ruSMART 2012. LNCS, vol. 7469, pp. 63–74. Springer, Heidelberg (2012)
14. Sprint Geofence API, http://developer.sprint.com/dynamicContent/geofence/ (retrived: February 2013)

15. Open API platform, `http://www2.alcatel-lucent.com/application_enablement/` (retrieved: March 2013)
16. Namiot, D.: Context-Aware Browsing–A Practical Approach. In: 2012 6th International Conference on Next Generation Mobile Applications, Services and Technologies (NGMAST), pp. 18–23. IEEE (2012), doi:10.1109/NGMAST.2012.13
17. Namiot, D., Schneps-Schneppe, M.: About location-aware mobile messages. In: NGMAST 2011, pp. 48–53 (2011), doi:10.1109/NGMAST.2011.19
18. Namiot, D., Sneps-Sneppe, M.: Proximity as a Service. In: 2012 2nd Baltic Congress on Future Internet Communications (BCFIC), pp. 199–205 (2012), doi:10.1109/BCFIC.2012.6217947
19. Rekimoto, J., Miyaki, T., Ishizawa, T.: LifeTag: WiFi-Based Continuous Location Logging for Life Pattern Analysis. In: Hightower, J., Schiele, B., Strang, T. (eds.) LoCA 2007. LNCS, vol. 4718, pp. 35–49. Springer, Heidelberg (2007)
20. Azizyan, M., Constandache, I., Roy, R.: SurroundSense: mobile phone localization via ambience fingerprinting. In: Proceedings of the 15th Annual International Conference on Mobile Computing and Networking (MobiCom 2009), pp. 261–272 (2009), doi:10.1145/1614320.1614350
21. Chen, Y., Chawathe, Y., LaMarca, A., Krumm, J.: Accuracy characterization for metropolitan-scale Wi-Fi localization. In: ACM MobiSys (2005)
22. Stuart, A.: The correlation between variate-values and ranks in samples from a continous distribution. British Journal of Statistical Psychology 7(1), 37–44
23. Kjaergaard, M., Wirz, M., Roggen, D., Troster, G.: Mobile sensing of pedestrian flocks in indoor environments using WiFi signals. In: 2012 IEEE International Conference on Pervasive Computing and Communications (PerCom), pp. 95–102 (2012)
24. Janga, W., Healyb, W.: Wireless sensor network performance metrics for building applications. Energy and Buildings 42(6), 862–868, doi:10.1016/j.enbuild.2009.12.008
25. Chandra, R., Padhye, J., Ravindranath, L., Wolman, A.: A Beacon-Stuffing: Wi-Fi without Associations. In: Eighth IEEE Workshop on Mobile Computing Systems and Applications, HotMobile 2007, pp. 53–57 (2007)
26. Smartphones movement track, `http://www.technologyreview.com/view/427687/if-you-have-a-smart-phone-anyone-can-now-track/` (retrieved: April 2013)
27. Meshlium Xtreme, `http://www.libelium.com/products/meshlium` (retrived: April 2013)
28. Hansen, J., Gronli, T., Ghinea, G.: Cloud to Device Push Messaging on Android: A Case Study. In: 2012 26th International Conference on Advanced Information Networking and Applications Workshops (WAINA), pp. 1298–1303 (2012)
29. Sneps-Sneppe, M., Namiot, D.: Spotique: A New Approach to Local Messaging. In: Tsaoussidis, V., Kassler, A.J., Koucheryavy, Y., Mellouk, A. (eds.) WWIC 2013. LNCS, vol. 7889, pp. 192–203. Springer, Heidelberg (2013)

An Integrated Smart System
for Ambient-Assisted Living

Thato E. Foko, Nomusa Dlodlo, and Litsietsi Montsi

CSIR-Meraka Institute, Box 395, Pretoria 001, South Africa
{tfoko,ndlodlo,lmontsi}@csir.co.za

Abstract. Ambient-assisted living (AAL) is an initiative to extend the time the elderly can live in their home environment by increasing their autonomy and assisting them carry out their daily activities. AAL systems exploit information and communication technologies (ICT) in the assistance to carry out daily activities, health and activity monitoring, enhancing safety and security and getting access to social, medical and emergency systems. These ICTs are in the form of smart systems, which are physical objects that are augmented with sensing, processing and network capabilities, enabling them not only to intercommunicate with one another, but also to exchange information with people and react to their environment. This paper is on a low-cost technology customised to the South African environment to support ambient assisted living. The technology takes advantage of South Africa's digitalisation programme to provide broadband access in the support of AAL. Digital television as a gateway to internet access, wireless mesh networks for communication, motes for machine to machine communication and smart phones are the technologies supported in this architecture. A survey of AAL technologies was conducted and features of these systems that would be useful in defining our architecture identified. These features contribute to the development of an architecture for AAL. This research feeds into extending the body of knowledge on AAL technologies.

Keywords: smart systems, ambient-assisted living, intelligent environment, emergency case prediction.

1   Introduction

The life expectancy of people is increasing and related to that is an increase in the population of the elderly. The idea is to ensure that the elderly stay longer in their homes, and only moving to nursing homes when they can no longer care for themselves.

Ambience is about creating a relaxing environment in the immediate surroundings. Ambient-assisted living (AAL) fosters the provision of equipment and services for the independent or more autonomous living of elderly people via the seamless integration of ICT within homes and residences, thus increasing their quality of life and autonomy and reducing the need for entering in residences or aiding it when it happens [1]. Ambient-assisted living (AAL) technologies will ensure that the elderly are monitored remotely. This paper is on a low-cost technology customised to the

S. Balandin et al. (Eds.): NEW2AN/ruSMART 2013, LNCS 8121, pp. 128–138, 2013.

South African environment to support ambient assisted living. The technology takes advantage of South Africa's digitalisation programme to provide broadband access in the support of AAL. Digital television as a gateway to internet access, wireless mesh networks, motes and smart phones are the technologies supported in this architecture. A survey of AAL technologies was conducted and features of these systems that would be useful in defining our architecture identified. These features contribute to the development of an architecture for AAL. This research feeds into extending the body of knowledge on AAL technologies.

Section 2 is the problem statement. Section 3 is an introduction to ambient-assisted living. The smart AAL technologies identified in Section 4 are classified under intelligent environments and emergency case prediction. Section 5 is on components and architecture of the system. Section 6 is on the conclusion.

2   Problem Statement

The need to extend the time that the elderly can live in their home environment by increasing their autonomy, security and assisting them carry out their daily activities calls for appropriate low-cost technologies to support ambience. The aim of this research is to come up with the design of an architecture to support AAL to both urban and remote areas of South Africa. There are many such systems in place, but each of these plays a limited role. Therefore we are moving towards designing an integrated system that incorporates the features of a number of such systems. The architecture takes advantage of South Africa's TV digitalisation programme which is set to be in full swing by the end of 2013 when analog switch-off occurs. The digitalisation programme will see an S.A brand of low-cost appropriate technologies including digital TV, TV white spaces, set-top-boxes and wireless mesh networks in use. The digitalisation programme supports broadband access through freeing, towards data carriage, of TV white spaces which were previously held by analogue signals.

The question that this papers answers is: "what architecture of an AAL system that incorporates the features of a number of such systems can we come up with that would take cognisance of South Africa's digitalisation programme?"

The approach used here is:

- The identification of AAL systems through a literature survey
- Analysis of the systems to identify features that would benefit our architecture
- Design of an architecture based on the identified features

3   Ambient-Assisted Living

Among the prominent goals of AAL solutions are:

- Improve the quality of life of care giving persons
- Reduce the need for external assistance

- Reduce health care costs for the individual and society
- Avoid stigmatisation of the elderly

AAL covers a hybrid product [23]: (1) a basic technical infrastructure in the home environment (sensors, actuators, communication devices) and (2) services provided by a third party with the aim of independent living at home with assistance in the following areas of communication, mobility, self-sufficiency and life at home. AAL systems include activity monitoring which includes a multi-sensor data acquisition in the home to monitor a person's behaviours as well as the processing of this raw sensor data in order to make high-order inferences about the person's activities and daily life patterns. In-home activity monitoring is essential in a whole range of potential applications such as falls monitoring and the development of a new generation of smart environments to support frail and disabled people living at home [25].

The technical components of an AAL system are [25]: (1) Fitting the user's home with ambient sensors and actuators, (2) Mobile components carried by the user (sensors on/near the body, mobile terminal devices) offering assistance in the home and when out and about, (3) The service providers' computing centre that performs external computing jobs and offers services such as remote maintenance, remote configuration, back-up or also an app-store with rechargeable software module on AAL systems, and (4) Third parties, i.e., those offering electronic services or services used by the AAL system without actually being part of the AAL system. Examples are IT systems from service providers on the healthcare sector that are integrated into an AAL system.

AAL includes several categories of support applications, for example (Hiroko, 2004): (1) Emergency treatment services aim at the elderly prediction of and recovery from critical conditions that might result in an emergency situation and the safe detection and alert propagation of emergency situations, (2) Autonomy enhancement service enables an independent living of the assisted persons, in case of lacking capabilities of the assisted persons and (3) Comfort services. The following application fields of AAL have been identified in [2]: (1) Information assistance, e.g., medication instruction in the morning, a key-finder, an all-in-one remote control, (2) Intelligent environment behaviour, i.e., learn human actions and offer the right action at the right time, (3) Emergency case prediction, i.e., predict future possibly dangerous situations, (4) Security – e.g., unlocked car detected and (5) Privacy.

Enabling technologies to develop AAL applications in these domains are [21]: (1) Sensing – anything, anywhere, in-body or on-body, in-appliances or non-appliance or in the environment (home, outdoor, public spaces), (2) Reasoning – collecting, aggregating, processing and analysing data, transforming them into knowledge in different and often across connected spaces, (3) Acting – automatic control through actuators and feedback, (4) Communication – sensors and actuators are connected to one or more reasoning systems which in turn might be connected to other reasoning systems or actuators and (5) Interaction – intelligent interaction of people within systems and services.

4   Smart Systems in Ambient-Assisted Living

According to [9], "smart objects are autonomous physical/digital objects augmented with sensing, processing and network capabilities. They carry chunks of application logic that let them make sense of their local situation and interact with human users. They sense, log and interpret what's occurring within themselves and the world, act on their own, intercommunicate with each other and exchange information with people". The smart systems in AAL in this section are classified under intelligent environments and emergency case prediction as follows:

4.1   Intelligent Environments

SenseWear armband is a wearable body monitor designed for both normal subjects and patients with chronic pulmonary disease and allows capturing body movement and energy expenditure [20]. Vitaphone is a mobile phone with integrated heart monitoring functions [26]. Dr. FeelGood is a combination of a PDA and mobile phone which measures various physiological functions [13]. SmartPillow is an electronic monitoring device in the form of a traditional pillow, which checks the user's basic vital parameters such as respiration, pulse and body temperature and in the case of an emergency or illness immediately and notifies medical personnel [19]. The SmartVest is a wearable physiological monitoring system that monitors various physiological parameters such as ECG, heart rate, blood pressure, body temperature and transmits the captured physiological data along with geographic location of its wearer to remote monitoring stations [18]. Different types of sensors are integrated into the design of the Smart Shirt, which allow monitoring of a variety of vital parameters, including heart rate, electrocardiogram (ECG), respiration, temperature of vital functions [6].

In the diabetes system [24], patients take measurements which are fed into the biometric devices and then read by a mobile device. The mobile device captures the vital signs, interprets them and then compares them independently to the measurement ranges established for the patient's disease. The rest are interpreted and transmitted by the control modules embedded in the mobile device. Vital signs such as the glucose level, blood pressure, temperature are measured and transmitted via Bluetooth technology between the mobile and biometric devices. Information is transmitted to a central database and advisory system for evaluation and monitoring by the medical server. The program is installed on the patient's mobile phone. The application is developed using the Android operating system with remote connectivity through MySQL.

4.2   Emergency Case Prediction

In the mobile approach to AAL [8], communication and delivery services is in a location-based manner using built in GPS, WiFi and 3G mobile connectivity. Bluetooth-compatible blood pressure and body-weight measurement devices are complemented with a body-mounted wireless physiological sensor to monitor activity, body temperature and stress. Telemetric data is continuously recorded on a local host computer while simultaneously being also sent to a central database, where rule-based systems or monitoring health personnel may make emergency assessment. The communications platform virtual human interface was designed to bridge the gap between people and computers by using virtual reality and animation.

In the ElderCare platform [12], elderly people are offered an interactive TV interface. Caretaking staff request and register information about their caring procedures through NFC mobiles and touchscreens. Relatives follow the elderly people's life logs through RSS feeds or micro-blogging services such as Twitter. The architecture is a low-cost and affordable, unobtrusive so that it can be integrated within a home, easily deployable so that relative can plug into the system through set-to-box hardware connected to the TV, and is usable and accessible to every user collective. The architecture centres around interactive TV (iTV) which combines OSGi middleware, RFID and NFC. iTV provides interaction with a device (remote control). Notification services to keep relatives up-to-date, e.g. email, SMS, RSS or Twitter. Care givers can access the ICT infrastructure to monitor elderly. OSGi-based service architecture allows easy service creation, context and knowledge intelligent management and the integration of some custom-built hardware such as RFID plugs to easily identify connecting devices. ElderCare's Central Server remotely manages the local systems, by collecting data at those installations, storing it and analysing it. The MobileCare Logging system is used by relatives and staff to record, through NFC mobiles in elderly people's RFID wristbands, events and caring procedures performed on them. RFID wristbands are also reported by mobiles regularly through Bluetooth to the local system, which forwards them to a Central Server which then notifies non-privacy invasive selections of them to relatives or friends.

A wireless system for fall detection is based on an accelerometer controlled by an FPGA (programmable logic) [11]. The AAL application communicates with the careholders and relatives of the assisted person through an ADSL-based gateway. The wearable system is a PCB with commercial three-axes, digital output MEMS accelerometer chip, a ZigBee transceiver and programmable logic (FPGA). The raw acceleration stream from the MEMS sensor is processed by suitable algorithms on the FPGA, whose aim is to detect potential falls and their level of confidence, deliver alarm information to the gateway by means of the ZigBee transceiver. The core of the system is the FPGA that controls the accelerometer, reads out its digital output and delivers the necessary information to the ZigBee radio module, in order to have it transmitted to the gateway.

Ambient Care System (ACS) [14] frameworks provide remote monitoring, emergency detection, activity logging and personal notifications dispatching services. The ACS system is built on top of a WSN composed of devices, external sensors and PDAs enabled with ZigBee technology. The middleware is integrated into an OSGi framework that processes the acquired information to provide ambient services and also enable smart network control. The devices continuously acquire health parameters, ambient information and other context data. Sensors are equipped with Bluetooth. The WSN is composed of traditional motes (small and resource-constrained wireless devices with sensing capabilities), but may also include personal mobile devices acting as full-enabled network nodes. The ACS consists of:

- A personal network which includes all devices a person must wear
- ZigBee tag to enable indoor positioning and a PDA equipped with ZigBee technology for outdoors
- A Home Network which includes home infrastructure sensors
- A Core Care Network which acts as a bridge of communication between the home and the service providers

CareLab's LifeStyle assistant [7] consists of a range of sensors distributed throughout the home to register activities carried out within it, which are then processed by a content-aware reasoning engine. The reasoning engine identifies critical incidents and can respond to them by alerting care centres or relatives.

VirtualECare [16] [5] is an intelligent multi-agent system to monitor, interact and provide its customers with healthcare services based on open standards. The system is interconnected with healthcare institutions, leisure centres, training facilities, shops and patients' relatives. The VirtualEcare architecture is a distributed one, interconnected through a network (LAN, MAN, WAN). Clinical data collected via sensors is sent to CallCareCentre which then redirects it to a Group Decision Support System. The rest of the data is sent to a CallServiceCentre. The CallServiceCentre analyses the data and makes decisions on actions to be taken according to it. On the other hand the CallCareCentre staff analyse the clinical data and act accordingly. The Group decision is in charge of all decisions taken at the VirtualECare platform.

Breathing problems such as chronic obstructive pulmonary disease, chronic bronchitis, etc., present the need to carry out home respiratory therapy. This requires the deployment of specific devices for supplemental oxygen therapy and monitoring of the status of the patient. The architecture [26] permits to connect wirelessly the biomedical sensors worn by the patient to the gateway deployed at the house and the safe and global communication and secure communication with remote monitoring centres and intelligent information systems. The sensors measure lung capacity, peak expiratory flow, breathing frequency, breathing amplitude, electrocardiogram, oxygen saturation and lung noise.

5   Components of the Architecture

This section consists of a subsection on components of the system and another on the architecture.

5.1   Components of the System

An analysis of the previous section identifies the following features that the system should have:

- Sensors for information acquisition and dissemination. These sensors should be wearable. Examples of such sensors are motes, RFID tags and QR codes.

The body parameters that these sensors should identify are: heartbeat, respiration rate, pulse, temperature, ECG, blood pressure, glucose levels, body movements, energy expenditure, lung capacity, peak expiratory flow, breathing frequency, breathing amplitude, oxygen saturation and lung noise

- Wireless communications for connected environments. These include Bluetooth, near field communication (NFC), Zigbee, WiFi, 3G, 6LowPan
- Mobile technology for remote monitoring, emergency detection, activity logging and remote control of smart devices
- Predictive and decision-making capabilities (intelligence) to interpret the data acquired
- Intelligent devices to support intelligent actions, e.g. smartphones
- Notification of medical personnel, caregivers and relatives, e.g. sms, mms, RSS, Twitter
- Geographic location of patient via GPS
- Interactive TV with set-top-box as a gateway to the internet
- Home Network

Important technologies of digital TV and white spaces, wireless mesh networks and motes are described in this subsection.

5.1.1   Broadband through Digital TV

Broadband internet access is a high rate connection to the internet. It is characterised by the ability to handle high volumes of data and high data rate content. Broadband access has not reached all corners of South Africa and yet the need to adopt it for improved health of the citizens does exist. This research is about taking advantage of South Africa's digitalisation programme to deliver health services. The architecture includes the integration of an internet TV platform and TV whitespaces technology for broadband access. TV whitespaces is about using TV frequencies that are not in use for communication. Digital Terrestrial TV (DTT) is a set of standards for transmission of TV signals in digital form. A set-top-box/adapter (STB) converts the signal used by DTT of older analogue models of TV receivers to digital. To extend DTT coverage a wireless mesh network is used in the architecture

5.1.2   Wireless Mesh Networks

A mesh is connectivity between two or more nodes in a network. Mesh nodes are small radio transmitters that function in the same way as a wireless router. Nodes use the common WiFi standards known as 802.11a, b and g to communicate wirelessly with other users. Nodes are programmed with software that tells them how to interact within the larger network. Information travels across the network from point A to point B by hopping wirelessly from one mesh node to the next. The node automatically chooses the quickest and safest path in a process known as dynamic routing. Between the source and destination nodes there is more than one relay. Characteristics of mesh networks that distinguish them from the linear chain are multi-routing capability, multi-hopping and multi-relays. On the software end the mesh network has to self-configure, self-heal and self-organise. Self-configuration is finding the route automatically without human intervention by locating the route, neighbour, topology and connectivity. It should set its own parameters. Self-healing means that if one node dies due to battery failure, theft, lightning strike, etc, the network picks up by using other routes. Self-organising means that if any new nodes are added to the network, the network should organise and recognise their IP addresses.

5.1.3   Motes

A sensor node, also known as a mote, is a node in a wireless network that is capable of performing some processing, gathering sensory information and communicating with other connected nodes in the network. A mote is a node but a node is not always a mote. Motes are low power wireless sensor network devices. Sensors have limited resources in terms of power/CPU/memory. They operate at low voltage and low current. A mote has a sensor unit, a power unit, a transceiver unit, an ADC unit and a processor. Motes have radio links, which enable them to communicate and exchange data with one another and to self-organise into ad-hoc networks.

5.2   Architecture of the System

The wearable sensors (motes) (as shown in Figure 1) detect various body parameters such as the body temperature, blood pressure, diabetes, heart rate, etc.

Each of these sensors has a valid internet protocol (IP) address which makes it part of the internet, and also has a GPS location. These wearable sensors (motes) communicate with the motes in the house via mote to mote communication. These house motes, known as border routers, are connected by wire to a digital TV, which can read the border router's IP address. A border router is a device that routes packets of data that reside on the edge or boundary of a network. The TV acts a gateway to the internet. The TV gateway's IP address is registered at the doctors', and so are the TV gateways for all the other patients' homes. The function of the digital TV can also be controlled via remote control. The set-top-box (STB) converts analogue signals into digital signals for the older models of the TV, which is not necessary in the newer digital TV models. In this case, the TV acts as a gateway to the computing centre via a TV base station. The computing centre has a database which records data sent to the doctor from the sensors (motes) and doctor's response to the data received. The intelligence management software in the computing centre extracts data from the database, analyses it against set rules to determine the state of the patient.

An SMS, MMS or Twitter message is sent via mobile network to the doctor's/caregiver's cell phone should the physiological parameters fall outside the norm. In this case the mobile network is a wireless mesh network. The benefits of wireless mesh networks are as described in section 5.1.2. Occasionally the doctor logs onto the computing centre server to check on a patient. If need be the doctor contacts emergency services to dispatch a service to the patient. The doctor may also contact

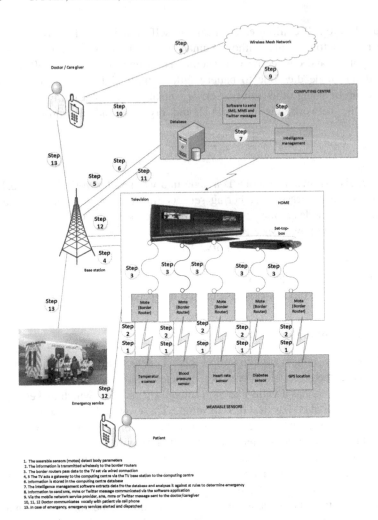

1. The wearable sensors (motes) detect body parameters
2. The information is transmitted wirelessly to the border routers
3. The border routers pass data to the TV set via wired connection
4, 5 The TV acts a gateway to the computing centre via the TV base station to the computing centre
6. Information is stored in the computing centre database
7. The intelligence management software extracts data fro the database and analyse it against st rules to determine emergency
8. Information to send sms, mms or Twitter message communicated via the software application
9. Via the mobile network service provider, sms, mms or Twitter message sent to the doctor/caregiver
10, 11, 12 Doctor communicates vocally with patient via cell phone
13. In case of emergency, emergency services alerted and dispatched

the patient directly via smartphone. The message goes from the doctor's smartphone, via the computing centre server, to the TV base station and then to the patient. The purpose of the GPS location for each wearable sensor (mote) is so that the position of the patient can be identified. Relatives can also plug into the TV to monitor the patient's state.

6   Conclusion

The research was on the design of an architecture to support AAL. There has been many systems in place to support AAL. Our research is an addition to the body of knowledge on these AAL technologies by trying out a combination of a number of them. This opens room for research into other possibilities to defining new AAL systems architectures. This has been achieved through an analysis of existing technologies, extracting features that we consider beneficial and designing a new architecture. It shows that there is still room for expansion in AAL research and development by building on what is on the ground and coming up with different combinations of technologies.

The shortcomings of this research would be that no single architecture would encompass all the features identified. It is also true that each of the technologies identified is a whole research on its own, encompassing various research areas, which takes many years of work by multidisciplinary teams. Constituting such teams is complex enough. The final results of this research though would be of use in the real world, in that a new technology is developed.

This paper reports on smart AAL technologies, to improve the quality of life of the elderly. These smart technologies have the ability to sense, reason, act, communicate and interact. The technologies identified in this paper fall under emergency case prediction and intelligent environments. There is room though for further research into smart technologies that offer information assistance to the elderly, e.g. reminders to take medication, security and privacy.

References

1. Ambient-assisted living joint programme (2009), http://www.aal-europe.eu
2. Andrushevich, A., Kistler, R., Bieri, M., Klapproth, A.: ZigBee / IEEE 802.15.4 Technologies in AAL applications, http://www.ihomelab.ch/fileadmin/Dateien/PDF/NewsEvents/EuZDC2009_paper.pdf
3. Beigl, M., Gellersen, H.W., Schmidt, A.: MediaCups: experience with design and use of computer-augmented everyday objects. Computer Networks 35(4), 401–409 (2001)
4. Cook, D.J., Das, S.K.: How smart are our environments? An updated look at the state of the art. Pervasive and Mobile Computing 3, 53–73 (2007)
5. Costa, R., Carneiro, D., Novais, P., Lima, L., Machado, J., Marques, A., Neves, J.: Ambient-assisted living
6. Demiris, G., Jan, J.: Rejuvenating home health care and tele-home care. In: Jan, J. (ed.) E-health Care Information Systems: an Introduction for Students and Professionals, pp. 267–290. Jossey-Bass, San Francisco
7. De Ruyter, B., Zwartkruis, J., Pelgrim, E., Aarts, E.: Economic and social implications: ambient-assisted living research in the Care Lab. Gerpsych 23(2), 115–119 (2010)
8. Hanak, D., Szijarto, G., Takacs, B.: A mobile approach to ambient-assisted living, http://cs.bme.hu/~dhanak/iadis_wac.pdf
9. Kortuem, G., Kawsar, F., Fitton, D., Sundramoorthy, V.: Smart objects as building blocks for the internet of things. In: Thiesse, F., Michahelles, F. (eds.) IEEE Internet Computing, pp. 30–37 (2010)
10. Lhotska, L., Havlik, J., Panyrek, P.: Systems approach to ambient-assisted living applications: a case study
11. Lombardi, A., Ferri, M., Rescio, G., Grassi, M., Malcovati, P.: Wearable wireless accelerometer with embedded fall-detection logic for multi-sensor ambient-assisted living applications. In: IEEE Sensors 2009 Conference, pp. 1967–1970 (2009)

12. Lopez-de-Ipina, D., Blanco, S., Laiseca, X., Diaz-de-Sarralde, I.: ElderCare: an interactive TV-based ambient-assisted living platform. In: 2nd International Workshop on Ambient-Assisted Living, IWAAL 2010, Valencia, Spain, September 7-10, pp. 75–82 (2010)

13. Marey, A., Buchner, M., Noehte, S.: Mobile monitoring – eine neve Chance fur die Diagnostik. In: Proceedings of Workshop "Mobiles Computing in der Medizin", Cologne, Germany, April 2, pp. 158–165 (2011)

14. Martin, H., Bernados, A.M., Bergesio, L., Tarrio, P.: Analysis of key aspects to manage wireless sensor networks in ambient-assisted living environments

15. Mattern, F.: From smart devices to smart everyday objects. In: Proceedings of Smart Objects Conference, SOC 2003, pp. 15–16 (2003)

16. Novais, P., Costa, R., Carneiro, D., Machado, L., Lima, L., Neves, J.: VirtualECare: Group support in collaborative network organisations for ambient-assisted living. In: Yogega, K., Bos, L., Gibbons, M.C. (eds.) Handbook of Digital HomeCare. Springer Series: Series in Biomedical Engineering, pp. 151–178 (2009)

17. Ovaska, E., Evesti, A.: Challenges and solutions of secure smart environments. VTT technical Research Centre, Finland (2011)

18. Pandia, P.S., Safeer, K.P., Gupta, P., Shakunthala, D.T., Sundersheshu, B.S., Padaki, V.C.: Wireless sensor for wearable physiological monitoring. Journal of Networks 3(5), 21–29 (2010)

19. Park, S.H., Won, S.H., Lee, J.B., Kim, S.W.: Smart home – digitally engineered domestic life. Personal Ubiquitous Computing 7(3&4), 189–196 (2003)

20. Pitta, F., Troosters, T., Probst, V.S., Spruit, M.A., Decrammer, M., Gosselink, R.: Quantifying physical activity in daily life with questionnaires and motion sensors in COPD. European Respiratory Journal 27(5), 1040–1055 (2006)

21. Siciliano, P.: Enabling technologies for ambient-assisted living, Institute for Microelectronics and Microsystems, CNR-IMM, Leece, Italy

22. Smart environment, http://en.wikipedia.org/wiki/Smart_environment (accessed December 05, 2012)

23. The German AAL Standardisation Roadmap, VDE Association for Electrical. Electronic and Information Technologies (January 2012)

24. Villarreal, V., Fontecha, J., Hervas, R., Bravo, J.: An architecture for development of ambient assisted living applications: a case study in diabetes, http://mami.uclm.es/ucami2011/Proceeding/papers/Long%20Paper/ucami2011_submission_51.pdf

25. Virone, G., Sixsmith, A.: Towards information systems for ambient-assisted living, http://www.gerontechnology.info/Journal/Proceedings/ISG08/papers/136.pdf

26. Vitaphone, Telemedizin aktuel, Medizin – Service institute fur Startegisches Marketing and Kommunikation, Wesel, Germany (August 2, 2008)

Discovery of Convoys in Network Proximity Log

Dmitry Namiot[1] and Manfred Sneps-Sneppe[2]

[1] Lomonosov Moscow State University,
Faculty of Computational Mathematics and Cybernetics,
Moscow, Russia
dnamiot@gmail.com
[2] ZNIIS,
M2M Competence Center,
Moscow, Russia
manfreds.sneps@gmail.com

Abstract. This paper describes an algorithm for discovery of convoys in database with proximity log. Traditionally, discovery of convoys covers trajectories databases. This paper presents a model for context-aware browsing application based on the network proximity. Our model uses mobile phone as proximity sensor and proximity data replaces location information. As per our concept, any existing or even especially created wireless network node could be used as presence sensor that can discover access to some dynamic or user-generated content. Content revelation in this model depends on rules based on the proximity. Discovery of convoys in historical user's logs provides a new class of rules for delivering local content to mobile subscribers.

Keywords: location, lbs, proximity, convoys, context aware.

1 Introduction

The term a trajectory refers to the movement of an object given by a continuous curve in the space. The past trajectory of an object is usually approximated by a collection of time stamped positions. For example, in our research we target mobile phones where positions usually could be obtained from a GPS device.

Convoy is a group of objects that travel together for more than some minimum duration of time. More probably, that the original task for discovery of convoys (groups of objects with coherent trajectory patters) was oriented to the military applications. As per nowadays research papers, a number of applications may be envisioned. The discovery of common routes among citizens may be used for the scheduling of public transport. The discovery of convoys for trucks may be used for throughput planning. The identification of cars that follow the same routes approximately at the same time may be used for creating carpooling, etc.

In our paper we will follow to convoy definitions from [1] and avoid restrictions on the sizes and shapes of the discovered trajectory patterns. Generic trajectory pattern of any shape and any extent will be based on the notion of density connection [2]. It enables the formulation of arbitrary shapes of groups.

S. Balandin et al. (Eds.): NEW2AN/ruSMART 2013, LNCS 8121, pp. 139–150, 2013.

Shortly, convoy is a group of moving object where included objects are in density connection the consecutive time points. Objects are density-connected if a sequence of objects exists that connects the two objects and the distance between consecutive objects does not exceed the given value. As it follows from this definition, convoy definition depends on the time, during which the objects in the density-connected group traveled together. As per our target area, we will consider relatively short traveling time and does not consider the distances between pairs of trajectories across all of time.

The next often used in this context terminology is moving cluster (or cluster of moving objects) [3]. The moving cluster exists if a shared set of objects exists across some finite time, but objects may leave and join a cluster during the cluster's life time. So, the semantic is different and moving clusters do not necessarily qualify as convoys (in the pure terms). Both the location and the set of objects of a moving cluster change over time. But sometimes, both definitions are mixed and moving cluster means the same as convoy. Some of authors define dynamic convoys and evolving convoys [4]. Dynamic convoys allows dynamic members under constraints imposed by some parameters (actually, by user-defined parameters). An evolving convoy captures the relationship between different stages of convoys, so that convoys in some stage has more (fewer) members than its previous stage. Another interesting term in this space is flock. Flock is a set of objects that travel within a range while keeping the same motion. Anyway, all patterns covering capturing "collaborative" or "group" behavior between moving objects. The difference between all the above mentioned patterns is the way they define the relationship between the moving objects.

In this paper we will investigate a special case for convoys (moving clusters) discovery. At the first hand, we have not location information in our database. We are working with some context-aware data discovery application, which lets mobile users get hyper-local content right on the mobile phones. This application (namely, SpotEx, first time described in [5]) based on the network proximity. Shortly, it defines logical rules (productions) that depends on the network proximity and lets mobile users discover local content via fired rules. Also, this application can record historical proximity data during the execution. This proximity log becomes the analogue of trajectory database. For example, this application, being executed indoor, records the track (in proximity terms, again) for the mobile user. Discovery of convoys (coherent motions) in such database let us define new class of rules. For example, in proximity marketing applications we can unveil a special kind of offers for those reached our point of sale (be nearby in proximity terms) in the group, etc.

The structure of our proximity log and the way we are getting measurements caused the need for the yet another definition of convoys. It is provided below.

The rest of the paper is organized as follows. Section 2 contains an analysis of existing approaches for discovery of convoys. It covers, at the first hand, the aspects we need for the future development. In Section 3, we describe our SpotEx approach. In Section 4 we describe discovery of convoys for our proximity logs.

2   The Discovery of Convoys and Network Measurements

This section contains the basic definition for the covered area. Analyzing research papers, we can list the following key issues in the discovery process [6]:

a) cluster related issues. How to define and described cluster for objects?

b) consistency related issues. The detected groups should be consistent enough to last for a given time.

c) group size related issues. Many applied tasks may have requirements on the cluster's size.

Let us give the basic definitions for this area.

Neighborhood: given a distance threshold e and a set of points S, distance operator D the e-neighborhood of a point p is given as $NH_e(p) = \{q \in S \mid D(p, q) \le e\}$.

Density-reach: given a distance threshold e and an integer m, a point p is directly density-reachable from a point q if $p \in 2\ NH_e(q)$ and $\mid NH_e(q)\mid \ge m$. A point p is density-reachable from a point q with respect to e and m if there exists a chain of points $p_1, p_2, \ldots, p_n$ in set S such that $p_1 = q$, $p_n = p$, and p_{i+1} is directly density-reachable from p_i.

Density connection: given a set of points S, a point $p \in S$ is density-connected to a point $q \in S$ with respect to e and m if there exists a point $x \in S$ such that both p and q are density-reachable from x.

The definition of density-connected elements is the basic formation for the definition of convoys. Figure 1 illustrates this.

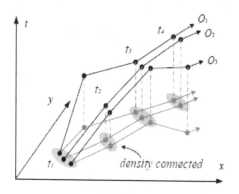

Fig. 1. An example of convoy [1]

Given the density-connected objects for consecutive time points, the convoy could be defined as follows: given a set of trajectories of N objects, a distance threshold e, an integer m, and an integer lifetime t, the convoy query returns all possible groups of objects, so that each group consists of a (maximal) set of density-connected objects with respect to e and m during at least t consecutive time points [1].

Another definition, which could be more interesting to our future development, is traveling company [6]. A group of objects form a traveling company, if members of the group are density-connected for themselves during some given time and the size of the group is not less than the given threshold.

In order to discover the traveling groups, most of the algorithms use the concepts of density-based clustering [7]. The simplest and, probably, most often used technique for discovering of convoy is to perform density-connected clustering on the objects at each time and then extract their common objects. Note, that our trajectories may have some missing time points due to non-regular sampling for locations. It prevents us from the checking the density-connection for all objects involved over those missing times. Some authors suggest linear interpolation for creating virtual points for missed times (e.g., Coherent Moving Cluster algorithm in [1]).

An idea to use network measurements for moving detection has been presented in many papers. For example, Locadio [8] uses Wi-Fi signal strengths from existing access points measured on the client to infer both pieces of context. For motion, authors measure the variance of the signal strength of the strongest access point as input to a simple two-state hidden Markov model (HMM) for smoothing transitions between the inferred states of "still" and "moving." This was based on the observation that when a Wi-Fi receiver (mobile phone) is moving, the signal strengths it receives are noisier than when it is not moving. For location, authors exploit the fact that Wi-Fi signal strengths vary with location.

Software based systems that provide location estimation based on the received signal strength indication (RSSI) of wireless access points are becoming popular nowadays. The main benefit of RSSI measurement based systems is that they do not require any additional sensors or actuators and can use existing infrastructure and already available communication parameters.

In general, RSSI based positioning includes two phases:

- the training phase where the wireless map of the environment is determined using field measurements
- the positioning phase where position estimation is performed based on the wireless map. Note that the training phase is an offline process and as such only needs to be redone if there have been major changes to the wireless propagation environment (e.g., relocation of access points) [9]

Note, the calibration phase could be very costly actually. Also, it does not support dynamic Wi-Fi nodes. What can we do if our Wi-Fi hot spot will be opened right on the mobile phones?

Technically, RSSI based measurement approaches can be divided into deterministic and probabilistic techniques.

In deterministic techniques our location area is subdivided into smaller cells and training phase readings are taken in these cells from several known access points. In the positioning phase the most likely cell (the cell that best fits the current measurement) is selected.

In probabilistic positioning techniques a probability distribution of the user's location is defined over the area of the movement. The goal of the positioning is to reach a single mode for this distribution, which is the most likely location of the tracked user. Probabilistic approaches to mobile node positioning from RSSI measurements rely on the precise estimation of a posterior probability distribution, $p(s_t \mid d_1, \ldots, d_t)$, of the likelihood of the node's state (location), s_t, given a history of the received measurements, $d_1, \ldots, d_t$ [9].

3   Spot Expert as a Network Proximity Service

Originally, the main idea of Spot Expert (SpotEx) comes as an extension for Wi-Fi based indoor positioning service (IPS). Spotex uses only a part of Wi-Fi based IPS. It stops process on the phase of detection Wi-Fi networks. Due to local nature of radio interfaces in Wi-Fi, this detection already provides some information about the location. More precisely, we can get information about proximity. As the second step, we add an external database with some rules (productions or if-then operators), related to the Wi-Fi access points. The typical examples for conditions in our rules are: Access point with SSID SomeCafé is visible for mobile device; time is within the given interval, signal strength is within the given interval, etc. And based on such conclusions, we will provide context-aware data retrieval and present some user-defined messages to mobile terminals. In other words, in SpotEx content's visibility depends on the network context (fingerprint for Wi-Fi network environment).

For the first time, SpotEx service [10], developed by Dmitry Namiot was described in article published in NGMAST-2011 proceedings [5]. You can see the latest state of SpotEx development in papers [11] and [12], for example.

SpotEx model does not require calibration phase as the most Wi-Fi based IPS do and based on the ideas of network proximity. Proximity based rules replace location information, where Wi-Fi hot spots work as presence sensors. SpotEx approach does not require from mobile users to be connected to the detected networks. SpotEx uses only broadcasted SSID for networks and any other public information.

Technically, SpotEx contains the following components:

- Server side infrastructure. It includes a database (store) with productions (rules), rules engine and rules editor. Rules engine is responsible for runtime data retrieval. Rule editor is a web application that lets work with rules database.
- Mobile application. This part is responsible for getting context info, matching it against productions (rules) and visualizing the output.

SpotEx could be deployed on any existing Wi-Fi network (or networks especially created for this service) without any changes in the infrastructure. Rule editor lets easily define some rules described context visibility to that network. Context here is just some text (HTML code) that should be opened (delivered) to the end-user's mobile terminal as soon as the appropriate rule is fired. For example, as soon as one of the above-mentioned networks is getting detected via our mobile application.

Existing use cases target proximity marketing, at the first hand. The whole process looks like an "automatic check-in" (by analogue with Foursquare, etc.) One shop can deliver proximity marketing materials right to mobile terminals as soon as the user is near some selected access point. Rather than directly (manually or via some API) check-in at the particular place (e.g., similar to Foursquare, Facebook Places, etc.) and get back deals info, with SpotEx mobile subscriber can collect deals info automatically. The prospect areas, by our opinion, are information systems for campuses and hyper local news delivery in Smart City projects. Rules could be easily linked to the public available networks.

One interesting use case could be based on the fact that most of the smart phones let users open Wi-Fi hot spots right on the phone. We can associate our rules with such mobile hot spot (hot spots). Another example of mobile hot spot is connected car. In this case our content becomes linked to the phones. It is a typical dynamic LBS. The available services are moving when phone is moving and hot spot is switched on/off. Services automatically follow to the phone.

Smart phone is all what we need for creating a new information channel. It is infrastructure-less approach.

This approach does not discuss security and connectivity issues. We do not need to connect mobile subscribers to our hot spot. SpotEx is all about using hot spot attributes as triggers in data discovery.

Each rule is a logical production (if-then operator). The conditional part includes the following data measured by the mobile application:

Wi-Fi network (SSID, mac-address)
RSSI (signal strength - optionally)
Time of the day (optionally)
ID for the client (mac-address)

In other words it is a set of operators like:

IF IS_VISIBLE('mycafe') AND FIRST_VISIT() THEN
{present the coupon info }.

Figure 2 presents use case for proximity marketing in retail area.

Because our rules present the standard production rule based system, we can use an old and well know Rete algorithm [13] for the processing.

Each rule looks like a production (if-then operator). The conditional part depends on the above mentioned measurements and logical functions (predicates). The predicates (in the current version) are:

IS_VISIBLE ( )
NOT_VISIBLE ( )
CLOSE_THAN( )
FIRST_VISIT( )
FOLLOW_UP_VISIT( )
TIME()
TIME_WITHIN()

Function IS_VISIBLE() or NOT_VISIBLE() accept as a parameter network ID (e.g., SSID or mac-address for access point) and returns a Boolean value depends on the current network's visibility.

Function CLOSE_THAN() accepts two parameters identified wireless networks (Wi-Fi access points) and returns Boolean value true if mobile terminal is close to the Wi-Fi access point described in the first parameter.

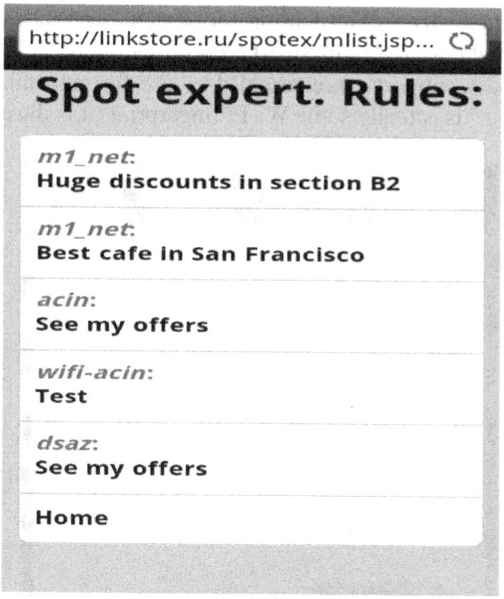

Fig. 2. SpotEx console snapshot

Two functions FIRST_VISIT() and FOLLOW_UP_VISIT( ) based on the simply fact that in Wi-Fi based system we have MAC-address for mobile terminal. The whole system does not require authorization. With SpotEx users can discover data anonymously. But in the same time we have some analogue of UUID, allowing us distinguish the users. It is MAC-address. We keep historical logs for vectors (MAC-address, wireless environment info) and use it for detecting new or retuned "visitors". For example, if for the same MAC-address we have at least two historical records where at least one Wi-Fi access point mentioned twice or more it is follow-up visitor.

Here is the starting point for our discovery of trajectories. Via the recorded track of networks snapshot environment we can try to discover how the current point (moment, when we are checking rules for the particular user) was reached. It let us define rules that depend on that track.

Let us describe another example. In the modern LBS applications that are mostly circling near the idea of "check-in", we lack the history of the movement almost completely. It is especially true on the micro-level (indoor). Suppose I have a new check-in in Foursquare. How do I come to this place? In the ordinary web browsing, any hyperlink click can have a referrer field. There are no references in LBS. In this paper we present our initial attempt to fill this gap. On practice, it means that we are going to add to our predicates a new logical function:

IN_GROUP_OF (n, t)

Here n presents some positive integer value and t describes a time (e.g., seconds). This function returns Boolean value true if mobile user traveled in the group of at least n people during at least t seconds. It is, by the SpotEx vision, of course, and those n people should be presented via own records in the proximity log. We think, that such a function (actually – qualification for context) could be useful in proximity marketing tasks, for data discovery in Smart City projects, etc.

For example, SpotEx supports an external database for customized check-ins. Any such checked "location" is actually some Wi-Fi fingerprint (it is illustrated on Figure 3):

Fig. 3. In-proximity check-in

This external database just keeps a temporal mapping between IDs in social networks (e.g. Facebook ID) and Wi-Fi fingerprints. And check-ins could be customized, as it is described in [14, 15], for example. In this case, the above-mentioned function IN_GROUP_OF() is a way to present, for example, some special offers to visitors. Group discount in retail is the simplest and obvious use case.

Our discovery process will use recorded wireless network environment snapshots (Wi-Fi fingerprints, actually). SpotEx application collects them from the moment user started the application. Of course, we can investigate historical logs too, but it is separate task. It is something that described in Reality Mining projects [16], for example. In the classical paper, authors perform cluster analysis for the previously collected data. A Hidden Markov Model conditioned on both the hour of day as well as weekday or weekend provided data separation for behavior patterns like "hone", "office", etc.

Of course, SpotEx is not the only approach uses phone as a sensor concept. We've tested the ability to implement our approach with the project Funf [17], for example. Funf Probes are the basic collection data objects used by the Funf framework. Each probe is responsible for collecting a specific type of information. These include data collected by on-phone sensors, like accelerometer or GPS location scans, etc. Actually, in Funf many other types of data (context info) can be collected through the phone. In other words, Funf is a rich data logger. We need only small part of it – collect information about wireless environment. And that log could be a source for data discovery too.

4   Trajectories in the Proximity Log

So, our context-aware browser collects wireless networks info during the execution. More specifically, our application collects snapshots that describe current Wi-Fi environment. This environment (it is an analogue of fingerprints used in Wi-Fi based indoor positioning) is a time stamped list of records. Each environment's record is a vector of triples. Each triple describes one Wi-Fi network:

Network ID (SSID)
mac-address
signal strength (RSSI)

and the whole environment could be described as a vector of triples:

$E = \{T_1, T_2, ..., T_n\}$

Our fingerprint is just a time stamped environment: [ti, Ei].

So, finally, we have a sequence of time stamped environment records. Technically our recording software (based on SpotEx or Funf) obtains data with regular time intervals. But technically again, not all our data could be available for the processing at the time of the calculation. For saving battery at the first hand, recorder can cache data and update central store in the batches (e.g., for every second, third, etc. cycle). This conclusion raises the important question about missing data. It is a common problem for discovery of trajectories. Of course, a robust system should be tolerant to such cases. There are several approaches for dealing with this problem. One possible solution is based on the introduced inactive period for candidates. This inactivity period is a threshold for the maximal allowed time interval between two position reports of the object. For any object is missing in a snapshot, as long as the inactive period is less than the selected inactivity threshold, the system still assumes that the object is traveling together with the companion in previous snapshot. In our current implementation the missing values simply ignored and appropriate object is deleted from snapshot candidate. Different strategies as well as their robustness are subject for future research.

The second important moment belongs to the general principles of our measurement. Suppose we have Wi-Fi access point with omni-directional antenna. As it is illustrated on Figure 4, having only proximity info we cannot distinguish two groups that actually reached our access point from the opposite directions. In this case, from the proximity point of view, groups 1 and 2 could be described (detected) as together moved objects. It means that in this paper we will use own definition for traveling groups. For our research it is a group of objects (mobile phones in this particular case) with the similar proximity track within the given time interval. It is consistent movement where the key metric is the relative proximity of an access point. In our research two proximity tracks (sequences of proximity records) are similar on some time interval if for the each sequential measurement in the first track we can get a sequential measurement from the second track for approximately the same timestamp where two networks snapshots have at least one pair of comparable Wi-Fi measurements.

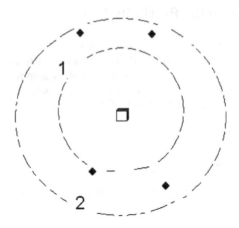

Fig. 4. Omni-directional Wi-Fi antenna

Lets us see some example. Suppose we have two tracks:

T_1 as $\{[t_{11}, E_{11}], [t_{12}, E_{12}], [t_{13}, E_{13}], \ldots\}$
and
T_2 as $\{[t_{21}, E_{21}], [t_{22}, E_{22}], [t_{23}, E_{23}], \ldots\}$

Here t_{ij} describes a time stamp and E_{ij} describes Wi-Fi environment. The similarity means that we can map measurements from the first track to the second one. And our mapping should keep the time sequence. So, for example, if we map a pair $[t_{11}, E_{11}]$ to $\{[t_{21}, E_{21}]$, then the next pair $[t_{12}, E_{12}]$ could be mapped to the time $t \geq t_{21}$.

Because each application (each mobile phone) executes and collects data independently, we can not warranty that for the given timestamp t_{1i} we will find exactly the same value t_{2j} in the second track. We will try to find approximately the same timestamp $t_{1i} \pm \Delta$ where Δ is some constantly selected threshold. Of course, it could be selected accordingly to the regular interval used for collecting measurements. In other words, comparing to the traditional trajectories discovery algorithms, this application does not restore (does not approximate or predict) missed values. If we cannot find some measurement within the given interval than we simply conclude that two tracks are not similar.

Citing absolutely the same reason (each phone collects data without the correlation with other possible participants) we can not warranty too, that two mobile phones record equal values for Wi-Fi signal strength on any selected place. It depends on battery level, external conditions, etc. It means that we will compare network measurements using the second given threshold (for signal strength). IDs for access points should be equal of course, where RSSI may vary within the given interval. That is why we mentioned above the comparable (approximately equal) Wi-Fi measurements.

Two networks measurements are comparable in this paper (it is the simplest metric) if they have at least one common access point with difference in signal strengths less than the given threshold.

Now we are ready to present our algorithm. We are calculating the Boolean value for function

IN_GROUP_OF( )

This function will be calculated in some of our predicates checked for the given mobile user. It means, that as a starting point we have data for the current wireless environment (Wi-Fi environment snapshot for this mobile user / mobile phone). Initial parameters are:

Δ - time threshold, Ω - RSSI threshold, E - an original network environment, T_0 – an original current time, T_{max} – argument for function

1. Initialize new candidate set R_1
2. Collect measurements within the time T_0- Δ $\rightarrow$ R_1;
3. **If** R_1 is empty **then** output *false*;
4. Remove from R_1 all measurements that are not comparable with E;
5. **If** R_1 is empty **then** output *false*;
6. **Set** $t = T_0$;
7. **While** $t > T_0$-T_{max}
        8. Find the previous measurement for the current user.   Update current settings $\rightarrow$ $\{t, E\}$;
        9. For the each measurement in R_1 find proximity data within $t \pm \Delta$ (update measurements with new data);
        10. Remove from R_1 elements without new data (not updated elements) ;
        11. Remove from R_1 elements that are not comparable with E;
        12. **If** R_1 is empty **then** break;
13. **End while**

The finally, R_1 presents the group we are looking for. Depends on the size of this array, we can calculate our function IN_GROUP_OF( ).

5   Conclusion

This article describes a new application (a new use case) for discovery of convoy task. This is an attempt to apply the known models for new use cases generated by the context-aware computing. Network proximity log is used here as a replacement for the classical trajectory database. This fact, as well as the technical aspects of how measurements are collecting, requires changes in the standard definitions and the corresponding modifications for the algorithms. The results of this research will be used for extending the functionality of a new service for context-aware data discovery. The future developments will include the analysis of the stability for the proposed algorithm to missing values and obtaining quantitative metrics for the speed and accuracy of recognition.

References

1. Jeung, H., Yiu, M., Zhou, X., Jensen, C., Shen, H.: Discovery of convoys in trajectory databases. Journal Proceedings of the VLDB Endowment 1(1), 1068–1080 (2008)
2. Ester, M., Kriegel, H.-P., Sander, J., Xu, X.: A density-based algorithm for discovering clusters in large spatial databases with noise. In: SIGKDD, pp. 226–231 (1996)
3. Kalnis, P., Mamoulis, N., Bakiras, S.: On discovering moving clusters in spatio-temporal data. In: Medeiros, C.B., Egenhofer, M., Bertino, E. (eds.) SSTD 2005. LNCS, vol. 3633, pp. 364–381. Springer, Heidelberg (2005)
4. Aung, H.H., Tan, K.-L.: Discovery of Evolving Convoys. In: Gertz, M., Ludäscher, B. (eds.) SSDBM 2010. LNCS, vol. 6187, pp. 196–213. Springer, Heidelberg (2010)
5. Namiot, D., Schneps-Schneppe, M.: About location-aware mobile messages. In: International Conference and Exhibition on Next Generation Mobile Applications, Services and Technologies (NGMAST), September 14-16, pp. 48–53 (2011), doi:10.1109/NGMAST.2011.19
6. Tang, L., Zheng, Y., Yuan, J., Han, J., Leung, A., Hung, C., Peng, W.: On Discovery of Traveling Companions from Streaming Trajectories. Microsoft Reserch, http://research.microsoft.com/pubs/156047/On%20Discovery%20o f%20Traveling%20Companions%20from%20Streaming%20Trajectories .pdf (retrieved: March 2013)
7. Sander, J., Ester, M., Kriegel, H.P., Xu, X.: Density-based clustering in spatial databases: The algorithm gdbscan and its applications. Data Mining and Knowledge Discovery 2(2), 169–194 (1998)
8. Krumm, J., Horvitz, E.: Locadio: Inferring motion and location from wi-fi signal strengths. In: Proceedings of International Conference on Mobile and Ubiquitous Systems: Networking and Services (MobiQuitous 2004) (2004)
9. Zàruba, G.V., Huber, M., Kamangar, F.A., Chlamtac, I.: Indoor location tracking using RSSI readings from a single Wi-Fi access point. Wireless Networks Archive 13(2), 221–235 (2007), doi:10.1007/s11276-006-5064-1
10. Namiot, D., Sneps-Sneppe, M.: Using Network Proximity for Context-aware Browsing. International Journal on Advances in Telecommunications 5(3&4), 163–172 (2012)
11. Namiot, D.: Context-Aware Browsing – A Practical Approach. In: 2012 6th International Conference on Next Generation Mobile Applications, Services and Technologies (NGMAST), pp. 18–23 (2012), doi:10.1109/NGMAST.2012.13
12. Namiot, D., Sneps-Sneppe, M.: Proximity as a Service. In: 2012 2nd Baltic Congress on Future Internet Communications (BCFIC), pp. 199–205 (2012) Print ISBN: 978-1-4673-1672-9, doi:10.1109/BCFIC.2012.6217947
13. Friedman-Hill, E.: Jess in action: rule-based systems in Java. Manning Publications Co., Greenwich (2003) ISBN: 9781930110892
14. Namiot, D., Sneps-Sneppe, M.: A new approach to advertising in social networks - business-centric check-ins. In: 2011 15th International Conference on Intelligence in Next Generation Networks (ICIN), pp. 92–96 (2011), doi:10.1109/ICIN.2011.6081110
15. Namiot, D., Sneps-Sneppe, M.: Customized check-in procedures. In: Balandin, S., Koucheryavy, Y., Hu, H. (eds.) NEW2AN 2011 and ruSMART 2011. LNCS, vol. 6869, pp. 160–164. Springer, Heidelberg (2011)
16. Eagle, N., Pentland, A.: Reality mining: sensing complex social systems. Personal and Ubiquitous Computing Archive 20(4), 255–268 (2006), doi:10.1007/s00779-005-0046-3
17. Funf Open Sensing Framework, http://funf.media.mit.edu/ (retrieved: April 2013)

FPGA Design and Implementation of MIMO-SDM Systems for Wireless Internet Communications Networks

Bui Huu Phu[1], Tran Van Tho[2], Tran Canh Vinh[2], Vu Dinh Thanh[3], and Nguyen Huu Phuong[4]

[1] DCSELAB, University of Technology,
Vietnam National University in Hochiminh City, Vietnam
[2] Hochiminh City University of Transport, Vietnam
[3] University of Technology, Vietnam National University in Hochiminh City, Vietnam
[4] University of Science, Vietnam National University in Hochiminh City, Vietnam
bhphu@dcselab.edu.vn

Abstract. Multiple-input multiple-output (MIMO) systems using spatial division multiplexing (SDM) technique have been considered a potential technology for high speed data transmission wireless internet networks, such as IEEE 802.11, 3GPP Long Term Evolution, WiMAX. Many studies have confirmed that the channel capacity is significantly increased and proportional to the number of transmit and receive antennas without additional power and spectrum compared with single-input single-output systems. Although a lot of technical papers have been considered and evaluated the MIMO SDM systems based on theory analyses and/or computer simulation, but just few ones investigated the systems based on hardware design, which is an important step before going to manufacture the integrated circuits. In the paper, we present our own design and implementation of three MIMO SDM systems on FPGA-based DSP Development Kit. Based on the design, we evaluate the bit-error rate performance of the systems and also consider the consumption of the FPGA elements in our design.

Keywords: MIMO, SDM, ZF, MMSE, FPGA.

1   Introduction

Multiple-input multiple-out (MIMO) systems have been considered as a high speed data transmission technology. Many studies have confirmed that the channel capacity of the systems can increase significantly and is proportionally to the number of transmit (TX) and receive (RX) antennas without additional power and bandwidth compared with single-input single-out systems. The systems have been standardized to be used in modern wireless internet networks such as IEEE 802.11, 3GPP Long Term Evolution, and WiMAX [1–3].

In MIMO systems, when channel state information (CSI) is not available at the transmitter, the spatial division multiplexing (SDM) technique is used to obtain maximum channel capacity by simultaneously transmitting multiple independent substreams with the same data rate and power level [4-7].

S. Balandin et al. (Eds.): NEW2AN/ruSMART 2013, LNCS 8121, pp. 151–161, 2013.

There have been a lot of technical papers considered about the MIMO SDM systems. They have been studied and evaluated the systems based on theory analyses and/or computer-based simulation [4-9]. However, there have just been few papers that investigated the systems based on their design and implementation on FPGA hardware, which is a very important step to evaluate the systems before going to design and manufacture integrated circuits of the systems [10-12]. In [10], authors present the design and implementation results of a digital 120 Mb/s MIMO orthogonal frequency-division multiplexing (OFDM) wireless LAN (WLAN) baseband processor based on the proposed decoding algorithms. In [11], the authors addressed the implementation in FPGAs of only a MIMO decoder embedded in a prototype of a 4G mobile receiver; in [12], authors design and implement on field programmable gate array (FPGA) board, reduced-complexity MIMO-maximum likelihood detection (MLD) system. In [10-12], authors have not considered the consumptions of FPGA elements in their system design, which evaluate the efficiency of system design.

In the paper, we present our own design and implementation of MIMO-SDM systems on the FPGA-based DSP Development Kit. A detailed design of full systems included transmit side, MIMO channel, and receive side is shown. The system performance is evaluated based on bit-error rate criteria, and the efficiency of our design is shown through the consumption of FPGA elements.

The remaining of the paper is as following. After a brief introduction, an overview of MIMO-SDM systems is described in section 2. In section 3, we present the design and hardware implementation of the MIMO SDM system. The results and discussion of our implementations are shown in section 4. Conclusions are presented in final part.

2   MIMO SDM Systems

Consider a MIMO system with N_{TX} TX antennas and N_{RX} RX antennas, as shown in Fig. 1. In the SDM technique, an input data stream is divided into N_{TX} different substreams and is then simultaneously transmitted with the same power level.

Received signals in narrowband fading channels are as follows:

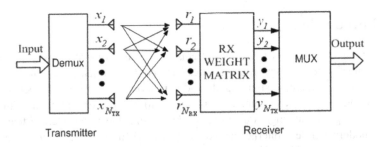

Fig. 1. Block diagram of MIMO-SDM systems

$$r(t) = Hx(t) + n(t), \tag{1}$$

Where $x(t) = \left[x_1(t) \; x_2(t) \cdots x_{N_{TX}}(t) \right]^T$ and $r(t) = \left[r_1(t) \; r_2(t) \cdots r_{N_{RX}}(t) \right]^T$ are transmit and receive signal vectors, $n(t) = \left[n_1(t) \; n_2(t) \cdots n_{N_{RX}}(t) \right]^T$ is AWGN noise, in which each $n_i(t)$ is an independent Gaussian process with zero mean and variance σ^2, H is MIMO channel matrix described as below:

$$H = \begin{pmatrix} h_{11} & h_{12} & \cdots & h_{1N_{TX}} \\ h_{21} & h_{22} & \cdots & h_{2N_{TX}} \\ \vdots & \vdots & h_{ij} & \vdots \\ h_{N_{RX}1} & h_{N_{RX}2} & \cdots & h_{N_{RX}N_{TX}} \end{pmatrix}, \tag{2}$$

Where h_{ij} is channel response from the jth transmit antenna to the ith receive antenna. Considering signal received at the ith receive antenna, we have:

$$r_i(t) = \sum_{j=1}^{N_{TX}} h_{ij} x_j(t) + n_i(t). \tag{3}$$

The equation shows that received signal at each antenna is a signal summation of sub-streams. Therefore, although total data transmit rate of MIMO SDM can be increased by transmitting multiple sub-streams; the performance of each sub-stream may be decreased by interference among the sub-streams. So, a detector to eliminate the interference is needed. In the paper, we just consider to use spatial filtering method to detect received signals. Other methods can be referred in [8].

2.1   Zero-Forcing (ZF) Algorithm

Received signals are detected as follows [8]:

$$y(t) = W^T r(t)$$
$$= W^T H x(t) + W^T n(t), \tag{4}$$

Where $y(t)$ is the detected signal and W is the receive weight matrix. The ZF weight can cancel all inter-stream interference. ZF weight matrix is chosen as:

$$W_{ZF} = H^* \left(H^T H^* \right)^{-1} \tag{5}$$

Note that $W_{ZF}^T = \left(H^H H \right)^{-1} H^H$ is Moore-Penrose inverse matrix. Substituting (5) into (4), we have:

$$y(t) = x(t) + W_{ZF}^T n(t). \tag{6}$$

The above equation shows that the ZF weight clearly eliminates all inter-stream interference. Therefore, it can maximize signal to interference ratio (SIR) at the output of the detector. However, the received power of the desired signal can be decreased when maximizing the SIR, so the ZF algorithm has a signal to noise ratio (SNR) lower than the other detectors.

2.2   Minimum Mean Square Error (MMSE)

The MMSE algorithm is almost the same as ZF. However, its detection uses filter matrix W_{MMSE} which is described as follows [8]:

$$W_{MMSE} = (H^H H + \sigma^2 I)^{-1} H^H \tag{7}$$

where σ^2 is the variance of Gaussian noise.

3   Design of System on Hardware

In the section, we will present our hardware design of a 2x2, a 2x3 and a 3x3 MIMO-SDM system using 4-QAM on Simulink platform, Altera DSP builder library and Altera Quartus II software for Verilog coding. Fig. 2 shows the block diagram of 2x2 MIMO-SDM systems.

3.1   Transmistter Side

At transmitter (block Data, 4QAM and Add training symbol in Fig. 3), two serial bit streams are modulated by using 4-QAM. The symbol after that is grouped into a pair of symbols and they are sent on two different antennas at the same time slot. Each of the signals includes two parts: in-phase (I) and Quadrature (Q) components.

Two bits data are created at first prior to entering 4QAM modulation module. The counter values are saved into Look Up Table (LUT). LUT carries out a random database based on the input data as shown in Fig. 3.

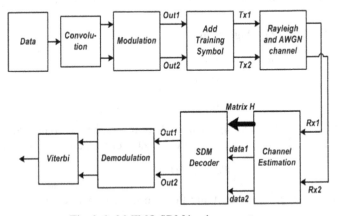

Fig. 2. 2x2 MIMO SDM hardware system

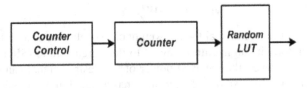

Fig. 3. Data created process

The modulation module uses a 4QAM modulation and applies the LUT method. The content in the LUT at the address includes values of -1or 1, The constellation of 4QAM is shown in Fig. 4.

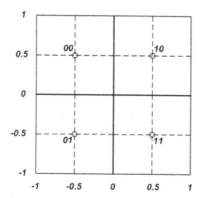

Fig. 4. Constellations of 4QAM

The preamble training data sequences are added into the original data for channel estimation at the receiver. At each transmit antenna, we use 8 orthogonal Hadamard bits for CSI estimation.

Fig. 5 is the hardware design of inserting data training sequence as showed above. After attaching the training symbol, they are fed into the MIMO channel prior to entering the receiver site.

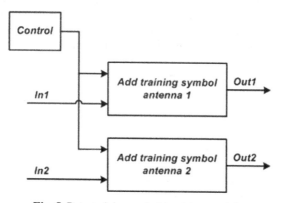

Fig. 5. Data training sysbol inserting module

3.2   MIMO Flat Fading Channel

In this paper, a flat fading channel model is considered. This means that the multipath channel has only one path (tap) as shown in Fig.6. Each of the impulse response h_{ij} of the channel corresponding to jth transmitter to i^{th} receiver antenna. It is considered as a Rayleigh channel, which have mean of 0 and variance of 1/2.

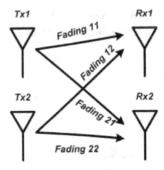

Fig. 6. Flat fading channel

Received signals are also affected by AWGN noise. Thus, each of them is passed through two modules: Rayleigh fading and AWGN noise. The hardware description is showed in Fig. 7.

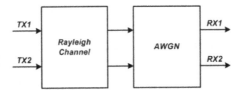

Fig. 7. Reyleigh and AWGN channel modeling

The Module of 2x2 MIMO Rayleigh fading channels is shown in Fig. 8. Values of **H** matrix are saved in LUT in fix-point format. In the design, the narrowband channels are assumed and their values are kept constant in a data frame of 192 data symbols. After a data frame, the channels are changed randomly.

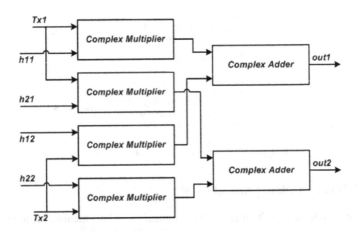

Fig. 8. Module of 2x2 MIMO Rayleigh channel

3.3   Receiver Site

At the receiver, the channel coefficients h_{ij} are estimated by using preamble training sequences which were created at the transmitter, as shown in Fig. 9.

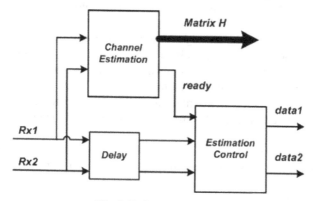

Fig. 9. Estimation module

The channel coefficients after that are used for extracting data.

As we can see in Fig. 2, Detector module of MIMO-SDM systems is placed at the receiver. Here we use ZF or MMSE weights to detect received signals.

After being detected, the complex data streams are demodulated "4QAM demodulation module" as shown in Fig. 10. After that, we compare the received bit stream to the transmitted bit stream to calculate BER of the systems.

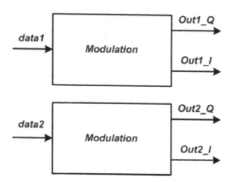

Fig. 10. De-modulation module

In addition, we have Viterbi decoder at Receiver to get data back to original. This module is designed with Register Exchange algorithm which has the best latency comparing with the others.

4   Experimental Results

In this section, we will show our results of the implementation of three 2x2, 2x3, 3x3 MIMO-SDM systems with both detection algorithms ZF and MMSE.

4.1   BER Performance of Designed Systems

BER performance of MIMO SDM systems with various numbers of antennas configuration on hardware is shown in Fig.11. When using ZF weights to detect received signals, the more antennas there are the worse the BER performance is. It can be seen that the degradation is about 2dB when comparing 3x3 antenna configuration with 2x2 case. This is because the ZF algorithm could not eliminate noise components, thus noise power increase proportionally to the number of antennas. For the 2x3 case, on the other hand, BER performance is much better than the cases of 2x2 and 3x3, as seen BER 6e-6 in 2x3 system compared to around 1e-3 in the other ones at SNR of 30dB. This is because of obtaining high diversity gain with the 2x3 system, meanwhile there is no diversity gain when the number of transmit antennas is equal to the number of receive ones.

When applying MMSE algorithm, it can be seen that its BER performance is better than the ZF algorithm. This is because the algorithm tries to suppress both the inter-sub stream interference and noise.

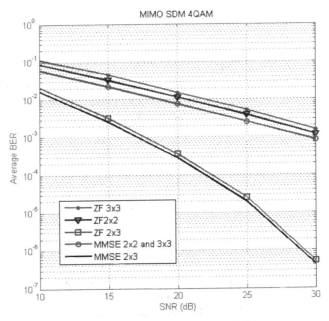

Fig. 11. Hardware performance of MIMO SDM systems

Fig.9 shows BER performance of MIMO SDM using Zero Forcing algorithm with channel coding modules. As we can see, at high SNR values, the channel coded systems are better than un-coded ones. However, at low SNR values, the result is opposite because the channel decoding with Viterbi module does not have high accuracy when data signals are effected by high noise.

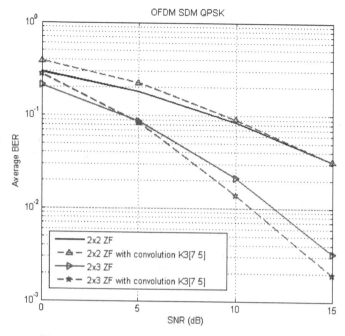

Fig. 12. Hardware performance of MIMO SDM systems

4.2   Hardware Consumption

Table 1 shows the total hardware consumption. Here, the used FPGA device was Stratix III 3SL150F1152C2. It can be seen from table I that hardware resource can be

Table 1. Consumption of 2x2 Model With Fixed-Point 10.22

Blocks	Consumption			
	The number in Model	Max Speed (MHz)	Combinational ALUTs Max:113,600	Dedicated Logic Registers Max:113,600
Modulation	2	420	0 (0%)	20 (<1%)
Convolution coding	1	416.32	11 (<1%)	4 (<1%)
Demodulation	2	420	18 (<1%)	2 (<1%)
Add training symbol	2	243.72	15 (<1%)	74 (<1%)
SDM decoder	1	162.60	22,519 (20%)	19,596 (17%)
Channel estimation	1	147.12	3,530 (3%)	7,505 (7%)
Estimation control	1	206.06	20 (<1%)	141 (<1%)
Viterbi	1	191.37	182 (<1%)	1552 (1%)
Total evaluation		147.12	<30%	<31%

free approximately 70% (only used about 30%). That is the reason we chose fixed point 32 bits (10.22) to increase the accuracy of system.

5   Conclusions

The MIMO system using the SDM technique is considered as a potential technology for high speed wireless internet communications. In the paper, we have presented our own design and implementation of three kinds of MIMO-SDM systems on the FPGA-based DSP Development Kit. A detailed design of full systems is shown. Based on the design, we have measured and evaluated BER performance of the systems. The results have shown that our design and implementation are good. Here, we also evaluated our design by considering the consumption of FPGA elements.

Acknowledgement. This research is funded by Vietnam National Foundation for Science and Technology Development (NAFOSTED) under grant number 102.02-2011.23. The authors would like to thank for the support.

References

[1] Prasad, R., Muoz, L.: WLANs and WPANs towards 4G Wireless. Artech House (2003)

[2] Andrews, J.G., Ghosh, A., Muhamed, R.: Fundamentals of WiMAX: Understanding Broadband Wireless Networking. Prentice-Hall (2007)

[3] Dahlman, E., Parkvall, S., Sköld, J., Beming, P.: 3G Evolution: HSPA and LTE for Mobile Broadband. Elservier (2007)

[4] Foschini, G.J.: Layered Space-Time Architecture for Wireless Communication in a Fading Environment When Using Multiple Antennas. Bell Laboratories Technical Journal, 41–59 (1996)

[5] Paulraj, A.J., Gore, D.A., Nabar, R.U., Bölcskei, H.: An overview of MIMO communications – A key to gigabit wireless. Proc. IEEE 92(2), 198–218 (2004)

[6] Hochwald, B.M., Ten Brink, S.: Achieving near-capacity on a multiple-antenna channel. IEEE Trans. Commun. 51(3), 389–399 (2003)

[7] Hedayat, A., Nosratinia, A.: Outage and Diversity of Linear Receivers in Flat-Fading MIMO Channels. IEEE Transactions 5, 5868–5873 (2007)

[8] Ohgane, T., Nishimura, T., Ogawa, Y.: Applications of space division multiplexing and those performance in a MIMO channel. IEICE Trans. Commun. E88-B(5), 1843–1851 (2005)

[9] Ban, K., Katayama, M., Yamazato, T., Ogawa, A.: Joint optimization of transmitter/receiver with multiple transmit/receive antennas in band-limited channels. IEICE Trans. Commun. E83-B(8), 1679–1704 (2000)

[10] Jung, Y., Kim, J., Lee, S., Yoon, H., Kim, J.: Design and implementation of MIMO-OFDM baseband processor for high-speed wireless LAN. IEEE Trans. on Circuits and Systems II 54, 631–635 (2007)

[11] Pacheco, A.J., Herrero, Á.F., Quirós, J.C.: Design and Implementation of a Hardware Module for MIMO Decoding in a 4G Wireless Receiver. Hindawi Publishing Corporation VLSI Design, Article ID 312614, 8 pages (2008), doi:10.1155/2008/312614
[12] Zahrani, Y.A., Marshed, S.A., Dhotyan, A.A., Ulyman, A.S., Dosari, S.A., Elnamaky, M., Shebeili, S.A.: Design and FPGA Implementation of Reduced Complexity MIMO-MLD Systems. In: IEEE International Symposium on Signal Processing and Information Technology (2010)

On Suitability of the Reinforcement Learning Methodology in Dynamic, Heterogeneous, Self-optimizing Networks

Milos Rovcanin, Eli De Poorter, Ingrid Moerman, and Piet Demeester

Ghent University - iMinds, Department of Information Technology (INTEC) Gaston
Crommenlaan 8, Bus 201, 9050 Ghent, Belgium
milos.rovcanin@intec.ugent.be

Abstract. An ever growing number of deployed wireless networks dictates a tempo with which the inter-network cooperation techniques are being developed. Cooperation, in this sense, can go far beyond a simple activation of an interference avoidance techniques. This paper describes and evaluates the performance of a reinforcement learning based reasoning engine, used in a self-learning, cognitively controlled cooperation between heterogeneous, co-located networks. Coupled with a concept of cooperation through the network service negotiation, this approach represents an efficient, yet scalable solution for the dynamic network self-optimization.

Keywords: Self-learning, reinforcement learning, network optimization, network service negotiation.

1 Introduction

There is a growing need for network solutions that dynamically support at run-time cooperation between devices from different networks [1]. Currently, the only way to support connectivity between co-located devices is to statically group them into different sub-nets, depending on their communication technology. This approach ensures that the same network policies are used for a sub-net, regardless of the characteristics of the devices. However, the process is usually quite complex and it might lead to a huge waste of available resources.

Dynamic, at run-time management is expected to improve many networking aspects [2]: decrease energy consumption, decrease interference, improve availability and bandwidth allocation etc. The major issue here is the fact that different sub-nets usually have different requirements (high level goals), that must not be neglected. Otherwise, this can lead to a misbehavior some of the participating sub-nets and, once again, an unacceptable waste of available resources.

Obviously, there is a need for an intelligent entity that controls and supervises the process of establishing an inter-network cooperation. Having it performed through a process of activating or deactivating network services [3], this reasoning entity will have to be able to determine the optimal set of services for each

S. Balandin et al. (Eds.): NEW2AN/ruSMART 2013, LNCS 8121, pp. 162–175, 2013.

one of the participating sub-nets. An additional dimension to a problem is given by the fact that activation of a service in one sub-net, consequently influences the performance of all the involved sub-nets.

Reasoning engine, proposed here, is based on an on-line, reinforcement learning methodology that does not require any sort of an a priori knowledge about the influences that different services pose on the network high level goals. Instead, it uses a number of network features to make the precise assessment of them. Collected features are used as an input for the Least Squares Temporal Difference (LSTDQ) [5] algorithm that generates numerical values for the system performance at each state. The highest value is given to the best performing service set, which is therefore considered to be the optimal one, considering the given high level network goals.

Performance is continuously evaluated using the most recent data. No deprecated values are used, as it was the case with many of the previous machine learning approaches. By constantly collecting measurements from the environment, our engine is capable of noticing sudden condition changes and adapting to them.

The following chapters of the paper are organized as follows: Section 3 describes the fundamentals of the LSTDQ algorithm and its strong points, in regards to other machine learning approaches. Section 4 discusses the suitability and prospects of using the LSTDQ in our use case. The following Section 5 gives a detailed overview of the experimentation set up, used for the proof of concept and testing. Section 6 presents results and discussion regarding each one of them. Section 7 proposes possible improvements and future course of development. Finally, Section 8 summarizes and concludes the paper.

2   Related Work

Work presented here is considered to be an improvement of an already used paradigm [3], which describes inter-network cooperation through an activation of network services in co-located networks. The process is initiated and controlled by a linear programming based reasoning entity - CPLEX ILPSolver [4], but requires design-time knowledge of the impact of a network service on the network performance.

3   Least Squares Temporal Difference Algorithm - LSTDQ

3.1   Fundamentals

When modeled using the reinforcement learning approach [6] [7], a problem transforms into a Markov Decision Process (MDP), with the continuous or discrete set of states $S = s_1, s_2, s_3, ..., s_n$. A decision maker passes from one state to another by selecting an action at every step. Set of actions can also be continuous or discrete, $A = a_1, a_2, a_3, ..., a_n$. A reward is given upon taking each action.

Learning, in this case, considers using actions, thus following the decision making paths that maximize rewards:

$$Q(s,a) = r(s,a) + \gamma \sum_{s'} P(s'|s,a) maxQ(s',a') \quad (1)$$

Above given is the well known Bellman equation, where $Q(s,a)$ represents the state-action function, also known as the Q function. It assigns numeric value to every state-action pair from the problem's state-action-space, based on the immediate reward $r(s,a)$, given for taking an action a at the state s and the future expected reward $\sum_{s'} P(s'|s,a) maxQ(s',a')$. The last argument includes the state-action transition probability, which is generally not known a priory.

3.2   Approximations and LSTDQ

LSTDQ was first introduced by M.G.Lagoudakis and R.Parr in [5] and further elaborated in [8], as a part of the well known Least Squares Policy Algorithm (LSPI). The algorithm is founded on the idea of representing the Q function as a linear combination of a certain number of problem features, *basis functions*:

$$Q(s,a;w) = \sum_k \phi_j(s,a)\omega_j \quad (2)$$

Argument ω_j is the weight parameter. Basis functions are arbitrary and generally non-linear functions of s and a. Their linear independence ensures no redundancy. Generally, the number of basis functions k is significantly smaller than the number $|S||A|$ of state/action pairs.

Combining equations (1) and (2) yields a system:

$$\omega = A^{-1}b$$
$$\text{where: } A = \Phi^T(\Phi - \gamma P^\pi \Phi)$$
$$b = \Phi^T R$$

Theoretically, in order to populate matrices A and b, one must acquire information regarding basis functions and rewards for each pair from the problem's state-action space, which is not practical in large problem spaces and impossible if the problem space is continuous. Therefore, approximations are used.

Samples $D = (s_{d_i}, a_{d_i}, s'_{d_i}, r_{d_i} | i = 1, 2, ..., L)$ are collected. Here, (s_{d_i}, a_{d_i}) is sampled from the state-action space and (s'_{d_i}) is sampled according to $P(s'_{d_i}|s_{d_i}, a_{d_i})$ that is defined by the used sampling policy (a random policy - uniform distribution). Argument L represents the number of gathered samples. Finally, here are the approximated versions of the above given matrices:

$$\hat{\Phi} = \begin{pmatrix} \phi(s_1,a_1)^T \\ ... \\ \phi(s_n,a_n)^T \end{pmatrix} \quad \widehat{P^\pi \Phi} = \begin{pmatrix} \phi(s'_1,\pi(s'))^T \\ ... \\ \phi(s'_n,\pi(s'))^T \end{pmatrix}$$

$$\hat{R} = \begin{pmatrix} r_1 \\ ... \\ r_2 \end{pmatrix}$$

Matrices A and b can now be approximated in the following way:

$$\hat{A} = \hat{\Phi}^T(\hat{\Phi} - \gamma P^{\hat{\pi}}\hat{\Phi}) \text{ and } \hat{b} = \hat{\Phi}^T \hat{R}$$

Originally, LSPI was introduced as an off-line learning technique that starts with an initial policy and iterates it, over the same training set of training samples, in order to converge to its optimal outlook. Every iteration considers using the LSTDQ to calculate the respective weight vector so that two iteration can be compared. An on-line version was introduced later, in []. In this form, the LSPI does not use an acquired training set to iterate on policies, but continuously acquires samples and adds them (updates) the already existing set to correct weight factors and yield better Q value approximations. The common part of both approaches is the underlying LSTDQ engine that uses whatever set of samples is given to it to calculate the corresponding weight vector.

Implementation complexity and processing power needed for the LSTDQ depends on the number of samples it has to deal with. Dealing with smaller number of samples will reduce complexity, but depending on the size of the entire state-action space, it might have an unacceptable precision. The consistency between approximated and true values A and b matrices are given by these claims:

$$E(\hat{A}) = \frac{L}{|S||A|}A \text{ and } E(\hat{b}) = \frac{L}{|S||A|}b$$

The following section will explain how the same LSTDQ based approach can be used for the purpose of inter-network self-optimization.

4 LSTDQ - Suitability and Prospects

At the beginning, being a core of the LSPI algorithm, LSTDQ was primarily used in an offline type of approach. An off-line method demands a training set of samples to be formed prior to the learning process. During each policy iteration, samples are picked from the training set and fed to the LSTDQ algorithm. The corresponding set of weights is calculated.

Within an on-line approach, samples are gathered during the learning process. In the extreme case, weights are recalculated after each new sample is collected. Obviously, an on-line mechanism does not iterate policies in a number of distinct trials, but continuously, sample by sample. Cross comparison between an on-line and off-line LSPI algorithms is elaborated in details and published in [9].

We utilize an on-line method, without defining stopping rules. Consequently, no optimal policy is proclaimed. This makes sense since we are dealing with a dynamic environment. Policy determined as the optimal one might yield sub-optimal results after the change a network properties occur. Therefore, continuous sample collection is performed for two reasons:

- Fine tuning of the weight parameters $W = w_1, w_2, w_3, ..., w_k$
- Detection of a network condition change

Markov decision process, in our case, is memoryless. In other words, performance of the network at the current state does not depend on previous states it has been through. This property is then used to update up to N_{states} Q values after each learning episode. In use cases with relatively small number of states, this leads to an extremely fast collection of all the samples from the state-action space.

LSTDQ is invoked for the first time once all the samples are collected. It calculates the vector of weight factors for this sample set and determines Q values for every state-action pair. The following samples are collected using the "ϵ greedy" exploration technique [10]. At each state, with ϵ probability, our reasoning engine picks the action with the highest Q value. With $1-\epsilon$ probability, the action is picked at random. LSTDQ gets invoked every time a new sample is obtained.

Obviously, the ϵ factor (exploration factor) has the major influence on the algorithm. It directly influences both the time that system spends in the optimal/sub-optimal states and engine's capability to quickly respond to the change of network conditions. Further elaboration through experimental results and discussion will be given in the following sections.

5   LSTDQ Howto

5.1   Use Case and Experimentation Set up

We consider a case of two co-located wireless sensor networks. In the reminder of the paper, they will be referred to as network A and network B (see Figure 1). Process of establishing a communication between networks A and B [3] is out of the scope of this paper. It is assumed that networks A and B are already able of communicating with each other.

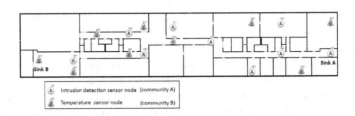

Fig. 1. Experimentation network topology. Part of the WiLab.t testbed [11], located at the iMinds research facilities, University of Ghent, Belgium.

At this early stage of research, live measurements are still unavailable. Therefore, we used fabricated, artificial measurements, affected by the influence of fictive network services. Network A provides a set of fictive services N_{S1} = Service A, Service B. Similarly, network B provides services N_{S2} = Service C , Service D. Given the high level goals of network A: *High Network Lifetime* and

Low Average Delay (both are assigned the same priority), the prime objective of the cognitive engine is to determine the optimal set of services in both networks so that the performance of network A, regarding its high level goals, is maximized.

5.2   Implementation Details

For this use case, the state is determined by the joint combination of active/non-active services in both networks. Given the set of four services N_{S1} and N_{S2}, the total number of states is $N_{states} = 16$. The states are: $S_t = A, B, C, D, AB, AC, AD, BC, ..., ABCD$. Performance of the network in each state is determined following the general rules set-up bu the system designer:

- Service A - always improves performance in a fixed manner
- Service B - improves performance when activated alone, degrades when combined with C,D,E
- Service C - improves performance when combined with D, degrades when combined with A and B
- Service D - best performing state when combined with C and A, minor improvement/degradation with others

Combined in this manner, state CB represents the worst possible service combination, while DCA is the optimal set of services.

To be able to calculate Q values for each state-action pair, two basis functions are used: ϕ_1 - *end-to-end delay* and ϕ_2 - *average number of re-transmissions*. As stated before, measurements regarding basis functions are fabricated, not obtained during the run-time. Q values are calculated in the following manner:

$$Q(s, a) = \phi_1 \omega_1 + \phi_2 \omega_2$$

Cognitive engine is allowed to switch between any of the two states, which means that the set of 16 actions ($A_c = a_1, a_2, a_3, ..., a_{16}$) is available at each state. The engine is ultimately expected to determine which state is the optimal one and force that respective service combination. Forcing a certain service combination in this context mean forcing transitions from any given state to the optimal one. If the system is already in the in the optimal state, the optimal decision would be to remain in it.

5.3   Exhaustive Exploration Phase

To take an advantage of the memoryless property of our set up, the first 16 episodes are used for an exhaustive exploration over the entire state-action space. All the possible states are investigated using a pseudo-random walk. Upon completion of this phase, LSTDQ is invoked using the gathered samples as an input data. The initial set of weight factors $W = \omega_1, \omega_2$ is calculated, followed by the calculation of the Q values for every state-action pair.

Exhaustive exploration will make an initial differentiation between "good" and "bad" states. Furthermore, it will pinpoint the best and the worst performing states. However, forcing this initial policy from then on will make our cognitive engine rigid and not capable of adapting to a dynamic environment behavior.

5.4   Exploitation Phase

Exploitation of the collected data begins after all the necessary samples are collected - upon completion of the exhaustive exploration phase. The process is conducted in the "ϵ greedy" fashion (Section 4). An appropriate values of the ϵ factor allow the reasoning engine to enforce the optimal set of services as much as possible, while still "being fair", up to an acceptable level, to sub-optimal states. Frequency with which the engine checks the sub-optimal states strongly affects its capability of noticing environmental disturbances. On the other hand, the system should not be kept "too long" in the states where it's performance is sub-optimal. Certain trade-offs are inevitable.

6   Results and Discussions

Experimental results are presented in three separate subsections:

- Behavior of the reasoning engine in a static environment
- Behavior of the reasoning engine after a condition disturbance
- Behavior after utilizing a simple optimization technique

Explanation about the obtained results are given at the end of each subsection.

6.1   Static Environment Conditions

Figure 2 depicts initial Q values, calculated per each state upon completion of the algorithm's *exploration phase*. Since values of our basis functions depend exclusively of the destination state(??), Q values for transferring from any state to a designated one are equal.

The best performing state, DCA, considers Service A, Service B and Service C to be activated, while Service B should be kept inactive. *Exploitation phase*, initiated at the end of the *exploration phase*, relays on these Q values .

The following Figure 3 depicts percentages of the number of learning episodes the system spent in each state during 100 learning episodes of the *exploitation phase*. Results are sorted in respect to values of the ϵ parameter.

As expected, the best performing state is determined to be CBA - services A and B activated in the Network A and service C activated in the Network B. Service D should be kept inactive. Corresponding percentages clearly depict the dependence on the value of the ϵ factor. After the exploration phase is over and the initial Q values are calculated, instances of the algorithm with the lower ϵ values tend to keep the system in the optimal state for as much as possible. The percentages vary from 77 down to only 15 percent, in cases when ϵ was set to

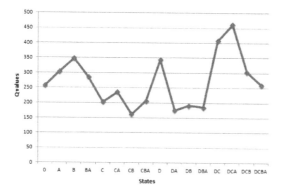

Fig. 2. Defined system states and their corresponding Q values upon completion of the *exploration phase*

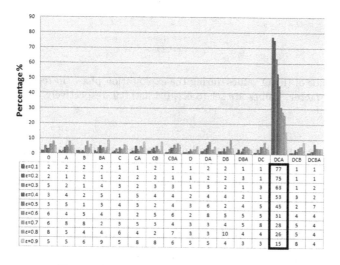

	O	A	B	BA	C	CA	CB	CBA	D	DA	DB	DBA	DC	DCA	DCB	DCBA
ε=0.1	2	2	2	2	1	1	2	1	1	2	2	1	1	77	1	1
ε=0.2	2	1	2	1	2	2	2	1	1	2	2	3	1	75	1	1
ε=0.3	5	2	1	4	3	2	3	3	1	3	2	1	3	63	1	2
ε=0.4	3	4	2	5	1	5	4	4	2	4	4	2	1	53	3	2
ε=0.5	3	5	1	3	4	3	2	4	3	6	2	4	5	45	2	7
ε=0.6	6	4	5	4	3	2	5	6	2	8	5	5	5	31	4	4
ε=0.7	6	8	8	2	3	5	5	3	4	3	3	4	5	28	5	4
ε=0.8	8	5	4	4	6	4	2	7	3	3	10	4	4	26	5	4
ε=0.9	5	5	6	9	5	8	8	6	5	5	4	3	3	15	8	4

Fig. 3. Percentages (Y - axis) are given in respect to the corresponding values of the ε factor, for every defined system state (X - axis)

0.1 and 0.9, respectively. It is worth noticing that, for the ε values of 0.4 and lower, our cognitive engine keeps the system in the optimal state for more than 50 of the time.

During the exploitation phase, ω factor are being constantly recalculated and q values reshaped. The following Figure 4 illustrate this process in three distinct cases, with the ε values set to 0.1, 0.5 and 0.9, respectively:

Algorithm instances (exploration policies) with lower ε values shape the Q values so that the optimal one is being increased at the highest rate. It appears as if the the sub-optimal ones are being repressed. This is an expected behavior since every visit to a certain state shapes the ω factors in its favor, thus

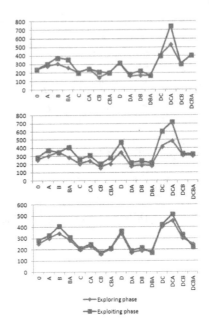

Fig. 4. Behavior of the Q values during the exploitation phase, in regards to different values of the ϵ factor. Values are set to 0.1, 0,4 and 0.9, respectively.

constantly increasing the corresponding Q values. As opposed, in the case with an almost completely random policy ($\epsilon = 0.9$) the states are picked almost uniformly, without favorizing any in particular. Therefore, Q values remain shaped similarly as after the exploration phase.

6.2   Reaction to a Network Condition Disturbance

Exploitation phase has two closely related objectives:

- Updating the sample set after each learning episode, thus reshaping the ω factors so that the respective Q values can be updated.
- Detects network condition disturbances and reshapes the decision making policy accordingly

The main indicator that the conditions in the network have changed are the Q values, calculated using re-shaped weight factors and respective basis functions. The following Figure 5 depicts the number of episodes needed to notice a condition change in the "worst case" scenario - best and worst performing states switch places (DCA - CB).

The results in Figure 5 are averaged over 10 trials of 250 learning episodes. As expected, the quickest response is achieved with $\epsilon = 0.9$ (13.2 episodes), since this mechanism checks system states at near-uniform manner. In the worst case,

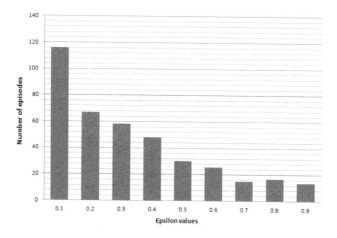

Fig. 5. Number of episodes needed to notice a network disturbance. The results are given in respect to different values of the ϵ exploration factor.

when $\epsilon = 0.1$, it takes almost 120 episodes, in average, to react on disturbances and re-shape Q values accordingly.

Results presented in sections 6.2 and 6.1 reveal one major issue of the taken approach - how to find a compromising solution for the ϵ value, so that the system is kept in the optimal (or the nearest-to-optimal states) for as long as possible, while being able to "quickly" react to environmental changes? For an illustration, with $\epsilon = 0.5$, system is kept in the optimal state for 45 percent of time, while engines ability to notice condition changes was limited to around 30 episodes, on average.

Fixing the ϵ to a certain value throughout the entire exploitation phase obviously does not provide acceptable results neither from the condition change versatility nor from the optimality point of view. One simple improvement of algorithm's efficiency during the exploitation phase is described and evaluated in the following sub-section.

6.3   A Simple Efficiency-Improving Procedure

In environments where the worst case scenario is impossible or rarely expected (moderate network dynamics), it is safe to restrain the exploitation phase to an optimal and a number of near-optimal states. This methodology can be applied in certain use cases where the traffic intensity does not change drastically over time. Figure 6 depicts the case where all the states with below the 50 percent of the optimal performance are discarded after the exploration phase.

In this case, an applied threshold will rule out six worst performing states - C, CB, CBA, DA, DB, DBA. The reduced set of ten remaining states is taken into account during the exploitation phase. The following Figure 7 describes the

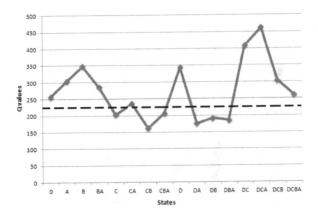

Fig. 6. Applying a threshold on Q values, prior to initiation of the *exploitation phase.* Threshold is set to 50 percent of the highest Q value.

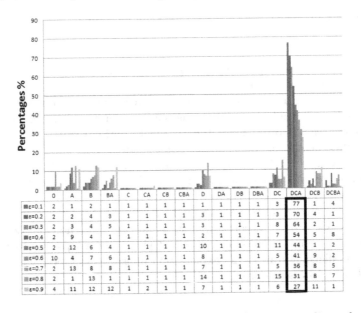

	O	A	B	BA	C	CA	CB	CBA	D	DA	DB	DBA	DC	DCA	DCB	DCBA
ε=0.1	2	1	2	1	1	1	1	1	1	1	1	1	3	77	1	4
ε=0.2	2	2	4	3	1	1	1	1	3	1	1	1	3	70	4	1
ε=0.3	2	3	4	5	1	1	1	1	3	1	1	1	8	64	2	1
ε=0.4	2	9	4	1	1	1	1	1	2	1	1	1	7	54	5	8
ε=0.5	2	12	6	4	1	1	1	1	10	1	1	1	11	44	1	2
ε=0.6	10	4	7	6	1	1	1	1	8	1	1	1	5	41	9	2
ε=0.7	2	13	8	8	1	1	1	1	7	1	1	1	5	36	8	5
ε=0.8	2	1	13	1	1	1	1	1	14	1	1	1	15	31	8	7
ε=0.9	4	11	12	12	1	2	1	1	7	1	1	1	6	27	11	1

Fig. 7. Percentages (Y - axis) are given in respect to the corresponding values of the ϵ factor, for every defined system state (X - axis), upon applying a simple efficiency-improving procedure

percentage of the number of episodes that are spent in each state during 100 episodes long exploitation phase, given different values of the ϵ factor:

Interestingly, there is no significant difference in the number of episodes spent in the optimal state for almost all the values of ϵ. However, by applying a threshold, the number of episodes spent in the worst performing states is significantly reduced. More importantly, the number of episodes needed to detect condition changes is reduced, as depicted on Figure 8:

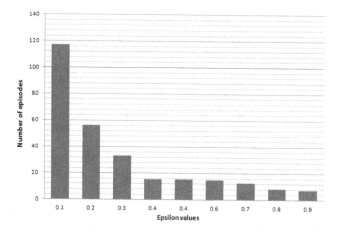

Fig. 8. Number of episodes needed to notice a network disturbance, while applying a simple speed up rule. The results are given in respect to different values of the ϵ exploration factor.

Except for the case when $\epsilon = 0.1$ a noticeable improvement is recorded during each run. Using a fixed value, $\epsilon = 0.5$, system is kept in the optimal state 44 percent of time, with an average condition change response time of around 15 episodes. If we take into account that an additional 33 percent of time is spent in nearest-to-optimal states (DC, D, B), as opposed to 12 percent when no threshold is applied. Conclusively, even such a simple speed up procedure yields a considerable performance improvement.

More advanced techniques will be presented as the part of the future work, in the 7 section.

7   Future Work

Future work implies improving the existing approach as well as expanding it on additional use cases. An at most attention needs to be paid on increasing the algorithm's efficiency during the exploitation phase, since efficiency is expected to become the major issue in use cases with the large state-action spaces.

One initiative is to develop techniques to "intelligently" reduce the number of system states that are being investigated. An already mentioned "performance threshold" approach is the simplest one. Others may also involve interpolation or prediction of a state performance. Both techniques aim at describing the performance of a system in states which have not been visited before. Some level of a priori knowledge can help defining a number of possibly forbidden states (e.g. certain services cannot be active at the same time). These techniques will have the major impact on efficiency once the duration of a learning episode becomes a factor.

Additional methods should focus on a dynamical change of the ϵ value during run-time. A critical network parameter, such as traffic intensity, is a good indicator of the network's overall behavior. Abrupt changes can be used to trigger an increase of the ϵ factor, thus stimulating an inspection of sub-optimal states. Depending on a use case, it might be reasonable to introduce a certain probability distribution over available actions, according to their respective Q values, according to which actions will be taken when the "ϵ greedy" is utilized.

An additional use cases this approach can be expanded to are:

- Optimizing a performance of a single network protocol
- Optimizing a negotiation process between multiple networks and coordinating the cooperation

The main goal of optimizing a single protocol would be to determine the optimal set of protocol settings. Setting combinations will represent system states. Therefore, complex services, with a large number of settings that need to be tweaked will imply a large number of state-action pairs. The above mentioned techniques for increasing the algorithm's efficiency will be reused here as well.

To optimize multiple co-existing networks, some compromises will be inevitable. Metrics that will precisely describe whether certain compromise is justified or not, from each sub-net's and the entire network's point of view, will have to be designed.

8   Conclusion

Optimizing multiple co-located networks, each with a variable number of network functionalities, influencing each other, is a complex problem that has not yet received a substantial attention in the research community.

This paper proposes and evaluates an application of the LSTDQ based, reinforcement learning approach. Our use case aims at discovering the optimal operating point of a single network, participating in a symbiotic cooperation with co-located, in general case, heterogeneous networks. Influences of the neighboring sub-nets are taken into account when calculating the optimal operational point. However, as opposed to most of the other approaches, no a priori knowledge about the influences is needed.

Use case described in this paper represents the proof of concept. Both strong and weak points of the implementation have been pointed out, along with a number of possible solutions and efficiency-improving techniques, all part of the future work. Most importantly, the final result of the learning process will be used as a starting point in the process of inter-network cooperation negotiation. Applying the same algorithm to all co-located sub-nets will yield an optimal, as well as the number of near-optimal operational points. They will represent a solid foundation in the following negotiation process.

We strongly believe that the problem of interfering co-located networks will only increase. As such, innovative cross-layer and cross-network solutions that take these interactions into account will be of a great importance to the successful development of efficient next-generation networks in heterogeneous environments.

Acknowledgment. This research is funded by the Institute for the Promotion of Innovation through Science and Technology in Flanders (IWTVlaanderen)through the IWT SymbioNets project, by iMinds through the QoCON project and by the FWO-Flanders through a FWO post-doctoral research grant for Eli De Poorter

References

[1] Wakamiya, N., Arakawa, S., Murata, M.: Self-Organization Based Network Architecture for New Generation Networks. In: 2009 First International Conference on Emerging Network Intelligence, October 11-16, pp. 61–68 (2009)

[2] Thomas, R.W., DaSilva, L.A., MacKenzie, A.B.: Cognitive networks. In: Proc. IEEE DySPAN 2005, pp. 352–360 (2005)

[3] De Poorter, E., Latre, B., Moerman, I., Demeester, P.: Symbiotic networks: Towards a new level of cooperation between wireless networks. Published in Special Issue of the Wireless Personal Communications Journal 45(4), 479–495 (2008)

[4] http://www-01.ibm.com/software/commerce/optimization/linear-programming

[5] Lagoudakis, M., Parr, R.: Model-free least-squares policy iteration. In: Proc. of NIPS (2001)

[6] Dietterich, T.G., Langley, O.: Machine Learning for Cognitive Networks:Technology Assessment and Research Challenges in Cognitive Networks: Towards Self Aware Networks. John Wiley and Sons, Ltd., Chichester (2007), doi:10.1002/9780470515143.ch5

[7] Kaelblign, L.P., Littman, M.L., Moore, A.W.: Reinforcement learning: A Survey. Journal of Artificial Intelligence Research 4, 237–285 (1996)

[8] Lagoudakis, M.G., Parr, R.: Least-Squares Policy Iteration. Journal of Machine Learning Research 4, 1107–1149 (2003)

[9] Busoniu, L., Babuska, R., De Schutter, B., Ernst, D.: Reinforcement Learning and Dynamic Programming Using Function Approximators, pp. 978–971. Taylor and Francis Group, LLC (2010) ISBN 978-1-4398-2108-4 (Hardback)

[10] White, J.M.: Bandit Algorithms for Website Optimization. O'Rilley Media, Inc.

[11] Tytgat, L., Jooris, B., De Mil, P., Latr, B., Moerman, I., Demeester, P.: UGentWiLab, a real-life wireless sensor testbed with environment emulation, https://biblio.ugent.be/publication/676545

Type II Hybrid-ARQ for DS-CDMA: A Discrete Time Markov Chain Wireless MAC Model

Francisco Ganhão[1,2], José Vieira[1], Luis Bernardo[1], and Rui Dinis[1,2]

[1] CTS, Uninova, Depto. de Eng. Electrotécnica, Faculdade de Ciências e Tecnologia - Universidade Nova de Lisboa, 2829-516 Caparica, Portugal
[2] Instituto de Telecomunicações, Lisboa, Portugal
fjsg@campus.fct.unl.pt

Abstract. Future wireless systems will need to cope with highly dispersive channels in order to support high data rates. A slotted Prefix-assisted DS-CDMA allows the multiplexing of various Mobile Terminals (MTs) at the uplink with appropriate Frequency Domain Equalization (FDE) to support the dispersive channels. However data packets can be received with errors due to channel interference or from a deep fade that persists for several slots; to cope with those errors a type II Hybrid Automatic Repeat reQuest (H-ARQ) protocol could be employed to re-use the signals from past packet copies to diminish errors. Most wireless CDMA models that employ H-ARQ assume a simplified characterization of the wireless channel, based on an average Signal to Interference-Noise Ratio (SINR), with simultaneous data transmissions from the MTs to a Base Station (BS). This paper proposes a DS-CDMA model that accounts the MTs' channel interference and channel noise simultaneously; packet reception is possible with the aid of a linear equalization method previously published by the authors. The wireless MAC model is characterized with Discrete-Time Markov Chain (DTMC), where the delay and throughput are obtained for a *Poisson* packet generator. The performance of the wireless model shows accurate results against the simulation values.

Keywords: Direct Sequence Code Division Multiple Access, Hybrid Automatic Repeat reQuest, Medium Access Control, Multipacket Reception.

1 Introduction

Wireless systems that need to multiplex several Mobile Terminals' (MTs) data at the uplink can use Code Division Multiple Access (CDMA) [1] to separate each MTs' data. CDMA Packet-data-oriented systems have been an hot research topic for the last decades, attracting contributions with the design and analysis of physical-layer receiver algorithms to improve Multi-Packet Reception (MPR) [1] or the Medium Access Control (MAC) performance, e.g. [2].

CDMA implements handles MTs' interference in two possible ways: frequency [3] or time [4]. In the frequency-domain, or Multi-Carrier CDMA (MC-CDMA)

S. Balandin et al. (Eds.): NEW2AN/ruSMART 2013, LNCS 8121, pp. 176–187, 2013.

[3]; data symbols are spread with an orthogonal spreading code for each user over different sub-carriers.

Concerning the time-domain, MTs can transmit data simultaneously using orthogonal spreading codes such as Direct Sequence CDMA (DS-CDMA) [4]. However, for a broadband channel, Frequency Domain Equalization (FDE) is necessary to cope with a highly variable dispersive channel [5]. Furthermore, the use of MC-CDMA has the cost of high envelope fluctuations, making DS-CDMA better suited for the uplink transmission when combined with FDE [6]. This work uses a linear MPR FDE technique from [7].

Despite the use of DS-CDMA to handle MTs' interference, packet errors may persist for several slots assuming a slotted system, either due to a deep fade or noise. Therefore there should be a way to cope with channel errors, where previous packet copies with errors are re-used for data equalization to improve the packets' reception. A known technique that re-uses packets with errors is type II H-ARQ, this technique ensures that packets are persistently retransmitted until they are correctly received at the Base Station (BS). Upon the reception of packets at the BS, the MTs' data will be verified for errors; if a packet has errors, the BS will not discard the actual packet and will ask for an additional transmission to cope with channel errors/interference [8]. The BS will therefore combine all transmissions from a given packet to retrieve its data, otherwise at the end of R retries, these are discarded. There are two techniques of type II H-ARQ: Code Combining (CC) [9] and Diversity Combining (DC) [8]. The first technique, employs a punctured error correcting code to the user's packets. The latter technique, DC, uses all transmitted copies to create a single packet with more reliable data symbols. This work considers an H-ARQ DS-CDMA solution with DC, due to its simple implementation when compared to CC.

Most works (e.g. [10–17]) attempt to model H-ARQ CDMA systems with simplified channel models where the receiver performance is modeled by a threshold value that defines the minimum SINR that satisfies a given Quality of Service (QoS). Approximate Bit Error Rate (BER) values are calculated, using average interference estimations (e.g. [11, 18] and references inside). However, the BER also depends on the number and power distribution of the packets being concurrently received [19].

This work proposes a broadband type II H-ARQ DS-CDMA slotted system, based on DC, employing the FDE linear equalization technique from [7] to cope with severe channel errors standing out from other works by accounting all the relevant interferences. In addition, a comprehensive modeling of the MAC protocol is shown characterizing the Delay and Throughput.

Regarding the paper's structure, Section 2 characterizes the DS-CDMA H-ARQ system; 3 describes the linear receiver design; Section 4 describes the Discrete-Time Markov Chain (DTMC) wireless MAC model; Section 5 shows the proposed model results; and Section 6 briefs the paper's conclusions.

As for the paper's notation for matrix computation, a matrix is denoted as bold such as $\mathbf{A}$; $\mathbf{A}^T$ is the matrix transpose of $\mathbf{A}$; $\mathbf{A}^H$ is the complex conjugate transpose of $\mathbf{A}$; and $\mathbf{A}^*$ is the complex conjugate of $\mathbf{A}$; an identity matrix of size x is denoted as $\mathbf{I}_x$.

2   System Description

The proposed model assumes a structured wireless system, with full-duplex communication, where the BS coordinates the wireless uplink access in a time slotted manner. In addition, perfect synchronization is assumed with perfect time advance mechanisms.

For a DS-CDMA spreading factor of K, each uplink slot should support up to K simultaneous transmissions. The wireless medium channel has $Q \leq K$ Mobile Terminals (MTs) contending the medium. The BS is capable of detecting the number of MTs transmitting in an uplink slot; a data packet at the uplink has the same size of an uplink slot.

The BS periodically signals the beginning of a new slot, so any MT with a non-empty queue transmits a packet; the MT stays idle if the queue is empty. At the end of the slot the BS verifies all packets for errors using the linear equalization technique from section [7], acknowledging at the next slot all packets that were successfully received; packets with errors will be re-transmitted in that slot for H-ARQ purposes. Any MT whose packet is retransmitted R times will be discarded from the queue, either with or without errors.

The BS performs an H-ARQ packet recovery by means of time diversity, i.e. for a given packet, the BS combines all of its transmissions to enhance packet reception. So the BS can store up to $R + 1$ failed transmissions from all MTs, controlling all MTs in the lth H-ARQ stage, where $0 < l \leq R$. If a MT succeeds transmitting a packet, then all of its transmissions are removed from the BS' database.

The number of idle users will be accounted as I and the number of users that are present on a given l-th H-ARQ stage is H_l, i.e. the number of MTs that have re-transmitted a packet l times. H_0 denotes the number of MTs that have only transmitted a single packet copy.

2.1   Descriptive Example

Figure 1 presents an example where MTs A, B and C transmit data packets to a BS, assuming a retransmission limit of $R = 1$. A packet from any MT, e.g A, is denoted in the figure as $A_i^{(r)}$, where i denotes A's packet identifier and r the rth transmission. The BS at the end of each slot acknowledges packets that were received with success and signals the beginning of the next slot.

At the first slot, the three MTs transmit data to the BS, where B and C packets were received by the BS with success, while A's packet, $A_1^{(1)}$, has errors. To recover from $A_1^{(1)}$'s errors, MT A will retransmit the same packet now denoted as $A_1^{(2)}$, and the BS will combine both $A_1^{(1)}$ and $A_1^{(2)}$ using the linear equalization technique. During the second slot, MT C also transmits a second packet, $C_2^{(1)}$,

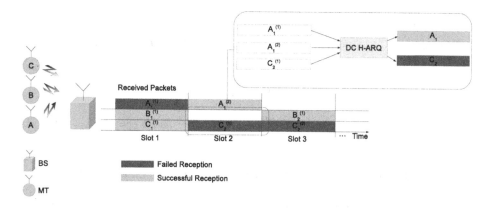

Fig. 1. CDMA H-ARQ Reception scheme

though received with errors; in similar fashion, MT C will retransmit the same packet, denoted as $C_2^{(2)}$. Since the second transmission from MT C fails, the BS and MT C will discard any information regarding the packet copies.

2.2   Physical (PHY) Channel Notations

This section briefs this paper's notations concerning the Physical (PHY) channel. A data block from MT p, where $0 < p \leq \sum_{r=0}^{R-1} H_r$, with M symbols in the time domain is $[a_{0,p}, \ldots, a_{M-1,p}]$, and its respective Discrete Fourier Transform (DFT) is $[A_{0,p}, \ldots, A_{M-1,p}]$.

For a spreading factor, K, the spreading sequence assigned to a MT p is $\mathbf{c}_p = [c_{0,p}, \ldots, c_{K-1,p}]$ and its respective DFT is $\mathbf{C}_p = [C_{1,p}, \ldots, C_{K,p}]$. For $k \in [0, \ldots, M-1]$ any transmitted symbol from MT p is $\mathbf{S}_{k,p} = \mathbf{C}_p^T A_{k,p}$ with size $K \times 1$; this is roughly equivalent of having K diversity replicas of $A_{k,p}$.

The channel realizations, concerning $\mathbf{S}_{k,p}$, from any lth transmission from MT p, where $0 < l < L$ and $L = max\{l; H_l > 0\}$, are

$$\mathbf{H}_{k,p}^{(l)'} = \begin{bmatrix} H_{k,p}^{(1,l)'} & 0 & \cdots & 0 \\ 0 & H_{k,p}^{(2,l)'} & \cdots & 0 \\ \vdots & \cdots & \ddots & \vdots \\ 0 & 0 & \cdots & H_{k,p}^{(K,l)'} \end{bmatrix}. \tag{1}$$

For simplicity it will be assumed that $\mathbf{H}_{k,p}^{(l)} = \left[H_{k,p}^{(1,l)'} C_{1,p}, \ldots, H_{k,p}^{(K,l)'} C_{K,p} \right]^T$,

so the spread channel realizations are $\mathbf{H}_{k,p} = \left[\mathbf{H}_{k,p}^{(0)T}, \ldots, \mathbf{H}_{k,p}^{(L)T} \right]^T$.

Note that for any MT p that is in the lth H-ARQ stage, its channel realizations $\left[\mathbf{H}_{k,p}^{(l+1)}, \ldots, \mathbf{H}_{k,p}^{(L)} \right]$ are nil, since MT p has only re-transmitted l copies.

The channel noise for the lth transmission is $\mathbf{N}_k^{(l)} = \left[N_k^{(1,l)}, \ldots, N_k^{(K,l)} \right]$.

Grouping the channel realizations from all $P = \sum_{l=0}^{L} H_l$ MTs, results $\mathbf{H}_k = [\mathbf{H}_{k,1}, \ldots, \mathbf{H}_{k,P}]$.

The received content is $[\mathbf{Y}_0, \ldots, \mathbf{Y}_{M-1}]$, where $\mathbf{Y}_k = \left[\mathbf{Y}_k^{(1)^T}, \ldots, \mathbf{Y}_k^{(L)^T} \right]^T$,

while the channel noise is $\mathbf{N}_k = \left[\mathbf{N}_k^{(0)^T}, \ldots, \mathbf{N}_k^{(L)^T} \right]^T$, so $\mathbf{Y}_k = \mathbf{H}_k^T \mathbf{A}_k + \mathbf{N}_k$.

3   Linear Receiver Design and the Packet Error Rate

This section lightly describes the linear equalization receiver design for a Quadrature Phase Shift Keying (QPSK) modulation, largely supported by the analytical PER model from [7], for the DS-CDMA context.

For any given time slot, an estimated data symbol from MT p at the output of the DFE receiver is

$$\tilde{A}_{k,p} = \mathbf{F}_{k,p}{}^T \mathbf{Y}_k, \tag{2}$$

where for a given number of re-transmissions $L = max\{l; H_l > 0\}$, $\mathbf{F}_{k,p}{}^T = \left[\mathbf{F}_{k,p}^{(1)}, \ldots, \mathbf{F}_{k,p}^{(L)} \right]$ are the feed-forward coefficients, used to remove channel interference, where $\mathbf{F}_{k,p}^{(l)}{}^T = \left[F_{k,p}^{(1,l)}, \ldots, F_{k,p}^{(K,l)} \right]$.

From the optimal $\mathbf{F}_{k,p}$ coefficients in [7], it is possible to compute the Mean Square Error for a given k-th symbol and user p, $\mathbb{E}\left[\left| A_{k,p} - \tilde{A}_{k,p} \right|^2 \right]$ (also in [7]). Defining

$$\sigma_p^2 = \frac{1}{N^2} \sum_{k=0}^{N-1} \mathbb{E}\left[\left| A_{k,p} - \tilde{A}_{k,p} \right|^2 \right], \tag{3}$$

and the Gaussian error function, $\Phi(x)$, the Bit Error Rate (BER) of user p at the ith iteration for a QPSK constellation is

$$BER_p \simeq \Phi\left(\frac{1}{\sigma_p} \right). \tag{4}$$

For an uncoded system with independent and isolated errors, the Packet Error Rate (PER) for a fixed packet size of K bits is

$$PER_p \simeq 1 - (1 - BER_p)^K. \tag{5}$$

4   DTMC MAC Model

The current section describes the DTMC model, based on the assumptions of section 2. Let us assume a network of Q MTs that transmit data to a single BS, and that the spreading sequences are enough for all MTs, i.e. $K \geq Q$.

MTs transmit data according to a λ rate *Poisson* distribution. The BS can hold up to $R + 1$ packet transmissions from each MT, otherwise it drops the packet copies from any MT whose data was re-transmitted R times after a(n) (un)successful reception.

4.1 Steady-State Probability Distribution

The universe of system states can be described by

$$\Omega = \left\{ \emptyset, \left\{ I, H_0, H_1, ..., H_R; (I, H_l \in \mathbb{N}_{[0,Q]}, \forall l \in [0, R]) \wedge \left(I + \sum_{l=0}^{R} H_l = Q \right) \right\} \right\}. \tag{6}$$

Once again, I stands for the number of idle MTs and H_l, $l \in [0, R]$, for the number of MTs in the lth H-ARQ stage; so any system state can be defined as $\chi \in \Omega$, where a non-empty state is

$$\chi = \left\{ I, H_0, H_1, ..., H_R; (I, H_l \in \mathbb{N}_{[0,Q]}, \forall l \in [0, R]) \wedge \left(I + \sum_{l=0}^{R} H_l = Q \right) \right\}. \tag{7}$$

Let us define a random process $\{\chi^1, \chi^2, ..., \chi^n\}, \forall n \in \mathbb{N}$, where $\chi^n \in \Omega$ and $\chi^n = \{I^n, H_0^n, H_1^n, ..., H_R^n\}$. Acquainted with the system's possible states, and admitting that

$$\chi^{n+1} = \left\{ I^{n+1}, H_0^{n+1}, H_1^{n+1}, ..., H_R^{n+1} \right\}, \tag{8}$$

$$\chi^n = \left\{ I^n, H_0^n, H_1^n, ..., H_R^n \right\}, \tag{9}$$

the state transition probability can be described as

$$\mathbb{P}\left[\chi^{n+1}|\chi^n\right] = \tag{10}$$

$$\begin{cases} 0, & \text{if } H_{l+1}^{n+1} > H_l^n, \quad \forall l \in [0, R-1] \\ \left(\prod_{l=0}^{R-1} bi\left(H_{l+1}^{n+1}, H_l^n, p_{err}^l(\chi^n)\right) \right) \times bi\left(I^{n+1}, I^n + H_R^n + \sum_{l=0}^{R-1} (H_l^n - H_{l+1}^{n+1}), p_{QE}\right), \\ & \text{otherwise.} \end{cases}$$

$bi(x, N, p) = \binom{N}{x} p^x (1 - p)^{N-x}$ is the binomial Probability Density Function (PDF); $p_{err}^l(\chi^n)$ is the average packet error rate of the MTs at the lth H-ARQ stage at slot n, computed from equation (5); p_{QE} is the probability of having an empty queue. Since it is assumed that MTs transmit with a *Poisson* rate λ, from [20] p_{QE} can be computed with the following approximation,

$$p_{QE} \approx 1 - min\left(\lambda \mathbb{E}\left[N\right], 1\right), \tag{11}$$

where $\mathbb{E}[N]$ is the expected number of packet transmissions whose calculus is referred below in equation (12).

$$\mathbb{E}[N] = \sum_{l'=1}^{R+1} l' \times \mathbb{P}\left[Tx = l' - 1|Tx \geq 0\right], \tag{12}$$

$\mathbb{P}[Tx = l | Tx \geq 0]$ is the probability of re-transmitting a packet l times, conditioned by the fact that there is at least one transmission, where $\mathbb{P}[Tx = l | Tx \geq 0] = \frac{\Xi(l)}{\Upsilon}$ and

$$
\Xi(l) = \begin{cases} \sum_{\forall \chi^n \in \Omega, H_l^n > 0} \mathbb{P}[\chi^n] \sum_{k=1}^{H_l^n} k \times bi(k, H_l^n, 1 - p_{err}^l(\chi^n)) & \text{if } l < R \\ \sum_{\forall \chi^n \in \Omega, H_{R-1}^n > 0} \mathbb{P}[\chi^n] \sum_{k=1}^{H_{R-1}^n} k \times bi(k, H_{R-1}^n, p_{err}^{R-1}(\chi^n)) & \text{if } l = R \end{cases} ;
$$

(13)

$$
\Upsilon = \sum_{l=0}^{R} \Xi(l).
$$

(14)

So, based on the aforementioned expressions, the steady-state probability distribution is computed below in equation (15).

$$
\pi_\chi = \mathbb{P}[\chi] = \lim_{n \to \infty} \mathbb{P}[\chi^n = \chi].
$$

(15)

The steady-state distribution is independent of the first state, χ^0, and can be computed in an iterative manner using (10).

4.2   Delay

The mean packet delay, $\mathbb{E}[D]$, can be computed, based on the Pollaczek-Khinchine formula from [20], where

$$
\mathbb{E}[D] = \mathbb{E}[N] + \frac{\lambda \mathbb{E}[N^2]}{2p_{QE}}.
$$

(16)

4.3   Average Throughput

The average throughput per MT, S, can be obtained based on the system's steady-state probability distribution, as the ratio of the number of successful packet receptions over the average number of transmissions from all MTs where

$$
S \approx \frac{1}{Q\mathbb{E}[N]} \sum_{\forall \chi \in \Omega} \mathbb{P}[\chi] \sum_{l=0}^{R} \mathbb{I}_{H_l \subseteq \chi \wedge H_l > 0} \sum_{k=1}^{H_l} k \times bi(k, H_l, 1 - p_{err}^l(\chi))
$$

(17)

5   Performance Results

All results throughout this section's figures use lines as the model's results and markers as the respective simulations. A time dispersive channel was considered, with rich multipath propagation and uncorrelated Rayleigh fading, but with channel correlation between each packet retransmission. To cope with channel correlation for each retransmission, the Shifted Packet technique from [22] is used, where each retransmission has a different cyclic shift. Terminals scattered inside the BS' coverage area transmit uncoded data blocks with $N = 256$ symbols selected from a QPSK constellation with Gray mapping for a $4\mu s$ transmission.

5.1   H-ARQ DS-CDMA Wireless Access: Variable E_b/N_0

The results within this section present the H-ARQ DS-CDMA MAC performance for a variable range of the bit energy over the noise ratio, E_b/N_0, with $Q = [4, 8]$ MTs and a fixed data rate of $\lambda Q = 0.4$, except the last figure that portrays the system saturated throughput.

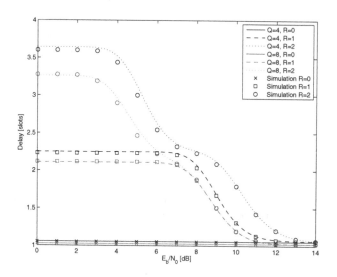

Fig. 2. System delay for $Q = [4, 8]$ MTs, considering an H-ARQ scenario up to two additional re-transmissions and $\lambda Q = 0.4$

Figure 2 displays the system delay for $Q = [4, 8]$ MTs, with a system load of $\lambda Q = 0.4$ and up to $R = 2$ H-ARQ transmissions. The simulations and model's results are practically matched, showing the accuracy of the proposed system model. As obvious the system delay does increase for lower E_b/N_0 values, but also with the increase of H-ARQ transmissions that ensure the correct transmission of the packets.

Figure 3 demonstrates the system throughput for $Q = [4, 8]$ MTs, with a system load of $\lambda Q = 0.4$ and up to $R = 2$ H-ARQ transmissions. The system model performs well, although the system model can be observed as pessimistic. Concerning the system performance, for an increasing number of transmissions, the H-ARQ behavior enhances the system throughput, decreasing the necessary E_b/N_0 ratio to attain a given system throughput for a fixed data rate.

Figure 4 demonstrates the saturated system throughput for $Q = [4, 8]$ MTs and up to $R = 2$ H-ARQ transmissions. Once more the model results are close to the simulation results. Concerning the system performance, once more for an increasing number of transmissions, the H-ARQ behavior enhances the throughput performance for the same E_b/N_0.

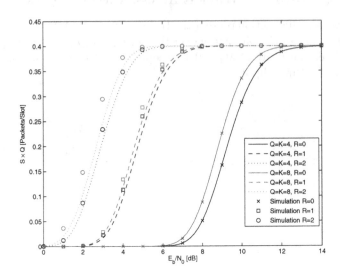

Fig. 3. System throughput for $Q = [4, 8]$ MTs, considering an H-ARQ scenario up to two additional re-transmissions and $\lambda Q = 0.4$

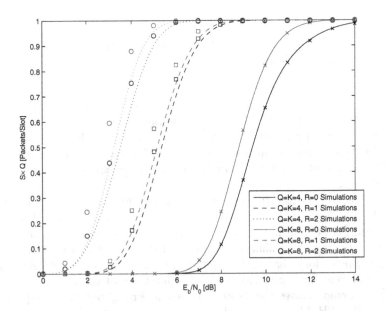

Fig. 4. Saturated system throughput for $Q = [4, 8]$ MTs, considering an H-ARQ scenario up to two additional re-transmissions

5.2   H-ARQ DS-CDMA Wireless Access: Variable Load

The results within this section present the H-ARQ DS-CDMA MAC performance for a variable range of the system load for $Q = [4, 8]$ MTs and $E_b/N_0 = [4, 6, 8]$ dB.

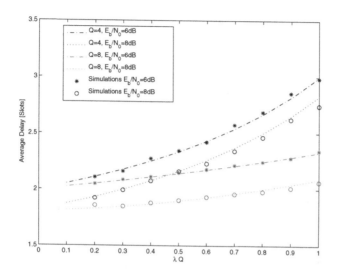

Fig. 5. Average system delay for $Q = [4, 8]$ MTs, considering an H-ARQ scenario for one additional re-transmissions and $E_b/N_0 = [4, 6, 8]$ dB

   Figure 5 portrays the average delay per packet according to a variable range of the system load for $E_b/N_0 = [4, 6, 8]$ dB, $Q = [4, 8]$ and $R = 1$. The model performance almost matches the simulation results. As for the MAC performance, the average delay increases as the system load increases; the delay also decreases as the E_b/N_0 increases, though for 4dB and 6dB the performance is practically the same. The performance of $Q = 8$ is better than $Q = 4$, in general, due to the fact that there are less transmitted packets per MT, therefore the lower interference and expected delay.

   Figure 6 portrays the system throughput according to a variable range of the system load for $E_b/N_0 = [4, 6, 8]$, $Q = [4, 8]$ dB and $R = 1$. The model's performance is illustrated by the lines while the simulation results are portrayed by markers. The model performance almost matches the simulation results, though being slightly pessimistic. Regarding the MAC performance, and as expected, the throughput increases as the system load and E_b/N_0 increase.

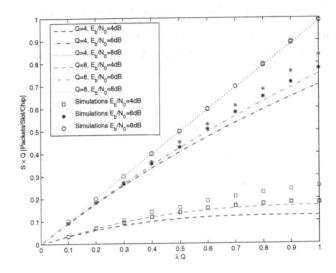

Fig. 6. System throughput for $Q = [4, 8]$ MTs, considering an H-ARQ scenario for one additional re-transmissions and $E_b/N_0 = [4, 6, 8]$ dB

6   Conclusion

This paper proposes a wireless MAC DS-CDMA model with H-ARQ, based on a previously published linear equalization receiver by the authors, that accounts the MTs' channel interference and channel noise simultaneously. The wireless access was characterized with DTMC, where the delay and throughput were extracted. The model's performance has a good accuracy when compared with simulation values. Results show that the H-ARQ DS-CDMA system has a good performance when compared with a regular DS-CDMA system without H-ARQ. For future work, the authors intend to extend the DTMC model with back-off mechanisms.

Acknowledgments. This work was supported by the FCT/MEC projects CTS PEst-OE/EEI/UI0066/ 2011; IT PEst-OE/EEI/LA0008/ 2013; MP-Sat PTDC/EEA-TEL/099074/2008; OPPORTUNISTIC-CR PTDC/EEA-TEL/115981/ 2009; ADCOD PTDC/EEA-TEL/099973/2008 and Femtocells PTDC/EEA-TEL/120666/2010; as well as grant SFRH/BD/66105/2009.

References

1. Verdu, V.: Multiuser Detection. Cambridge University Press (1998)
2. Costello, D., Forney, G.: Channel Coding: The Road to Channel Capacity. Proceedings of the IEEE 95, 1150–1177 (2007)
3. Hara, S., Prasad, R.: Overview of multicarrier CDMA. IEEE Communications Magazine 35, 126–133 (1997)

4. Moshavi, S.: Multi-user detection for DS-CDMA communications. IEEE Communications Magazine 34, 124–136 (1996)
5. Adachi, F., Garg, D., Takeda, K.: Broadband CDMA techniques. IEEE Wireless Communications 12, 8–18 (2005)
6. Benvenuto, N., Dinis, R., Falconer, D., Tomasin, S.: Single Carrier Modulation With Nonlinear Frequency Domain Equalization: An Idea Whose Time Has Come—Again. Proceedings of the IEEE 98, 69–96 (2010)
7. Ganhão, F., Dinis, R., Bernardo, L., Oliveira, R.: Analytical BER and PER Performance of Frequency-Domain Diversity Combining, Multipacket Detection and Hybrid Schemes. IEEE Transactions in Communications 60, 1–8 (2012)
8. Sindhu, P.: Retransmission error control with memory. IEEE Transactions on Communications 25, 473–479 (1977)
9. Hagenauer, J.: Rate-compatible Punctured Convolutional Codes (RPCP Codes) and Their Applications. IEEE Transactions on Communications 36, 389–400 (1998)
10. Levorato, M., Zorzi, M.: On the Performance of Ad Hoc Networks with Multiused Detection, Rate Control and Hybrid ARQ. IEEE Transactions on Wireless Communications 8 (2009)
11. Lu, B., Wang, X., Zhang, J.: Throughput of CDMA Data Networks With Multiuser Detection, ARQ, and Packet Combining. IEEE Transactions on Wireless Communications 3 (2004)
12. Souissi, S., Wicker, S.: A Diversity Combining DSICDMA System with Convolutional Encoding and Viterbi Decoding. IEEE Trans. on Vehicular Technology 44, 304–312 (1995)
13. Bigloo, A., Gulliver, T., Bhargava, V.: Maximum-Likelihood Decoding and Code Combining for DS/SSMA Slotted ALOHA. IEEE Trans. on Communications 45, 1602–1612 (1997)
14. Zhang, Q., Wong, T., Lehnert, J.: Performance of type-II hybrid ARQ protocol in slotted DS-SSMA packet radio systems. IEEE Trans. Commun. 47, 281–290 (1999)
15. Prakash, R., Veeravalli, V.: Analysis of code division random multiple access systems with packet combining. In: Proc. 33rd Asilomar Conf. Signals, Systems & Computers (2000)
16. Telatar, I., Gallager, R.: Combining queueing theory with information theory for multiaccess. IEEE J. Select. Areas Commun. 13, 963–969 (1995)
17. Win, M., Pinto, P., Shepp, L.: A Mathematical Theory of Network Interference and Its Applications. Proceedings of IEEE 97, 205–230 (2009)
18. Lai, K., Shynk, J.: Analysis of the Linear SIC for DS/CDMA Signals with Random Spreading. IEEE Trans. on Signal Processing 52, 3417–3428 (2004)
19. Das, S., Viswanathan, H.: A Comparison of Reverse Link Access Schemes for Next-Generation Cellular Systems. IEEE Journal on Selected Areas in Communications 24, 684–692 (2006)
20. Takács, L.: Introduction to the Theory of Queues. Oxford University Press (1962)
21. Cui, S., Goldsmith, A., Bahai, A.: Energy-constrained modulation optimization. IEEE Transactions on Wireless Communications 4, 2349–2360 (2005)
22. Dinis, R., Montezuma, P., Bernardo, L., Oliveira, R., Pereira, M., Pinto, P.: Frequency-domain multipacket detection: a high throughput technique for SC-FDE systems. IEEE Transactions on Wireless Communications 8, 3798–3807 (2009)

Erasure-Coding Based Data Delivery in Delay Tolerant Networks

Khalil Massri, Roberto Beraldi, and Andrea Vitaletti

Dipartimento Di Ingegneria Informatica Automatica e Gestionale
Università degli Studi di Roma "La Sapienza"
Roma, Italy
{massri,beraldi,vitale}@dis.uniroma1.it

Abstract. We consider the data delivery problem in delay tolerant networks, where a data content is located in a fixed source need to be delivered to a specific destination. We assume nodes have limited storage and computational capabilities. In this paper, we initially, explore the data delivery problem, for both unbiased and biased contact models. Based on our observations, we propose a data delivery scheme that can reduce both storage overhead and delivery delay. Our scheme combines erasure coding technique and the framework of simulated annealing optimization, in order to maximize the content delivery probability to the destination.

Keywords: Routing, Delay Tolerant Networks (DTNs), Erasure Coding.

1 Introduction

Delay-Tolerant Networks, DTNs [1], are intermittently connected mobile wireless networks that may suffer from frequent partitions, and thus a path connecting source and destination cannot be maintained over time. Routing in DTNs is one of the main challenges; the use of efficient routing schemes is a key element to performance of DTN networks. Therefore, the usability and applicability of DTN deployments are conditioned by efficient and optimized routing algorithms. Nodes exploit mobility in order to carry the message to the intended destination.

Due to the wide diversity in contexts in which DTN routing is applicable, many different protocols have been proposed in the literature. In our previous work [2], we classified most of the recently proposed protocols in the literature. In particular, we consider forwarding, replication and queue management as the main techniques for DTN protocols and we show how each protocol can be classified according to the techniques it adopts. Forwarding techniques control whether or not a node can forward a message to the other node when there is a contact. While replication techniques control the duplication of a message among nodes.

In this work, we explore the problem of data content delivery in DTNs, where relay nodes are limited-resources devices. Namely, when the delivery of the data

S. Balandin et al. (Eds.): NEW2AN/ruSMART 2013, LNCS 8121, pp. 188–200, 2013.

content to a given destination requires a number of packets. Our objective is to reduce the delivery delay for the whole content. At the same time, we want also to reduce storage overhead in the network. Two main factors mainly affect the delivery delay. The first one is the time needed by the source to disseminate the data content into the network. The second one is the technique used among the relay nodes to fasten the delivery of the data packets to the destination.

Erasure coding [3] is a powerful technique used in routing protocols for DTNs. It enables the source to inject redundant packets in the network, with fixed replication overhead. Our first concern is, to what extend would this replication overhead reduce the collecting time needed by the destination for the data packets required to restore the original content. Moreover, an interesting question arise here: how should these coded packets be forwarded or replicated among relay nodes in order to gain this reduction.

In a homogeneous environment, all node movements have the same stochastic characteristic, and thus, a good strategy to reduce the delivery delay is to replicate the same packet in the network. We call such scenario unbiased contact model. On the other hand, in a heterogeneous environment, some nodes may be "better" relay nodes for a given destination than others, because they meet the destination more often. In this case, called biased contact model, a good strategy to reduce the delay is to forward packets from one node to another with a higher probability of meeting the destination.

In this paper we study data delivery in both biased and unbiased cases, using real and synthetic mobility traces. For the biased case we adopted a Simulated Annealing (SA) algorithm that searches for the "best carriers" nodes available in the network. The algorithm is based on the framework reported in [4]. In our proposed scheme, the source exploits erasure coding technique as a mean of splitting the data content into packets suitable to be carried by the nodes, and to inject data packets into the network, with minimum replication overhead. We show that the amount of replication required to reduce the delay could be small and of the same order of magnitude of the original content.

To evaluate our scheme, extensive simulations were carried out. In our simulations we used both synthetic mobility, and real movement traces. Evaluation results confirm our observations about data delivery problem in DTNs, also they show that our data delivery scheme for biased contact model, reduces the delivery delay using fixed storage overhead.

2   Related Work and Background

Many different protocols for DTNs routing have been proposed in the literature. In our previous work [2], we evaluated a wide variety of them. We proposed a clear identification of the main techniques characterizing DTNs routing protocols in order to better understand and classify the solutions proposed in the literature.

Coding-based routing protocols are recently proposed to improve the delivery performance in DTNs. Depending on where the coding functionality is performed, they are divided into two families: erasure coding [3] and network coding based protocols [5]. In the erasure code family, coding operations are all done at the source, and coded packets are disseminated into the network. Well known examples of erasure coding include Raptor, Tornado and Reed-Solomon codes.

In [6] the authors explored the benefit of erasure coding based routing, showing that with coding the best worst-case performance can be achieved for a fixed overhead. Other important works include [7], that studied the problem of optimal routing in a DTN in the presence of path failures with different failure probabilities; [8] that proposed an adaptive protocol that estimates the Average Contact Frequency, used to regulate the spreading phase of a protocol that uses erasure coding; [9] that proposed an Inter-Coding protocol, where coded blocks are interleaved in order to cope with uncertainty about link failure probabilities prediction, and [10] that studied how the cost of erasure coding based routing protocols can be reduced, by leverage different spraying algorithms, right parameter selection and splitting spraying phase on the cost of message delivery.

Erasure-Coding Based Data Delivery. In erasure coding (EC) the source node splits the original content into G data packets, or fragments $E_1, E_2, ..., E_G$ of equal size, and it emits $K = rG$ packets. It first emits the original G packets followed by other $K - G$ linear combinations over them. r is the replication factor (r is such that K is an integer). An encoded packet x is a liner combination over $GF(2)$. $x = \kappa_1 E_1 \oplus \kappa_2 E_2 \oplus ... \kappa_G E_G$ where κ_i are random binary values, called the encoding vector. Assuming optimal erasure coding is used, the content can fully decoded in the destination when receiving any G out of K packets.

Optimization in Opportunistic Networks. In [4] the authors presented a distributed stochastic algorithm, based on the Markov Chain Monte Carlo method (MCMC) to search for the optimal solution in maximizing a utility function. Their method is based on the well-known simulated annealing algorithm. Their framework can be applied to many problems in opportunistic networking such as, routing, buffer management and content/service placement. Providing efficient local algorithms that can converge to a globally optimal solution. It aims at maximizing a distributed utility function (in their case study, sum of carriers node degree) based on using simulated annealing techniques.

3   Data Collection Model

In this section we present a mathematical model to estimate the impact of erasure coding replication overhead on the expected collection delay needed by the destination. Then, we estimate the impact on this delay when there are two classes of carriers, one of which is more frequently contacted by the destination.

We assume a source generates $K = rG$ packets using an ideal erasure coding protocol. The K packets are uniformly disseminated in the network (in the next section we show how). Each node has a buffer space sufficient to allocate only one data packet. The destination can retrieve the content as soon as it collects any G out of the K packets.

Our first concern is the effect of K on the delivery delay under the unbiased case. The destination node undergoes a sequence of contacts with nodes that come in-range. These are opportunities for the destination to pull a data packet. Let Ω_i be the number of *new packets* the destination collects during the i-th contact ($i \geq 1$), and let $\mathbb{E}[\Omega_i]$ be the expected number of *new packets* the destination gets at contact i. Therefore, $\mathbb{E}[\Omega_1] = 1$, because the destination always gets a new packet in the first contact. Then we have:

$$\mathbb{E}[\Omega_i] = \frac{K - \sum_{j=1}^{i-1} \mathbb{E}[\Omega_j]}{K} \tag{1}$$

From which we also have:

$$\mathbb{E}[\Omega_{i+1}] = \frac{K - \sum_{j=1}^{i} \mathbb{E}[\Omega_j]}{K} \tag{2}$$

Now, by subtracting equation (1) from (2) we get:

$$\mathbb{E}[\Omega_{i+1}] = \mathbb{E}[\Omega_i](1 - \frac{1}{K})$$

From which we get:

$$\mathbb{E}[\Omega_i] = (1 - \frac{1}{K})^{i-1} \tag{3}$$

Then, the number of contacts (delay) d_G required in collecting G packets is:

$$d_G = \min_i \{i | \sum_{j=1}^{i} \mathbb{E}[\Omega_i] \geq G\} \tag{4}$$

In Fig.1 we see the delay as a function of K, for different values of G. We see how the delay sharply decreases with K, until it becomes pretty close to the minimum at a replication factor $r = 2$ i.e., $K = 2G$.

Lets now consider a simplified biased scenario, where two different type of nodes exist characterized by different meeting probabilities. The first type belong to class C_1 and the other, to class C_2. At each contact, the destination has a probability β_1 to make a contact to C_1, and $\beta_2 = 1 - \beta_1$ to make a contact to C_2. Let's assume that class C_1, carries α_1 *different packets*, and class C_2 carries α_2 *different packets*. where $\alpha_1 + \alpha_2 = K$.

Let $\mathbb{E}[\Omega^{c_1}{}_i]$ and $\mathbb{E}[\Omega^{c_2}{}_i]$ be the average number of *new packets* the destination gets from C_1 and C_2 respectively at contact i. Then, the probability of getting a useful packet at contact i from C_1 is: $p_i = \frac{\alpha_1 - \sum_{j=1}^{i-1} \mathbb{E}[\Omega^{c_1}{}_j]}{\alpha_1}$, from which:

$$\mathbb{E}[\Omega^{c_1}{}_i] = \beta_1 \left(\frac{\alpha_1 - \sum_{j=1}^{i-1} \mathbb{E}[\Omega^{c_1}{}_j]}{\alpha_1} \right) \tag{5}$$

As we did in equation (3):

$$\mathbb{E}[\Omega^{c_1}{}_i] = \beta_1 (1 - \frac{\beta_1}{\alpha_1})^{i-1} \tag{6}$$

As well, for class C_2:

$$\mathbb{E}[\Omega^{c_2}{}_i] = \beta_2(1 - \frac{\beta_2}{\alpha_2})^{i-1} \qquad (7)$$

Then, the average number of packets the destination gets at each contact i is:

$$\mathbb{E}[\Omega_i] = \mathbb{E}[\Omega^{c_1}{}_i] + \mathbb{E}[\Omega^{c_2}{}_i] \qquad (8)$$

From which we can get the delay d_G as in equation (4).

We confirm the validation of this model using numerical simulation reported in Fig.1 and Fig.3. Fig.3 shows the delay as a function of α_1 for $G = 20$ and $K = 40$. As expected, the delay reduces with α_1. However, it worth to note that the delay becomes small even when we are able to allocate a slightly more than G packets to class C_1. These results will drive us to design our proposed protocol.

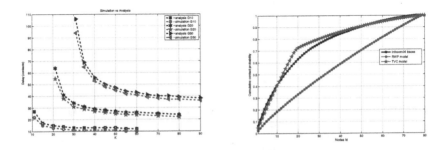

Fig. 1. Delay(contacts) vs K. (blue) theo-retical model. (red) simulating the model

Fig. 2. cumulative contact probability. b/w the destination and other nodes.

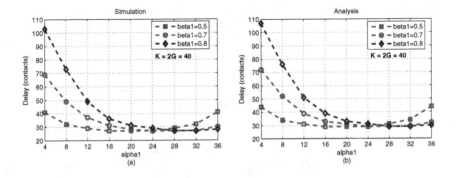

Fig. 3. Effect of β and α on the Delay, given $K = 2G = 40$. (a)is simulated, (b)is the theoretical model.

4   Forwarding and Replication

In this section we discuss the techniques most suitable for data delivery in unbiased and biased contacts cases. We start by analyzing three types of movements traces to see their contact characteristics.

The first movements are synthetic traces generated by simulating 80 nodes moving according to the random way-point model (RWP) [11]. These traces supposed to produce non-biased contacts among the nodes. Simulation area is a square of dimension 1000 by 500 meters. Nodes transmission range is 20 meters with movement speed uniformly selected from the range of [1.0, 2.0] meters per second (walking speed), and maximum pause time of 100 seconds. Time duration is 4-days.

The second movements are also synthetic traces, using similar simulation settings as above. But nodes are moving according to the time-variant community mobility model (TVC) [12], which defines communities that are visited often by the nodes to capture skewed location visiting preferences. In particular, we define 4-communities each of 20 nodes. Nodes undergoes two epoch stages, of duration 100 and 30 seconds respectively. For each stage a different aggregation value is used to aggregate the nodes around the community center (we used 0.5, 0.0 respectively). The more this value approximates to 1, the nodes will be more aggregated and closer to the group center. With aggregation 0, the nodes are randomly distributed in the simulation area. Both RWP and TVC contacts are averaged over 50 runs.

The last movements are from Infocom06 [13] real traces extracted from CRAW-DAD [14]. These traces were logged using devices carried by 78 users, during 4-days experiment. More details about this traces are in section 6.

In Fig. 2 we plot the cumulative contact probability between a selected destination and each other node. The x-axis represents the contacted node Ids, sorted by their contact probabilities, while the y-axis is the cumulative contact probability between the destination and these nodes.

Infocom06 traces and TVC model clearly show how the cumulative contact probability rapidly increased up to specific group of nodes, then it increased slowly. This indicates that there is a *small* group of nodes in the network which the destination is more biased to contact than the rest of other nodes. On the other hand, the cumulative contact probability generated by RWP increases linearly with respect to the contacted nodes, which indicates that the destination has no biasing in contacting any node.

4.1   Unbiased Contact Model

In the unbiased contact case, such as in RWP, all nodes have the same probability to reach the destination. Hence, the optimal approach to maximize the probability for the destination to get any coded packet at each contact, is to uniformly replicating the K packets among all relays. This can be achieved by Binary Spraying protocol (BSW) with parameter $L = \frac{N}{K}$, where N is the total number of nodes. Indeed, in [15], its proven that, binary spraying optimally minimizes the dissemination time. As discussed in the previous section, also shown in Fig. 1, using $K = 2G$ is sufficient to reduce the collecting time by the destination, also its a good compromise to reduce the time consumed by the source to inject the data packets.

4.2   Biased Contact Model

In real environments, things are different. Some nodes may be "better" relays, such, for example, could be nodes that tend to see the destination more often, or have similarity in their movements pattern, as discussed in Infocom06 and TVC. We approximate this behavior to our model of two classes of nodes. For example in Fig. 2 we see, in both infocom06 and TVC the destination, approximately, contacts a node from the first 20 ones with probability $\simeq 0.7$.

If we adopt the replication based approach (used in biased contact case), where each packet is replicated $\frac{N}{K}$ times, then class C_1 will carry $\alpha_1 = \frac{N_1}{N} \times K$ different packets, where N_1 is the number of nodes in C_1. Now, to have at least G packets in C_1, assuming $r = 2$, then N_1 must be $\geq N/2$, however, this is not realistic for the destination to be biased or to have similar movements to half of network nodes.

An alternative approach is to use a forwarding scheme that search for the optimal K carriers that maximize the delivery probability to the destination in the whole network.

5   Proposed Data Delivery Scheme

In this section we design a scheme for data delivery under biased contact model. We aim at reducing the delivery delay and storage overhead. We assume that each node has a storage buffer size of one packet. This assumption is motivated by our goal of providing a lower bound for the use of erasure coding in a network of resource-constrained devices.

Initially, the source erasure-codes the data content and generates $K = 2G$ coded packets, then forwards them to the first K nodes. The source needs to exploit every contact to inject the data packets for two main reasons: 1) since each node can carry only one packet, the source need several contacts to disseminate the whole content, 2) the "best carriers" nodes not necessarily have high contacts probability with the source. Since they may have different movement patterns.

5.1   Estimation of Node's Contact Probability

Each node estimates its contact probability to the destination. We adopt a simple and effective approach, used in many previous DTNs routing algorithms [16], namely: exponentially weighted moving average (EWMA). More specifically, each node i maintains its contact probability μ_i to the destination, which is updated every time slot t, according to the following formula:

$$\mu_{i,t} = \begin{cases} (1-\delta)\mu_{i,t-1} + \delta & \text{meeting occurs} \\ (1-\delta)\mu_{i,t-1} & \text{no meeting} \end{cases}$$

Where δ is a constant parameter between 0 and 1. Clearly, this is a dynamic process, and thus μ_i doesnt necessarily equal to the actual contact probability P_i . However, In [16] it's shown that if nodes i and j have a probability of P_{ij} to meet in every time slot, then, the mean of EWMA converges to P_{ij}. Applying it to our case, yields:

$$\mathbb{E}[\mu_{i,1}] = (1-\delta)\mu_{i,0} + \delta P_i$$

$$\mathbb{E}[\mu_{i,2}] = (1-\delta)^2\mu_{i,0} + \delta P_i(1 + (1-\delta))$$

$$\mathbb{E}[\mu_{i,t}] = (1-\delta)^t\mu_{i,0} + \delta P_i \sum_{j=1}^{t}(1-\delta)^{j-1}$$

From which we get:

$$\lim_{t \to \infty} \mathbb{E}[\mu_{i,t}] = \delta P_i \frac{1}{\delta} = P_i$$

5.2   Forwarding Technique

By our forwarding technique we aim at finding the subset C_K of K nodes that maximize the utility function $U(C_K) = \sum_{\forall j \in C_K} \mu_j$. In [4], a well done framework is proposed where simulated annealing (SA) algorithm is employed for optimization problems for opportunistic networks. They showed that SA algorithm can be employed in DTNs for globally selecting a subset of nodes that maximizes a utility function.

SA employs randomization in searching for the optimal solution, which not only accepts states that increase the utility function but also some changes that decrease it, to avoid becoming trapped at local maxima. The latter are accepted with a probability $p = \exp(\frac{-\Delta U}{T})$, where ΔU is the decrease in U, and T is a control parameter (temperature). This randomization is controlled by the parameter T. The way in which the temperature is adapted is called a "cooling schedule": starts with a relatively high T, so that more randomization is used in searching for the high utility states, and gradually cool down the system, in order to converge to the maximum utility. In our implementation we used an empirical exponential cooling schedule.

As proved in [4], and since our utility function is evaluated locally, this enables a fully distributed implementation of the optimization algorithm. This implies that the marginal contribution of each node is independent of the of other nodes. Accordingly, when a relay node r carrying a packet pkt contacts an empty node e, the technique explained in Algorithm 1.

Algorithm 1. SA Forwarding Technique

$U_r = \mu_r$
$U_e = \mu_e$
if $U_e \geq U_r$ **then**
    forward *pkt* to node e
else
    $\Delta U = U_e - U_r$
    $p = \exp(\frac{-\Delta U}{T})$
    if $p \geq rand(0,1)$ **then**
        forward *pkt* to node e
    end if
end if

---

6   Evaluation

To evaluate data content delivery schemes, we create two simulating scenarios. For the first scenario, which reflects the biased case, we use Infocom06[13] real traces, which were logged by iMote devices (with wireless range around 30 meters) carried by 78 volunteers, joined a 4-days experiment conducted at Infocom 2006. The source and the destination were selected from two of the 20 static long range (around 100 meters) iMote devices, that were placed at various locations of the conference venue. We parsed these traces to be injected into the ONE simulator [11]. For the second scenario, which reflects the unbiased case, we use synthetic mobility generated by RWP model with the simulation settings as discussed in section 4. Two fixed nodes are selected to be the source and the destination. Data content is generated in the source after a reasonable warm-up time.

6.1   Discussion

Initially, we compare our delivery scheme (SA), against Binary Spray and Wait (BSW) with replication $L = N/K$ (number of packet copies). Recall that BSW with $L = N/K$ will uniformly disseminate the K packets among all the nodes. Fig. 4, depicts the delivery delay for these two protocols in both scenarios. In plot (a), (Infocom06 scenario), not only SA has less delay than BSW in all cases of G, but also after $K = 2G$ the delay is almost constant. This indicates that SA succeed in finding the high utility relay nodes, thus, injecting more packets has no impact on the delivery delay. On the other hand, BSW has the least delay when $K = 2G$, while after this point the delay increases. This is because, larger K will require the source to contact more nodes to inject them into the network. Also since $L = N/K$ the spraying time for all the K packets among nodes will be increased.

However, things are different with RWP scenario, in plot (b), BSW performs better than SA. This is because there is no biasing, thus, SA can't find the "best carriers". As a consequence, the delay just reduced by increasing K.

It is important to note that storage overhead, which is the total number of packets in the network, is an important factor. In particular when there are more than one content to be delivered. SA generates, in total, just the K packets. While BSW, to gain its optimal performance, floods the K packets among all nodes. Fig. 5 depicts in (a) the delay, when there is one or two data contents to be delivered. Its clear that the delay of SA almost doesn't affected when there are two data contents. In plot (b) the number of total transmissions needed by SA is almost similar to BSW, even its less with small G.

The other objective, is to evaluate our scheme against a protocol uses not only replication but also forwarding for data delivery. We select Binary Spray and Focus (BSF) [17] which has two phases, it binary sprays L copies for each packet in the spray phase, then starts the focus phase, in which, packets are forwarded to nodes having lower age of last contact time with the destination.

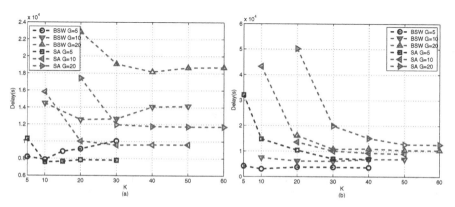

Fig. 4. Delivery delay vs K for different G. (a) using Infocom06 traces, (b) using RWP model.

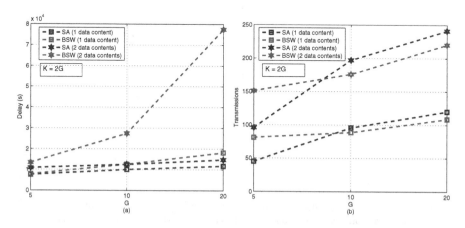

Fig. 5. SA and BSW protocols under Infocom06 contacts where the source generates one data content, or two data contents. (a) Delivery delay, (b) Number of transmissions.

To illustrate the impact of erasure coding, we run BSF with different L values in two cases: 1) when the source generates just the G packets, as in Fig.6, 2) when the source generates $K = 2G$ packets(we call it EC-BSF), as in Fig.7. In both figures we see how SA outperforms BSF or EC-BSF in terms of delivery delay and total transmissions. In addition, SA doesn't employ any replication among nodes while in BSF or EC-BSF each packet is replicated L times. If we compare BSF with EC-BSF as in Fig.6 (a) and Fig.7 (a) respectively. We notice that the delay is reduced using erasure coding; for example the delay in EC-BSF with $L = 2$ is less than the delay in BSF with $L = 4$.

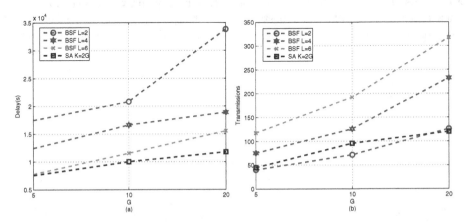

Fig. 6. SA with $K = 2G$, and BSF with $K = G$ packets each replicated L times. Under Infocom06 contacts. (a) Delivery delay and (b) Number of Transmissions.

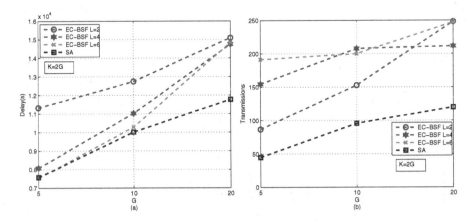

Fig. 7. SA with $K = 2G$, and EC-BSF with $K = 2G$ packets each replicated L times. Under Infocom06 contacts. (a) Delivery delay and (b) Number of Transmissions.

7   Conclusion

In this paper we have studied data delivery in DTNs for both biased and unbiased contact models. The source exploits erasure coding as a mean of splitting the data content that doesn't fit in a single packet, into a number of encoded packets suitable to be carried by the nodes. We showed that the erasure coding replication factor required to reduce the decoding delay could be small and of the same order of magnitude of the original content.

This small replication overhead reduces the time consumed by the source to inject the data packets into the network. Then, the underlying movement patterns of the nodes must be taken into consideration when designing the routing technique. For the unbiased contact model reducing the delivery time is proportional to replicating every packet. While in the biased case selecting the high utility group of nodes as data carriers is the key factor. For this issue, we have designed a scheme for data delivery, which is based on Simulated Annealing (SA) algorithm to searches for the best carrier nodes available in the network.

References

1. Fall, K.: A delay-tolerant network architecture for challenged internets. In: Proceedings of the 2003 Conference on Applications, Technologies, Architectures, and Protocols for Computer Communications, SIGCOMM 2003, pp. 27–34. ACM (2003)
2. Massri, K., Vernata, A., Vitaletti, A.: Routing protocols for delay tolerant networks: a quantitative evaluation. In: PM2HW2N 2012, pp. 107–114. ACM (2012)
3. Lin, W.K., Chiu, D.M., Lee, Y.B.: Erasure code replication revisited, pp. 90–97 (2004)
4. Picu, A., Spyropoulos, T.: Distributed stochastic optimization in opportunistic networks: the case of optimal relay selection, pp. 21–28 (2010)
5. Fragouli, C., Le Boudec, J.-Y., Widmer, J.: Network coding: an instant primer. SIGCOMM Comput. Commun. Rev. 36(1), 63–68 (2006)
6. Wang, Y., Jain, S., Martonosi, M., Fall, K.: Erasure-coding based routing for opportunistic networks, pp. 229–236 (2005)
7. Jain, S., Demmer, M., Patra, R., Fall, K.: Using redundancy to cope with failures in a delay tolerant network, pp. 109–120 (2005)
8. Liao, Y., Tan, K., Zhang, Z., Gao, L.: Estimation based erasure-coding routing in delay tolerant networks, pp. 557–562 (2006)
9. Tang, X., Yang, P., Tian, C., Peng, L., Yan, Y.: Inter-coding: An interleaving and erasure coding based stable routing scheme in multi-path dtn, pp. 446–451 (2010)
10. Bulut, E., Wang, Z., Szymanski, B.K.: Cost efficient erasure coding based routing in delay tolerant networks, pp. 1–5 (2010)
11. Keränen, A., Ott, J., Kärkkäinen, T.: The ONE Simulator for DTN Protocol Evaluation. ICST Institute for Computer Sciences, Social-Informatics and Telecommunications Engineering (2009)
12. Hsu, W.J., Spyropoulos, T., Psounis, K., Helmy, A.: Modeling time-variant user mobility in wireless mobile networks, pp. 758–766 (2007)
13. Scott, J., Gass, R., Crowcroft, J., Hui, P., Diot, C., Chaintreau, A.: CRAWDAD trace cambridge/haggle/imote/infocom2006 (May 29, 2009)

14. Dartmouth. A community resource for archiving wireless data,
    http://crawdad.cs.dartmouth.edu/
15. Spyropoulos, T., Psounis, K., Raghavendra, C.S.: Spray and wait: an efficient rout-
    ing scheme for intermittently connected mobile networks, pp. 252–259. ACM (Au-
    gust 2005)
16. Dang, H., Wu, H.: Clustering and cluster-based routing protocol for delay-tolerant
    mobile networks. IEEE Transactions on Wireless Communications, 1874–1881
    (June 2010)
17. Spyropoulos, T., Psounis, K., Raghavendra, C.S.: Spray and focus: Efficient
    mobility-assisted routing for heterogeneous and correlated mobility, pp. 79–85
    (March 2007)

Efficient Clustering of Cabinets at FttCab

Frank Phillipson

TNO, P.O. Box 5050, 2600 GB Delft, The Netherlands
frank.phillipson@tno.nl

Abstract. At this moment consumers want an internet connection with
20-50 Mb/s speed and around 100 Mb/s in the near future. Rolling out
Fibre to the Curb networks quickly will be the only way for telecom operators in some countries to compete with cable tv operators. This requires
a fibre connection to the cabinets. When the telecom operator wants to
connect the cabinets in a ring structure, he has to decide how to divide
cabinets over a number of circuits, taking into account a maximum number of customers per circuit. This we call the cabinet clustering problem.
In this paper we formulate this problem, present the heuristic approch
we developed and show the results of our extensive testing that shows
the method is accurate and fast. Finally we demonstrate the method on
a real life case.

Keywords: FttCab planning, VDSL, Clustering.

1 Introduction

More digital services come available to customers every day and even more important, these services are asking more bandwidth. This is mainly due to the
integration of video, high definition, 3D, into numerous services which are used
next to each other. We see the bandwidth demand grow approximately 30% to
40% per year between now and 2020 on fixed connections. Telecom operators
have to make their access networks ready for this. Therefore they have to make
the costly step to Fibre to the Cabinet (FttCab), where the services are offered
using the VDSL technique from the street cabinet, or, even more costly, the step
to Fibre to the Home (FttH). The roll out of FttH will be taking too long to
compete with the cable TV operators, who can offer the required bandwidth at
this moment. For many telecom operators bringing VDSL in the next two years
will be the only way to survive.

1.1 Migration to FttC

When migrating from ADSL to VDSL, a choice can be made between two options. The first option is to offer VDSL from the Central Office. This requires
a relatively small investment, as the location is already connected to a fibre
optic network; only the modems need adjusting. This is a viable option for residences which are situated less than 1 km copper distance from the Central Office.

S. Balandin et al. (Eds.): NEW2AN/ruSMART 2013, LNCS 8121, pp. 201–213, 2013.
© Springer-Verlag Berlin Heidelberg 2013

This 1 km is the cut-off for the added value of VDSL. For residences situated farther from the Central Office, it will be necessary to extend the fibre optic further into the direction of the houses. This is the second option, in which the cabinet is selected as the next logical active point. This is also called Fibre to the Cabinet (FttCab). Connecting the cabinet to fibre optic and installing the necessary hardware into it will be referred to as activating a cabinet from this point onwards. When we look at the second option, an operator does not want to activate all cabinets but only a selection. The operator wants to reach as many customers with as little investment as possible; usually the choice is made for a minimal penetration of, for example, 85%. The operator will therefore look for a minimal cost selection of activated cabinets, that collectively have more than 85% of the customers within 1 km. These cabinets are connected to the Central Office via new fibre optic cables. In this paper we discuss the connection by circuits. The fibre optic circuits have a maximum capacity in the number of cabinets that can be connected. The cabinets that are not activated will be connected to an activated cabinet using existing copper connections and are called 'placed in cascade'. Still, customers connected to these cabinets can be within 1 km from the activated street cabinet and hence use VDSL.

1.2   Planning of FttC

When the operator wants to migrate to FttCab, he has to design and plan the new network, starting with the available equipment and cables from the existing network. This is too complex to solve mathematically in one step. Therefore we divide the complex problem into three simpler sub-problems. These three sub-problems in our approach are:

1. Which cabinets must be activated in order to reach the desired percentage of households at minimal costs?
2. Which cabinet is served by which fibre optic circuit?
3. How will each fibre circuit run? Each circuit that is constructed should be edge and node disjoint. In other words, it is not allowed that an edge or node is used/passed twice within one ring.

The last two steps form a cluster-first-route-second approach as is well known in VRP literature. The restriction of the edge disjointness of the circuit makes an other approach difficult. Next to this, the approach was chosen due to the fact that this problem was in practice part of the tactical planning process of an operator. In our practice we learned that planners are interested in a good starting point, generated in a very short time (seconds) for their planning made by a simple tool, not in an optimal solution created in some expensive, difficult to understand application. There are several reason for this attitude:

— Methods based on 'rules of thumb' are better to understand and are closer to the way they work, what helps the credibility of the tool.
— The real life situation is much more complex then can be modelled in the tool. The geographical conditions need site surveys that influences the final

planning, think of questions like: 'can we place the drilling rig in that street?'.
These modifications to the 'optimal' planning have a major impact on the
final planning that void the performance gain.
- Big companies have a strict IT policy that restrict from using special applications. Sometimes the only way to run a tool is using Excel and VBA.

In this article we focus only on the second problem of the three mentioned
above. Figure 1 shows schematically the starting point. All cabinets (Cab) are
connected through copper to the Central Office (CO). Several residences are
connected to the cabinet; this is only shown for one cabinet in the illustration.
Now a subset of the cabinets has been chosen to be activated in order to reach the
intended number of households over copper from an activated cabinet within the
set distance, e.g., 85% of the customers within 1 km. How should these cabinets
be clustered?

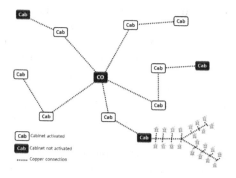

Fig. 1. Which activated cabinets should be clustered together?

1.3   Outline of the Paper

In this paper we will focus on the second problem described earlier, the cabinet
clustering problem: how to divide the cabinets in a number of clusters, where all
cabinets in one cluster are connected in one circuit. Mathematically this problem
can be seen as a clustering problem, more specifically Centroid-based clustering.
In centroid-based clustering, a centre point for every cluster is created, and points
(cabinets in this case) are appointed to the closest centre point, in which the
quadratic distance is minimized. If a number of k clusters is searched for, the
method is called 'K-means' clustering. In our problem we also have a maximum
number of cabinets that can be placed in one circuit. This gives a constrained
K-means clustering problem.
   In the remainder of this paper we will give an overview of the literature on
both related network planning theory and K-means clustering problems. Next
we will formulate the cabinet clustering problem and propose a heuristic approach. Finally we will test the proposed heuristic on a number of generated
test instances and present the solution to the real life case Amstelveen.

2   Literature Review

In this section we will give an overview of the literature on both related network planning theory and K-means clustering problems.

2.1   Network Planning

With respect to network planning we see a lot of related work, most of them having different network views in terms of the number of layers and the way of connecting. For the latter we see design problems consisting of determining graphs that represent network topology for stars, double stars and double connected trees, as also stated by [1]. We give here a short overview of other publications regarding network planning, but none of the methods are applicable directly to our case where ring structures are used, the nodes are capacitated, there is a special distance requirement and we use a minimal penetration rate.

Kalsch et al. [2] developed a mathematical model and a heuristic for embedding a ring structure in a fibre network. They developed a mathematical model that takes into account the following restrictions: ensuring a ring structure, a maximum number of nodes in a ring, each node in exactly one ring, and that the ring uses each edge only once. Very similar to our problem. However, they report only one test result: in a graph consisting of 13.246 cable nodes and 22.116 cable edges, 135 rings were constructed in 4.8 hours using a computer with decent specifications. It is, however, hard to draw conclusions on the performance of their approach, since no further information is given on the data than is mentioned over here. The goal of our work is, however, to come up with a method that is much faster: being able to solve real-life cases in the order of magnitude of seconds instead of hours. Another important disadvantage of their method is that no real attention is paid to the clustering of the nodes to the rings. They indicate clustering is part of the problem, but do not really treat it in there article; it is not clear how this is done. That part is the main topic of our article.

Gollowitzer et al. [3] present the Two Level Network Design (TLND) problem for greenfield deployments and roll-out mixed strategies of Fibre-To-The-Home and Fibre-To-The-Curb, i.e., some customers are served by copper cables, some by fibre optic lines. For two types of customers (primary and secondary), an additional set of Steiner nodes and fixed costs for installing either a primary or a secondary technology on each edge, the TLND problem seeks a minimum cost connected sub-graph obeying a tree-tree topology, i.e., the primary nodes are connected by a rooted primary tree; the secondary nodes can be connected using both primary and secondary technology. In the paper an important extension of TLND is presented in which additional transition costs need to be paid for intermediate facilities placed at the transition nodes, i.e., nodes where the change of technology takes place.

In an other article [4] Gollowitzer gives an overview of MIP models for connected facility location problems. Here also only tree structures and uncapacitated nodes are considered.

Mateus et al. [5] describe the network design problem of locating a set of concentrators which serves a set of customers with known demands. The uncapacitated facility location model is applied to locate the concentrators. Then, for each concentrator they analyse a topological optimization of its sub network based on a simple heuristic. In a third phase, they solve the upper level sub network connecting the concentrators to a root node in a tree structure.

In [6] Mitcsenkov et al. address broadband PON access network design minimizing deployment costs using a heuristic solution. The questions here are: how to form groups of customers that share a PON splitter, where is the splitter placed, what is the best path from the customer to its splitter unit and how to connect splitters to the central office. The first problem is regarded as a clustering problem which they solve heuristically by combining nearby shortest paths from the customer to the Central Office. The second problem is solved optimally by just calculating the optimal location from the (small) set of possible locations. The last two problems are solved by a Steiner Tree problem, using a 2-approximation heuristic, the Distance Network Heuristic as presented in [7] by Kou and Berman.

In [8] the topology design of hierarchical hybrid fibre / VDSL access networks is presented by Zhao et al. as an NP-hard problem. A complete strategy is proposed to find a cost-effective and high-reliable network with heuristic algorithms in a short time. The Ant Colony Optimization (ACO) has been implemented for a clustering problem. The network structure they look at is a two layer street cabinet solution with Branch Micro Switches (BMS) and Lead Micro Switches (LMS) where the BMS is connected with the CO with two paths and the LMS is connected to two BMSs. The users are connected in a star with one LMS. The major planning problem here is to build up the intermediate BMS level.

In [9] Gódor and Magyar look at the access planning problem from the side of mobile network planning. Here the facility location problem is seen as basis for solving this. They use a two-phase heuristic planning algorithm. In the first phase, they construct an initial network-solution with a K-means algorithm and in the second phase, they improve it by moves or swaps.

2.2   K-means Clustering

In literature much has been published about this problem, 'Centroid-based' or 'K-means' clustering, mostly in the applications of data-clustering. The name goes back to 1967, where MacQueen presents the problem in his article [10]. An nice overview of 50 years of 'K-means' clustering was presented by Jain [11]. That this optimization problem is NP-hard was proven by Aloise et al., see [12]. Usually these problems are therefore solved via an approximation algorithm. In special cases the problem can be solved in polynomial time, see for examples the work that Inaba et al. [13] did. For other cases a known method is Lloyd's algorithm named after the creator as described by Lloyd [14]. However, this method is not guaranteed to find a global optimum, but usually finds a local optimum. In the paper of Bradley et al. [15] an extension to this algorithm is created to handle constrained K-Means Clustering. The constraint here is a minimum number of items in a cluster.

3   Problem Description and Approach

In this section we formulate the cabinet cluster problem. As stated before this problem is NP-hard. To solve the problem we present a heuristic that is inspired on the methods of both Lloyd and Bradley, mentioned earlier.

3.1   Problem Description

We start in the situation that we have a collection of cabinets that have to be divided in groups, where all cabinets in one group are connected in one circuit. Mathematically this can be formulated as follows. Given is a set $D = \{x^i\}_{i=1}^m$ of m points in R^2, the location of the cabinets. We want to divide these cabinets in k ($k \leq m$) groups which each form a circuit. The cluster centres are centers of gravity, $C^1, C^2, \ldots, C^k$. We define for $i = 1, \ldots, m$ and $h = 1, \ldots, k$:

$$T_{i,h} = \begin{cases} 1 \text{ if data point } x_i \text{ is closest to center } C_h, \\ 0 \text{ otherwise.} \end{cases}$$

The cluster centres are calculated as follows:

$$C^h = \frac{\sum_{i=1}^m T_{i,h} x^i}{\sum_{i=1}^m T_{i,h}}$$

The problem that we have to solve is:

$$\min_T \sum_{i=1}^m \sum_{h=1}^k T_{i,h} (\|x^i - C^h\|_2^2) \tag{1}$$

This means that we look for the T that minimizes the sum of the quadratic distances, expressed in the sum of the squared 2-norm distance: $\|x\|_2 = (x_1^2 + x_2^2 + \ldots + x_n^2)^{\frac{1}{2}}$. In practice in this problem each circuit has a maximum number of households that can be connected. We model this by assuming each cabinet has a weight (a natural number) u_i and there is a limit τ to the total weight on one circuit. This gives the constraint:

$$\sum_{i=1}^m T_{i,h} u_i \leq \tau \quad h = 1, \ldots, k \tag{2}$$

Each point is assigned to only one circuit:

$$\sum_{h=1}^k T_{i,h} = 1 \quad i = 1, \ldots, m$$

$$T_{i,h} \in \{0, 1\} \quad i = 1, \ldots, m \quad h = 1, \ldots, k$$

As stated before, in [12] is shown that this problem is NP-hard. We will therefore solve the problem with a heuristic that has to perform well in practical problems.

3.2   Existing Methods

In the literature review we already mentioned the papers of Lloyd [14] and Bradley [15]. We present their approaches here as we use their methods later on. The algorithm of Lloyd is an iterative algorithm where T^t represents the assignment of the cabinets to the clusters in iteration $t \in \mathbb{N}$, resulting in cluster points $C^{t,1}, \ldots, C^{t,k}$ in iteration t. The steps in the algorithm are:

1. Start with a random distribution of the cabinets T^0. This results in an initial value of C^0.
2. Repeat for all x_i, $i = 1, \ldots, m$: assign x_i to k such that centre $C^{k,t}$ is nearest to x_i.
3. Update $C^{h,t+1}$ as follows:
   If $\sum_{i=1}^m T_{i,h}^t > 0$: $C^{h,t+1} = \frac{\sum_{i=1}^m T_{i,h}^t x^i}{\sum_{i=1}^m T_{i,h}^t}$
   otherwise $C^{h,t+1} = C^{h,t}$ $h = 1 \ldots, k$
4. Stop when $C^{h,t+1} = C^{h,t}, h = 1, \ldots, k$ else increment t by 1 and go to step 2.

An often named disadvantage of this algorithm is that you need to specify value k in advance. However, this is not a disadvantage in our clustering problem, here we know k on beforehand. Also, it has been shown that the worst case running time of the algorithm is super-polynomial in the input size and the approximation found can be arbitrarily bad with respect to the objective function compared to the optimal clustering. Finding a good start solution can be helpful here, as shown in [16]. This will be the case mainly in big data clustering problems, not in the limited problems we discuss here. Another often mentioned disadvantage is that the algorithm tries to create clusters of approximately similar values. For us, this is not a problem either; on the contrary, it is a requirement. The cabinet cluster problem has an upper limit to the number of cabinets, or total weight of these cabinets, in a circuit, as described in the constraint in equation (2).

Bradley gives an extension to this algorithm to handle constrained K-Means Clustering. The constraint they present is a minimum number of items in a cluster, instead of a maximum as we have due to equation (2). The algorithm they propose is:

1. Cluster assignment. Let $T_{i,h}^t$ be a solution to the following linear program with $C^{h,t}$ fixed:

$$\min_T \sum_{i=1}^m \sum_{h=1}^k T_{i,h}(\|x^i - C^{h,t}\|_2^2) \tag{3}$$

subject to

$$\sum_{i=1}^m T_{i,h} \geq \tau_h \quad h = 1, \ldots, k \tag{4}$$

$$\sum_{h=1}^k T_{i,h} = 1 \quad i = 1, \ldots, n$$

2. Cluster update. Update $C^{h,t+1}$ as follows:

If $\sum_{i=1}^{m} T_{i,h}^{t} > 0$: $C^{h,t+1} = \frac{\sum_{i=1}^{m} T_{i,h}^{t} x^{i}}{\sum_{i=1}^{m} T_{i,h}^{t}}$

otherwise $C^{h,t+1} = C^{h,t}$ $h = 1, \ldots, k$

Stop when $C^{h,t+1} = C^{h,t}, h = 1, \ldots, k$ else increment t by 1 and go to step 1.

Replacing equation (4) by (2) gives a solution method for our problem. However, we will show later on that our proposed heuristic performs better.

3.3 New Heuristic

We propose a algorithm that starts with Lloyd's algorithm. The found solution is adjusted such that no more than weight τ is placed in one circuit, due to equation (2), while Lloyd's algorithm doesn't have this check or constraint. The final step is trying to improve the solution using a $2 - opt$ improvement, like originally presented in solving the TSP, see [17]. This is depicted in Fig. 2. In more detail the proposed heuristic is:

1. Initial allocation according to Lloyds algorithm, with $k = \lceil \sum_{i=1}^{n} u_i / \tau \rceil$. [1]
2. If a circuit h is present with higher allocated weight than allowed, $\sum_{i=1}^{n} T_{i,h} u_i > \tau$, then choose the cabinet which can be moved to another with the least expenses. The cabinet i that needs to be moved follows from:

$$\min_{\{(i|T_{i,h}=1),l\}} \|x^i - C^l\|^2,$$

under the constraint

$$\sum_{j=1}^{m} T_{j,l} u_j \leq \tau - u_i$$

which means that only that circuit l can be used that has assigned less than the maximum weight minus the weight of cabinet i. If no solution can be found to remove the constraint using a single swap, a double swap must be found: which combination of cabinets can be swapped in the cheapest way. That means we have to find a combination of cabinets i in circuit h and i' in circuit h', such that

$$\min_{\{(i|T_{i,h}=1),(f|T_{i',h'}=1)\}} \|x^i - C^{h'}\|_2^2 + \|x^{i'} - C^h\|_2^2,$$

---

[1] Note that the ratio between the values of u_i and k can lead to insolvable problems, as a simple example shows. Let us say that we have 5 cabinets with weight 4. We want to create circuits with maximum weight 10. The definition of k says that 2 circuits are needed ($\frac{4 \times 5}{10}$). Only we cannot divide the 5 cabinets over 2 circuits, not violating the constraint. However, if the maximum value of u is small related to k and there is a big variation in number and place of the size of the cabinets (as in our practice) this problem never occurs.

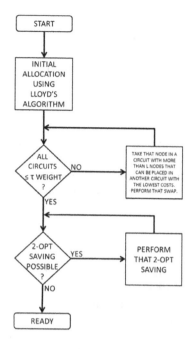

Fig. 2. Flow diagram proposed heuristic

under the constraints:

$$\sum_j (T_{j,h'} u_j) - u_{i'} + u_i \leq \tau$$

$$\sum_j (T_{j,h} u_j) - u_i + u_{i'} \leq \tau$$

3. Try to improve the solution by pair swapping until no improvement can be found any more. Determine for every combination of cabinets within two separate circuits the current score (total sum of quadratic distances) and the score after exchanging these two combinations. If the exchange results in an improvement, apply this exchange to the solution. This gives the following steps:

(a) For every combination (i, i') where $T_{i,h} = 1$, $T_{i',h'} = 1$ and $h' \neq h$:
   i. Define $T' = T$ and change the elements $T'_{i,h'} = 1$, $T'_{i',h} = 1$, $T'_{i,h} = 0$ and $T'_{i',h'} = 0$.
   ii. Calculate the centre points $C_T^1, \ldots, C_T^k$ based on solution T and $C_{T'}^1, \ldots, C_{T'}^k$ based on solution T'.
   iii. Define the score

$$S(T) = \sum_{j=1}^m \sum_{l=1}^k T_{j,l}(\|x^j - C^l\|_2^2)$$

and calculate $S(T)$ and $S(T')$.

  iv. If $\sum_{j=1}^{m}(T'_{j,h}u_j) \leq \tau$ and $\sum_{j=1}^{m}(T'_{j,h'}u_j) \leq \tau$ and $S(T') < S(T)$ then swap i and i': $T := T'$.

(b) If no swap occurred stop, else go to step 3a.

4   Performance

In this section we discuss the performance of the proposed heuristic. First we show the effect of the swapping operation, step 3 in the heuristic. Next we show the performance against the method of Bradley. Finally we present the application of the method to the case Amstelveen.

4.1   Swapping Operation

The swapping operation, step 3 in the proposed heuristic, results in a clear improvement of the solution and a quicker convergence to the best solution. To illustrate this, we generated 1000 cases, in which there are 40 cabinets which need to be allocated to 4 circuits. The maximum number of cabinets per circuit is 10. The xy-coordinates are arbitrarily drawn from the range (0,100) per cabinet. We solve each case 1000 times, each time with an arbitrarily generated starting solution (like Lloyds algorithm prescribes). For each case, we note the iteration for which the best solution is found, and what this solution is. We have applied the algorithm with and without step 3 (the swap).

Figure 3 (left) shows the frequency of the iteration at which the best solution was found. The heuristic with swap finds the solution far quicker; in 452 of the 1000 cases the best solution is already found within the first iteration. The remaining 999 iterations offer no improvement. Without swap this is the case only 228 times.

The final solution found with swap is better than the heuristic without swap in 312 cases. In about 200 cases this improvement is 1%, in 60 cases 2% and in two cases even 10% and 11%, see Fig. 3 (right).

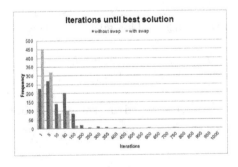
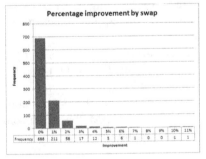

Fig. 3. Performance of swapping

The question arises whether the swap method also results in an improvement of the classic clustering problem without the constraint. In the tested 1000 cases this shows to be only a minor improvement. In only a few cases the number of iterations is smaller and only in 38 cases an improvement of 1-2% is realized.

4.2   Performance of the Heuristic

To show the performance of the heuristic we tested it against the LP-based method proposed by Bradley. To illustrate this, we generated again 1000 cases, in which there are 40 cabinets which need to be allocated to 4 circuits. The maximum number of cabinets per circuit is 10. The xy-coordinates are arbitrarily drawn from the range (0,100) per cabinet. Per case we generated 20 starting solutions at random and looked at the solution found by each method for this starting solution. So we got 20000 results per method. Per case we determined the best solution found by both methods[2] and the relative deviation of the other solutions. Those results are depicted in Fig. 4 (left). We see that in 37% of all problems the heuristic found, starting with a arbitrary solution, the best solution against 11% in case of the LP-based method. The method of Bradley was implemented using AIMMS that solved the linear program using the solver CPLEX 12.4. The test version of the heuristic was implemented in Delphi.

 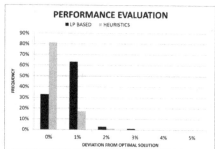

Fig. 4. Performance of the heuristic

Per case, having 20 starting solutions, the heuristic found in 81% the optimal solution against 33% in case of the LP-based method, as shown in Fig. 4 (right).

4.3   Real Life Case

The method was implemented in the tool GIANT-PlanXS[3], created by TNO. We show here the results of the case Amstelveen. Amstelveen is a Dutch city, close at

---

[2] Note that this is not necessarily the global optimum, but only the best solution found by both methods.

[3] Based on Matlab.

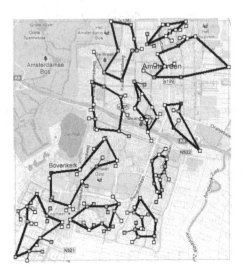

Fig. 5. Example of result

the south to Amsterdam, with approximately 80,000 inhabitants. In Amstelveen we have 180 cabinets, belonging to 1 Central Office, serving 38868 subscribers. The first step, activation, leaves 87 activated cabinets that have to be distributed over, in this case, 11 circuits. A result of the clustering of the Case Amstelveen is shown in Fig. 5. We mentioned before that calculation time is important in the real life use of the method presented. The implementation we made in Matlab and uses a lot of time for updating the cluster centres, especially due to the third step of the heuristic. The calculation time of the case Amstelveen is 0.9 seconds. However, in the case of Amsterdam, where 235 activated cabinets of a total of 513 cabinets have to be clustered, which is one of the biggest central office of the Netherlands, it still can be calculated within 14 seconds.

5   Summary and Conclusions

In this paper we presented the clustering problem when designing next generation telecommunication networks: how can we divide access points, cabinets in our example, over a number of circuits, taking into account a maximum weight per circuit. This is a NP-hard problem. To solve it we presented heuristic that uses Lloyd's algorithm, and added a re-clustering method to stay under the maximum weight and a 2-opt improvement method. Finally we presented the results of our extensive testing on the effect of the swapping operation, step 3 in the heuristic on the total heuristic against the method of Bradley. We showed that the method is accurate and fast.

Acknowledgement. The author wishes to thank prof. dr. Rob van der Mei and dr. Sandjai Bhulai, Vrije Universiteit Amsterdam, for their valuable comments and suggestions

References

1. Chardy, M., Costa, M.C., Faye, A., Trampont, M.: Optimizing splitter and ber location in a multilevel optical ftth network. European Journal of Operational Research (222), 430–440 (2012)
2. Kalsch, M.T., Koerkel, M.F., Nitsch, R.: Embedding ring structures in large ber networks. In: XVth International Telecommunications Network Strategy and Planning Symposium (NETWORKS) (2012)
3. Gollowitzer, S., Gouveia, L., Ljubić, I.: A node splitting technique for two level network design problems with transition nodes. In: Pahl, J., Reiners, T., Voß, S. (eds.) INOC 2011. LNCS, vol. 6701, pp. 57–70. Springer, Heidelberg (2011)
4. Gollowitzer, S., Ljubić, I.: Mip models for connected facility location: A theoretical and computational study. Computers and Operations Research 38, 435–449 (2011)
5. Mateus, G.R., Cruz, F.R.B., Luna, H.P.L.: An algorithm for hierarchical network design. Location Science 2(3), 149–164 (1994)
6. Mitcsenkov, A., Paksy, G., Cinkler, T.: Topology design and capex estimation for passive optical networks. In: Proceedings of BROADNETS 2009 (2009)
7. Kou, L., Markowsky, G., Berman, L.: A fast algorithm for steiner trees. Acta Informatica 14, 141–145 (1981)
8. Zhao, R., Liu, H., Lehnert, R.: Topology design of hierarchical hybrid ber-vdsl access networks with aco. In: Proceedings of Fourth Advanced International Conference on Telecommunications (2008)
9. Gódor, I., Magyar, G.: Cost-optimal topology planning of hierarchical access networks. Computers & Operations Research 32(1), 59–86 (2005)
10. MacQueen, J.: Some methods for classification and analysis of multivariate observations. In: Proceedings of 5th Berkeley Symposium on Mathematical Statistics and Probability, pp. 281–297 (1967)
11. Jain, A.K.: Data clustering: 50 years beyond k-means. Pattern Recognition Letters 31, 651–666 (2010)
12. Aloise, D., Deshpande, A., Hansen, P., Popat, P.: Np-hardness of euclidean sum-of-squares clustering. Machine Learning 75, 245–249 (2009)
13. Inaba, M., Katoh, N., Imai, H.: Applications of weighted voronoi diagrams and randomization to variance-based k-clustering. In: Proceedings of 10th ACM Symposium on Computational Geometry, pp. 332–339 (1994)
14. Lloyd, S.P.: Least squares quantization in pcm. IEEE Transactions on information theory IT 28(2) (1982)
15. Bradley, P., Bennet, K., Demiriz, A.: Constrained k-means clustering. Technical report, Microsoft Research MSR-TR-2000-65 (2000)
16. Arthur, D., Vassilvitskii, S.: k-means++: The advantages of careful seeding. In: Proceedings of the Eighteenth Annual ACM-SIAM Symposium on Discrete Algorithms, pp. 1027–1035 (2007)
17. Croes, G.: A method for solving traveling salesman problems. Operations Research 6, 791–812 (1958)

Proxy Mobile IPv6-Based Seamless Handover

Jari Kellokoski, Joonas Koskinen, Tuomas Rusanen,
Pasi Kalliolahti, and Timo Hämäläinen

Department of Mathematical Information Technology, University of Jyväskylä,
FI-40014 Jyväskylä, Finland
{jari.k.kellokoski,joonas.a.koskinen,tuomas.t.rusanen,
pasi.k.kalliolahti,timo.t.hamalainen}@jyu.fi

Abstract. A prospective next generation wireless network is expected to integrate harmoniously into an IP-based core network. It is widely anticipated that IP-layer handover is a feasible solution to global mobility. However, the performance of IP-layer handover based on basic Mobile IP (MIP) cannot support real time services very well due to long handover delay. The Internet Engineering Task Force (IETF) Network-based Localized Mobility Management (NETLMM) working group developed a network-based localized mobility management protocol called Proxy Mobile IPv6 (PMIPv6) to reduce the handoff latency of MIPv6. Moreover, PMIPv6 provides the IP with the mobility to support User Equipments (UEs) without it being required to participate in any mobility-related signaling. This was one of the reasons why the 3rd Generation Partnership Project (3GPP) chose PMIPv6 as one of the mobility management protocols when defining the Evolved Packed System (EPS). One of the key features of the standard is its support for access system selection based on a combination of operator policies, user preference and access network conditions. Although Android, which is one of the most popular "mobile" operating system, is not officially supporting IPv6 nor PMIPv6, this paper analyzes the required challenges for IPv6 and PMIPv6 usage and handover performance with real-time services. The analysis and measurement results show that, with the modifications presented IPv6 and PMIPv6 may be supported and utilized for localized mobility management and seamless handovers. ...

Keywords: Proxy Mobile IPv6, Handover.

1 Introduction

The demand for increased network capacity has been unrelenting. Network operators have responded to this challenge by increasing the density of their networks. At the same time, consumers expect a solid user experience regardless of the underlying network. The effective handover mechanism must keep the ongoing communications active. Users expect to maintain their connection without any disruptions when they move from one network to another. This process is called handover. In a wireless heterogeneous environment, effective handover is needed to achieve seamless mobility management support. MIPv6 [1] is a IP

S. Balandin et al. (Eds.): NEW2AN/ruSMART 2013, LNCS 8121, pp. 214–223, 2013.

mobility management technology that provides seamless mobility service by allowing a home of agent (HA) and a corresponding node (CN) to share the UE's home of address (HoA) and care of address (CoA) when the UE moves across an adjacent service area. Although this technology has been extensively studied and shows reliable performance, it has not been adopted frequently for commercial purposes because it requires the involvement of the UE in the mobility process, which is challenging for the existing devices and results in a shorter batter life in UE side. Furthermore, it results in high handover latency and large signaling overheads for registrations. PMIPv6 as a network-based mobility management protocol should level out UE signaling and make path for a wide adaption of the protocol as all the functionalities are carried out by the network itself. The aim of the present research is twofold: first, to analyze and test how the Android operating system could operate with IPv6 addresses and whether the PMIPv6 protocol poses any other requirements for the Android UE; second, to make performance measurements of handovers with real-time-service. The rest of the paper is organized as follows. Section 2 introduces the background and related work. Section 3 analyzes the IPv6 and PMIPv6 challenges in the Android operating system. This is followed by a test environment description and the handover performance results section. Section 6 finalizes the paper with Conclusion and Future Work.

2   Background and Related Work

Over the years there have been a number of investigations on host-based versus network-based mobility management. PMIPv6 substantially reduces the handoff latency of MIPv6 since its handoff procedure takes over the movement detection and duplicate address detection process from the handoff procedures. The analytical models as in [2], [3] and [4] show that the handoff latency of PMIPv6 is much lower than that of the MIPv6, Hierarchical Mobile IPv6 Mobility Management [5] (HMIPv6) and Mobile IPv6 Fast Handovers [6] (FMIPv6) host-based mobility management protocols. Similar results are shown by Guan et al. [7] in their implementation of protocols: their results show that PMIPv6 has lower handoff latency than the other schemes. This fact has inspired researchers to further improve PMIPv6. A Fast Handoff Scheme in PMIPv6 [8] and Optimized-PMIPv6 [9] presents analytical models that perform better than plain PMIPv6. Despite these analytical results [10], the real numerical and user experience data on handover delay in PMIPv6 managed networks with real UEs is minimal.

2.1   Evolved Packed System (EPS)

EPS, formerly known as the System Architecture Evolution (SAE), was standardized by 3GPP and consists of a radio part: (Evolved URTAN (E-UTRAN)) and a network part: (Evolved Packet Core (EPC)) [11]. The EPC architecture that interconnects 3GPP and non-3GPP access networks consists of several main subcomponents, namely eNodeB, Mobility Management Entity (MME), Serving Gateway (SGw) and Packet Data Network Gateway (PDN-Gw). The PDN

Gateway provides connectivity from UE to one or multiple external PDNs simultaneously. The PDN-Gw acts also as a mobility anchor, but between 3GPP and non-3GPP technologies and in addition performs packet filtering and charging. The Access Network Discovery and Selection Function (ANDSF) is a new EPC element in Release 8 and performs data management and controls the functionality to assist the UE on the selection of the optimal access network in a heterogeneous scenario via the S14 interface, the logical interface between ANDSF and UE [12].

Today's mobile devices are not yet at the 3GPP Release level 8 although most if not all of the lacking functionality is software based [13]. That research shows that one of the biggest challenges is lack of simultaneous connections for example in the Android environment. Existing Android solutions utilize only one access at a time and when the access is changed the existing connection is torn down. The network performance for off-the-shelf Android devices is around one second [14].

The 3GPP TS23.402 Architecture enhancements for non-3GPP accesses [15] and 3GPP TS access to the 3GPP Evolved Packet Core (EPC) via non-3GPP access networks [16] are specifying requirements toward UE. One of the basic definitions is IP Flow Mobility (IFOM) and Multi Access PDN Connectivity (MAPCON) where the packed data connections are going through different access networks. The IP Mobility Management Selection (IPMS) principles state, naturally, that UE and network must support the same mobility management mechanism. The two main choices are network based mobility (NBM) and host based mobility (HBM). In case of HBM the defined protocols are Dual Stack Mobile IPv6 (DSMIPv6) [17] and Mobile IPv4 (MIPv4) (RFC 5944). Upon initial attachment to 3GPP access, no IMPS is necessary since connectivity is always established with network based mobility. Upon initial attachment to a trusted non-3GPP access or handover to it, IMPS is performed before IP address is allocated and provided to UE. When applying NBM, session continuity can take place according to PMIPv6 or by legacy GPRS tunneling protocol (GTP) [18]. However, GTP is proprietary legacy protocol that is not supported in large scale outside 3GPP access.

2.2   IPv6 and Proxy Mobile IPv6

IP version 6 (IPv6) is a new version of the Internet Protocol, designed as the successor to IP version 4 (IPv4) [19]. The major changes are: expanded addressing capabilities, header format simplification and improved support for extensions and options.

IPv6 increases the IP address size from 32 bits to 128 bits to support more levels of addressing hierarchy and to provide, a much greater number of addressable nodes, and simpler auto-configuration of addresses. The scalability of multicast routing is improved by adding a "scope" field to multicast addresses. The following scopes and scope field values are defined: 1 Node-local, 2 Link-local, 8 organization-local, E global. For example, traffic with the multicast address of FF02::2 has a link-local scope. An IPv6 router never forwards this traffic beyond the local link. A new type of address, "anycast address" is defined. It is used to send a packet to any one of a group of nodes. An anycast address identifies multiple interfaces, and is used for one-to-one-of-many communication, with delivery to a single interface. With the appropriate routing topology, packets addressed to an anycast address are delivered to a single interface. In terms of routing distance, a packet addressed to an anycast address is delivered to the nearest interface identified by the address. That is, anycast addresses are used only as destination addresses and are assigned only to routers.

The Internet Engineering Task Force (IETF) proposed a host-based mobility management protocol, called the Mobile IPv6 (MIPv6) protocol [1], for UE to maintain continuous service when moving among different foreign networks. However, MIPv6 does not provide good service for real-time applications because it causes long disruptions during handoff. Improvements such as HMIPv6 and Mobile IPv6 Fast Handovers [6] introduced new host-based schemes to improve the performance of MIPv6.

PMIPv6, A network-based localized mobility management protocol, reduces the handoff latency compared to the MIPv6. Moreover, PMIPv6 provides IP with the mobility to support UEs without requiring its participation in any mobility-related signaling. Fig. 1 shows the network architecture of PMIPv6, which contains two network entities: the mobile access gateway (MAG) and the local mobility anchor (LMA). The MAG is responsible for detecting the movements of an MN and performs mobility-related signaling with the LMA in place of the MN. The LMA acts in a way similar to the home agent (HA) in MIPv6 and maintains the binding cache entries for currently registered UEs.

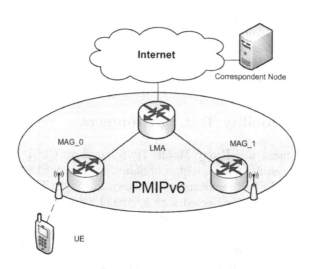

Fig. 1. Proxy Mobile IPv6 topology showing the network entities mobile access gateway (MAG) and local mobility anchor (LMA)

3   IP Version 6 and Proxy Mobile IPv6 Support in Android Framework

Android is a Linux-based operating system, mostly for portable devices such as smartphones and tablets [20]. At the moment it is the most popular smartphone operating system. However, IPv6 is not officially supported by Android. Despite this, the Linux kernel in Android is supporting IPv6. This allows an IPv6 traffic if the network provides IPv6 address and the application on top of Android can support it. Some challenges remain, however. The Android network manager can only manage IPv4 addresses. If the network is not providing an IPv4 address, the network manager considers the address faulty even when a valid IPv6 address is provided. Another challenge relates to stateless autoconfiguration of IPv6 addresses [21]: IPv6 addresses that use MAC address hiding might not be routed correctly by the Android system. Finally, by default the Android kernel is lacking mobility management protocol support such as PMIPv6.

Workarounds for these challenges are available. The IPv6 address can be manually setup to 0.0.0.0 as a non-routable address. After this, the Android network manager is contented and can keep the access connected and IPv6 addresses can be used in communication. The routing problems with a MAC-address based IPv6 address can be avoided by prohibiting the use of temporary addresses. This can be done with sysctl (which is an interface for examining and dynamically changing parameters in Linux). By the prohibition of temporary addresses the network interface obtains only the global and link-local address that are suitable for communication. The PMIPv6 itself does not require any action from the applications using it. On the other hand, the kernel needs to support basic mobility related options such as: Mobility, IPsec, Multiple Routing Tables, and source address based routing. These need to be included into the UE kernel. Further details about how to compile and install the kernel can be found from Cyanogen MOD 9 (Android version 4.0.4) setup pages. With these modifications the Android UE can be put to handover tests.

4   Seamless Mobility Test Environment

The test environment was Proxy Mobile IPv6 based on UMIP OpenAirInterface [22]. The user equipment consisted of Samsung Galaxy SI and II and Linux Lubuntu laptops. The test environment is shown in Fig. 2. The MAGs and the LMA are Linux desktops connected with a virtual network by HP a ProCursw 2650 switch. Cisco 120 Aironets were used as the basestations. In the test setup, the mobility is managed by the LMA based on the MAG information. When a PMIP device contacts a WLAN basestation the Media Access Control (MAC) information of the UE is used in the authentication. The Freeradius-based authentication server where a syslog service informs the MAG when the a UE connects the WLAN basestation. This information is then forwarded to the LMA which checks if the UE is known or not. If known, it returns the network prefix to the UE that is used in IPv6 address creation.

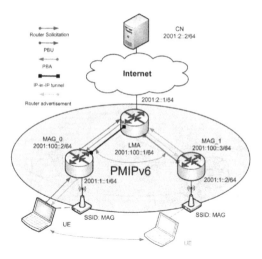

Fig. 2. PMIPv6 test environment

5 Performance Results

Two test scenarios were defined for handover tests. The First scenario was UE, where an address outside of a PMIPv6 domain was pinged while handover between two WLAN networks was taking place. Real-time protocol (RTP) [23] traffic between CN and UE was tested during the handovers. In this scenario, the traffic had the following characteristics 64 byte packets with a 0.5 second interval. The ping was initiated stationary close to WLAN basestation connected to the MAG_0 as Fig. 2 shows. From there, the UE was moved closer to the MAG_1 WLAN basestation causing the MAG_1 to notify the LMA that the UE had moved. As a result of this, the existing tunnel was torn down and a new tunnel created between the LMA and the MAG_1. The second scenario concerned RTP protocol traffic similar to that in the first scenario but with the different buffer sizes and flavors of the streaming protocol. The environment for this is depicted in the Fig. 3.

5.1 Challenges with the Android Operating System

During the test, in addition to IPv6 configuration challenges, there was lack of functionality with the Internet Control Message Protocol (ICMPv6 RFC4443) and WLAN roaming. An ICMPv6 challenge was encountered when connecting to a PMIPv6 network via the MAG and authenticating with the LMA by sending a router solicitation message to the MAG, the responses being a router advertisement message. After that, the MAG and the LMA created an IP-in-IP tunnel between them for UE traffic. The UE and the MAG kept the connection and the tunnel alive with the help of neighbor solicitation (NS) and neighbor advertisement (NA) messages. The tunnel and the connection were

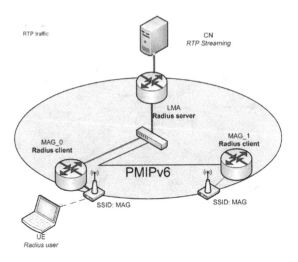

Fig. 3. PMIPv6 Test environment with RTP traffic

kept alive while producing traffic, e.g. pinging to the correspondent node. In other case the tunnel was torn down. There is a relevant issue reported at http://code.google.com/p/android/issues/detail?id=32662.

Another challenge was a WLAN roaming problem related to the fact the Android UE did not change its WLAN accesspoint as expected. Instead, the UE kept the existing connection until the connection was dropped due to poor signal level. The UE did not change from the MAG_0 to the MAG_1 until it was too late to achieve a seamless handover. There is an issue reported about it at http://code.google.com/p/android/issues/detail?id=12649. The implication from this becomes apparent in Fig.4. The results are from the case where the bitrate is 96 kbits/second with 200 ms buffer 16-bit monaural audio stream. Fig. 4 shows one complete session drop (starting around 50 second) that lasted around 20 seconds. Other handovers are visible when the bitrate drops below 40 kbits/second. The second problem is the high packet loss of 23.8 % during the measurements (275 seconds). The average jitter was 20.44 ms. Based on these results, the Android UEs may operate with IPv6 addresses and under PMIPv6 mobility management. However, there are challenges to its suitability for real-time communication.

5.2   Linux Results

Both of these scenarios were run also under a Linux environment to compare them with the results from the Android environment. The Linux environment consisted of a Linux Lubuntu distribution with kernel 3.4.4, the IPv6 options including mobility, IPsec and multiple routing tables. The network used the Zyxel USB wireless card as its interface card. The ping scenario results are shown in Table 1. The packet loss is visible, only about 2 %, and the average round-trip-time is higher where handovers occur. The test was repeated 50 times. The RTP

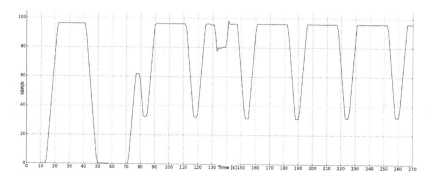

Fig. 4. UE (Samsung Galaxy II) RTP traffic during handovers

Table 1. Handovers with Linux UE

Sent/Received Packets	12800/12527
Packet Loss	2.1 % 273
RTT Handover (min, avg, max)	1.27 / 7.69 / 202.00 ms
RTT No Handover (min, avg, max)	1.27 / 5.40 / 20.00 ms

streaming results are depicted in Fig. 5. As with the Android UE, the handover points are clearly visible when the transfer bitrate drops significantly from 96 kbits/second to 20 or less kbits/second. Unlike with the Android UE, there are no total connection drops. The buffer size is 100 ms (16-bit monaural audio stream). The packet loss is 8.1 % and the jitter average is 20.99 ms. Due to the fact that the handovers were successful with the Linux UE, additional tests with RTP streaming were conducted. The tests were similar but conducted with different stream codecs such as MPEG-1 Audio and MPEG-2 Audio, and with buffer

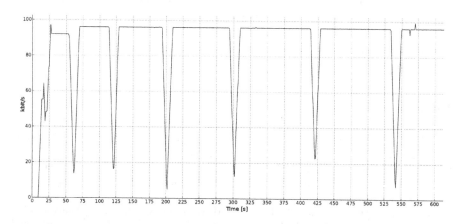

Fig. 5. RTP traffic with 100 ms buffer in Linux during handovers

values from 100 to 300 ms. These results show zero jitter as it was eliminated by the buffer and lower bitrate (16 - 24 kbits/second). The packet loss was in a range of 4.3% to 6.6%. Video testing was done purely for user experience where 480p video was streamed from CN to UE while executing handovers. With a 200 ms buffer there was no visible or audible indication that the handover would have occurred.

As a summary, the Linux UE was able to achieve seamless handovers in a PMIPv6 controlled network environment. The packet loss was there as expected but its effect on real-time communication remained non-existent with the executed bitrates and content.

6   Conclusion and Future Work

This paper dealt with seamless handovers in a Proxy Mobile IPv6 managed network environment with Android and Linux User Equipment. Although the IPv6 is not officially supported by the Android operating system, enabling it was possible. The limitations were that the network manager requires an IPv4 address to keep connection alive, routing of IPv6 addresses was limited, and, as the framework was not supporting the IPv6, it is up to the particular application to support IPv6. The PMIPv6 support in the Android and Linux UE is coming from the kernel support to the various IPv6 options that can be included into UE. The paper presented a test-bed for a PMIPv6 based mobility management. With the help of two user scenarios handover performance and its effect on user experience were evaluated in an Android UE. Due to the challenges in the Android environment, the same measurements were repeated in a Linux UE as well. The measurements did show some challenges with the existing Android UEs for example dormant WLAN-roaming there resulted in bigger packet loss than with the Linux UEs. On the other hand, with the Linux UE, measured handovers were successful and packet loss had minimum or no effect on the user experience. These facts defend the use of PMIPv6 on mobility management in EPC, and Android may support it once the challenges presented have been solved.

In future work, the aim is to study multibearer and multihomed solutions for further reduction of packet loss during handover and support for heterogeneous network environment.

References

1. Johnson, D., Perkins, C., Arkko, J.: Mobility Support in IPv6. RFC 3775 (Proposed Standard), Obsoleted by RFC 6275 (June 2004)
2. Kong, K., Lee, W., Han, Y.H., Shin, M.K., You, H.: Mobility management for all-IP mobile networks: mobile IPv6 vs. proxy mobile IPv6. IEEE Wireless Communications 15(2), 36–45 (2008)
3. Kong, K., Lee, W., Han, Y.H., Shin, M.K.: Handover Latency Analysis of a Network-Based Localized Mobility Management Protocol. In: IEEE International Conference on Communications (ICC), pp. 5838–5843 (May 2008)

4. Lei, J., Fu, X.: Evaluating the Benefits of Introducing PMIPv6 for Localized Mobility Management. In: Wireless Communications and Mobile Computing Conference (IWCMC 2008), pp. 74–80 (August 2008)
5. Soliman, H., Castelluccia, C., ElMalki, K., Bellier, L.: Hierarchical Mobile IPv6 (HMIPv6) Mobility Management. RFC 5380 (Proposed Standard) (October 2008)
6. Camarillo, G., Niemi, A., Isomaki, M., Garcia-Martin, M., Khartabil, H.: Referring to Multiple Resources in the Session Initiation Protocol (SIP). RFC 5368 (Proposed Standard) (October 2008)
7. Guan, J., Zhou, H., Xiao, W., Yan, Z., Qin, Y., Zhang, H.: Implementation and Analysis of Network-based Mobility Management Protocol in WLAN Environments. In: ACM Mobility Conference - MobiWorld Workshop (September 2008)
8. Chuang, M.C., Lee, J.F.: FH-PMIPv6: A Fast Handoff Scheme in Proxy Mobile IPv6 Networks. In: International Conference on Consumer Electronics, Communications and Networks (CECNet), pp. 1297–1300 (April 2011)
9. Rasem, A., St-Hilaire, M., Makaya, C.: A comparative analysis of predictive and reactive mode of optimized PMIPv6. In: International Wireless Communications and Mobile Computing Conference (IWCMC), pp. 722–727 (August 2012)
10. Chuang, M.C., Lee, J.F.: Comparative Handover Performance Analysis of IPv6 Mobility Management Protocols. IEEE Transactions on Industrial Electronics 60(3), 1077–1088 (2013)
11. 3GPP: Policy and charging control architecture; V10.8.0 Release 10. TS 23.203, 3rd Generation Partnership Project (3GPP) (September 2012)
12. 3GPP: Access Network Discovery and Selection Function (ANDSF) Management Object (MO); V10.6.0 Release 10. TS 24.312, 3rd Generation Partnership Project (3GPP) (July 2012)
13. Kellokoski, J.: Challenges of the Always-Best-Connected Enablers for User Equipment in Evolved Packet System. In: Ultra Modern Telecommunications and Control Systems and Workshops (ICUMT) (October 2012)
14. Kellokoski, J., Koskinen, J., Hmlinen, T.: Real-life Performance Analysis of Always-Best-Connected Network. In: New Technologies, Mobility and Security (NTMS) (May 2012)
15. 3GPP: Architecture enhancements for non-3GPP accesses; V10.4.0. TS 23.402, 3rd Generation Partnership Project (3GPP) (July 2011)
16. 3GPP: Access to the 3GPP Evolved Packet Core (EPC) via non-3GPP access networks, Stage 3. TS 24.302, 3rd Generation Partnership Project (3GPP) (March 2012)
17. Soliman, H.: Mobile IPv6 Support for Dual Stack Hosts and Routers. RFC 5555 (Proposed Standard) (June 2009)
18. 3GPP: GPRS Tunnelling Protocol (GTP) across the Gn and Gp interface. TS 29.060, 3rd Generation Partnership Project (3GPP) (December 2011)
19. Postel, J.: Internet Protocol. RFC 791 (Standard) (September 1981), Updated by RFCs 1349, 2474
20. Google Inc.: Android developers (January 2013), http://developer.android.com/
21. Thomson, S., Narten, T., Jinmei, T.: IPv6 Stateless Address Autoconfiguration. RFC 4862 (Draft Standard) (September 2007)
22. Eurecom: Openairinterface proxy mobile ipv6 (January 2013), http://www.openairinterface.org/ openairinterface-proxy-mobile-ipv6-oai-pmipv6
23. Schulzrinne, H., Casner, S., Frederick, R., Jacobson, V.: RTP: A Transport Protocol for Real-Time Applications. RFC 3550 (Standard) (July 2003), Updated by RFCs 5506, 5761, 6051, 6222

Influence of Buffer Size on TCP Performance in Heterogeneous Wired/Wireless Networks

Ivan Vujović[1] and Maroje Delibašić[2]

[1] CKB, IT Infrastructure Department, Podgorica, Montenegro
[2] MTEL, Network Development Department, Podgorica, Montenegro

Abstract. Data packets from different TCP (Transmission Control Protocol) flows, which are transferred over IP (Internet Protocol) networks, pass through the various links and through the different capacity router buffers on the way to the receivers. Most authors consider the size of the output buffer on the router that is connected to the rest of the network via bottleneck link, equal to bandwidth-delay product. In more recent studies, based on the assumption that TCP flows are desynchronized, much lower values are suggested. In this paper we analyse wireless last hop links with errors occurring during signal transmission. Established TCP flows are considered both synchronized and desynchronized. Also, reverse traffic is present. Simulations are made with different number of flows originating from different protocols. The obtained results show that proposed buffer size values do not correspond to the optimal ones. That is why we determine the optimal buffer sizes.

Keywords: buffer size, desynchronized flows, fairness, friendliness, goodput, synchronized flows, TCP.

1 Introduction

Transmission Control Protocol (TCP) is connection oriented protocol that provides reliable data transfer, flow control, connection management and congestion control. Critical point for the occurrence of congestion is the buffer of the router that is connected to the bottleneck link.

The rule of thumb for buffer sizing [1] is:

$$B = C \cdot AvgRTT \ , \tag{1}$$

where C is the bandwidth of the bottleneck link, and $AvgRTT$ mean value of RTT (Round Trip Time). Some authors have questioned the practicality of applying above formula, based on hypothesis that TCP flows in the real networks are not synchronized. In [2] is suggested that buffer size should be calculated according to the:

$$B = C \cdot AvgRTT / \sqrt{N} \ , \tag{2}$$

where N is the number of TCP flows sharing a bottleneck link. The authors of [3] showed results which suggest that the optimal value of the buffer size is:

$$B = 0.63 \cdot C \cdot AvgRTT / \sqrt{N} \ , \tag{3}$$

S. Balandin et al. (Eds.): NEW2AN/ruSMART 2013, LNCS 8121, pp. 224–235, 2013.
© Springer-Verlag Berlin Heidelberg 2013

while in [4] is concluded that optimal buffer size is 10 - 20 packets.

The buffer size can be defined as function of decrease parameter b in congestion avoidance phase [5]:

$$B = [(1 - b)/b] \cdot C \cdot AvgRTT . \tag{4}$$

Decrease parameter b has value 0.5 for Reno and Vegas [5], [6], [7], while for Veno and Feno it has value 0.8 [5], [8], [9].

Real networks contain the wireless links where errors occur during signal transmission. Scenarios in this paper are based on a network that contains a wireless links as a last hop to the receivers, which is the most common case where the wireless links appear. Errors that occur during signal transmission are usually bursty. Another important issue that is usually ignored in most of the analytical and simulation based studies is the fact that ACKs could also experience packet losses induced by both congestion as well as transmission errors [10]. We consider the case where TCP flows are synchronized as well as the case of desynchronized flows [2].

Chosen protocols for simulations are Reno [6], Vegas [7], Veno [8] and Feno [9]. They have similar congestion control algorithms, but have different performance depending on the buffer size, as will be shown through simulations results.

There is no standardized way to perform buffer size dimensioning. In this paper we will analyse the influence of buffer size on TCP performance in heterogeneous network environment. Simulation model includes wired links, last hop wireless links and traffic in reverse direction. All simulations will be done for two cases: synchronized and desynchronized TCP flows. Simulation results will show that optimal buffer sizes for different scenarios do not correspond to those obtained by [1]-[5]. Also, the optimal buffer sizes depend on the applied protocol. This topic is important since buffer sizes greatly influence the performance of the network and even though memory has become cheap, bigger is not always better.

The rest of the paper is organized as follows. Section 2 presents network topology and simulation scenarios. Simulation results are presented and discussed in Section 3. Finally, Section 4 concludes the paper with discussion about future work in the area of dimensioning router buffers.

2   Network Topology and Simulation Scenarios

Network topology used in simulations is shown in Fig. 1. There are n TCP sources that are connected to the router R1 through 100 Mb/s wired link. Link between routers R1 and R2 has capacity of 10 Mb/s, and it represents a bottleneck. All receivers are connected to the router R2 via 100 Mb/s wireless links. Therefore, the congestion may occur on the link between routers R1 and R2, while errors occur on wireless links between router R2 and receivers. Performances of the considered protocols are analyzed in relation to the router R1 output buffer size.

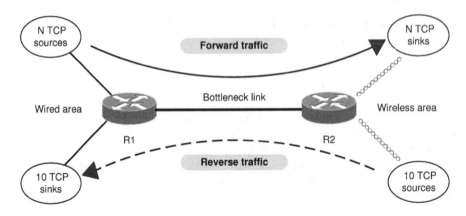

Fig. 1. Network topology

We do not include specific wireless links, but for the simplicity, more general wireless errors model, in order to obtain more general results.

We consider wireless channels between router R2 and receivers affected by bursty segment losses in both directions, and use Gilbert two state Markov chain to model the loss process [11]. In particular, we assume a segment loss probability equal to 0 when the channel is in the *Good* state, and equal to 0.1 when the channel is in the *Bad* state. The permanence time in the *Good* state is assumed deterministic and equal to 1 s whereas the permanence time in the *Bad* state is assumed also deterministic but this time we consider value 100 ms. When the permanence time in a state elapses, the state can transit to a *Good* or *Bad* state with a probability $p = 0.5$ [12].

Acknowledgements (ACKs) compression is achieved by introducing 10 TCP Reno flows in the reverse direction (from receivers to senders). Those flows are noted as reverse traffic. Total delay in this direction is 40 ms, and delay on the link from R2 to R1 is 20 ms. These flows are present during the entire simulation time. The TCP sinks implement delayed ACK option. Data segments are 1440 Bytes long. All routers use Drop Tail queuing mechanism. Duration of each simulation is 1000 s. All results are presented as mean values from 10 simulations and were obtained using ns-2 simulator [13], and supplements [14], [15].

The following simulation scenarios are analyzed:

- For every considered protocol $n = 10$ TCP flows are assumed.
- For every considered protocol $n = 100$ TCP flows are assumed.
- All protocols are present in the network, having 10 TCP flows each, alternately established ($n = 40$ flows in total).

Protocol performances will be analysed from the obtained simulation results for two cases. In the case A all flows are synchronized, while in the case B all flows are desynchronized.

In the case of desynchronized flows, data transfer of the first flow starts at 0 s, and ends when simulation ends. The second flow starts 1 s later and ends

1 s before the end of the first flow, while its RTT is 1 ms higher. This rule is applicable to every subsequent flow. Minimum value of RTT is 80 ms. When 10 flows are established, mean value of RTT is 84.5 ms, for 100 established flows this value is 129.5 ms and for 40 established flows is 99.5 ms.

In the case of synchronized flows, data transfer of all flows starts at the beginning of the simulation, ends at the end of simulation and all of them have the same value of RTT. In order to compare the results obtained for synchronized flows with those obtained for desynchronized flows, values taken for RTTs in scenario with synchronized flows are the same as mean values of RTT for desynchronized flows.

In all scenarios performances are presented for different buffer sizes. Based on (1), (2), (3) and (4), and taking the first major integer values, the router R1 output buffer sizes are:

- For $n = 10$ flows: 74, 24, 15, and 19 packets respectively.
- For $n = 100$ flows: 113, 12, 8, and 29 packets respectively.
- For $n = 40$ flows: 87, 14, 9, and 22 packets respectively.

Maximum considered buffer sizes Bm are obtained from (1).

In order to evaluate results for proposed buffer sizes, we will compare them to results obtained for some other buffer sizes, like 2, values from 10 to 20, and some integer multiples of 10. Furthermore, we will point out the values that are optimal.

Performances of the protocols will be observed through the values of goodput, fairness, friendliness and number of the first discarded flow. Goodput for a single flow is calculated as useful traffic measured in bits per second [16]. Mean goodput is calculated as the mean value of individual goodputs of all flows. Total goodput is calculated as sum of individual goodputs of all flows. Fairness is the rate of square of the total goodput and the sum of squares of the individual goodputs multiplied by the number of flows [16]. Friendliness referring to fairness between flows using different protocols [16]. Flow that is not established end-to-end, i.e. flow which packets are dropped, is noted as discarded flow.

3   Simulation Results

Choice of buffer size affects mean goodput and/or fairness values as well as friendliness behavior of considered protocols. In this section we will define optimal buffer size as value for witch goodput, fairness and friendliness are optimal.

3.1   Synchronized Flows

Scenario with $n = 10$ TCP Flows. In this case values of fairness for each protocol are very high (above 0.992) at all buffer sizes and practically the same as for Bm. This means that fairness is not strongly affected by the buffer size. Obviously, the values of goodput determine the optimal buffer sizes.

Mean goodputs for 10 flows of the considered protocols depending on the buffer size are shown in Fig. 2. It can be seen that goodput significantly increases with increase of buffer size to some particular value which we denote as optimal buffer size (Bo), and then remains almost constant. Goodput and fairness values for Bo and Bm are shown in the Table 1.

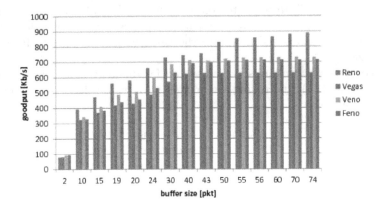

Fig. 2. Goodput values for 10 synchronized flows

Buffer sizes obtained from [1]-[5] do not match the optimal, except in the case of Reno where optimal value is the same as Bm.

Table 1. Goodput and fairness values in the case of 10 synchronized flows

Protocol	Reno	Vegas	Veno	Feno
Bo	74	43	55	56
Goodput [kb/s]	887.8	627.6	728.2	714.6
Fairness	0.9996	0.9998	0.9997	0.9997
Bm	74	74	74	74
Goodput [kb/s]	887.8	628.6	729.9	715.7
Fairness	0.9996	0.9998	0.9998	0.9997

Scenario with $n = 100$ TCP Flows. Mean goodput values for each protocol are almost constant for all considered buffer sizes and practically the same as for Bm. This means that goodput is not strongly affected by the buffer size. In this case, the values of fairness (see Fig. 3) determine the optimal buffer size (Bo). It is important to note that a number of flows, in this scenario, are not established end-to-end for low buffer sizes. Fig. 4 shows serial numbers of the first discarded flow. Low buffer sizes that cause rejection of one or more flows are not considered as optimal no matter on goodput or fairness. It can be seen that,

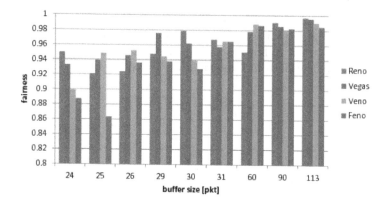

Fig. 3. Fairness values for 100 synchronized flows

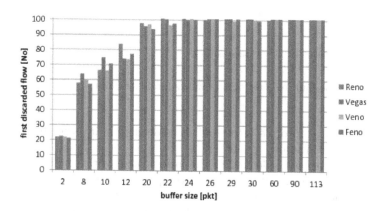

Fig. 4. First discarded flows for 100 synchronized flows

Table 2. Goodput and fairness values in the case of 100 synchronized flows

Protocol	Reno	Vegas	Veno	Feno
Bo	30	29	31	31
Goodput [kb/s]	89.5	89.2	90.1	89.7
Fairness	0.9789	0.9758	0.965	0.9647
Bm	113	113	113	113
Goodput [kb/s]	93.1	93.1	93.2	93.2
Fairness	0.9971	0.9954	0.9903	0.9847

for TCP Reno, optimal buffer size must be equal or greater than 24 packets, for Vegas 26, Veno 25 and Feno 26.

Goodput and fairness values for Bo and Bm are shown in the Table 2.

Results obtained in this scenario also show that the optimal buffer sizes do not correspond to those obtained from [1]-[5].

Scenario with $n = 40$ TCP Flows. This scenario investigates the influence of buffer size on TCP friendliness. Fig. 5 shows values of mean goodput for individual protocols, when 10 flows of each protocol are established in the same time (that is $n = 40$ flows in total) and values of friendliness between flows.

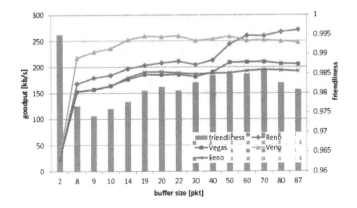

Fig. 5. Goodput and friendliness values for 40 synchronized flows

Flows of different protocols are established alternately in order to achieve balanced capacity utilization.

Optimal buffer size Bo is defined as the lowest value of the buffer size that provide friendliness and goodputs that are comparable to the values obtained for Bm. Those conditions are satisfied for $Bo=19$. It should be noted that some flows are discarded for buffer sizes from 2 to 7. Values of goodput for Bo and Bm are shown in the Table 3.

Table 3. Goodput values in the case of 40 synchronized flows

Protocol	Reno	Vegas	Veno	Feno
Goodput [kb/s], Bo	203.1	185.5	259.2	190.2
Goodput [kb/s], Bm	271.2	205.4	247.6	191.8

It can be concluded that optimal buffer size does not correspond to any value reported in [1][3], [5]. However, the optimal buffer size is in the range of those suggested in [4], where the proposed buffer sizes are from 10 to 20 packets.

3.2   Desynchronized Flows

Scenario with $n = 10$ TCP Flows. Values of fairness for every buffer size are practically the same as for Bm. Like for synchronized flows, the values of goodput determine the optimal buffer size. Goodput values for this scenario are shown in Fig. 6. For TCP Reno, goodput constantly grows with the increase of the buffer size to the Bm. Goodput values of other protocols do not significantly change with buffer size increase from Bo to the Bm. Optimal buffer size Bo is obtained by the same criteria as in the case of synchronized flows. Values of goodput and fairness for Bo and Bm are shown in the Table 4.

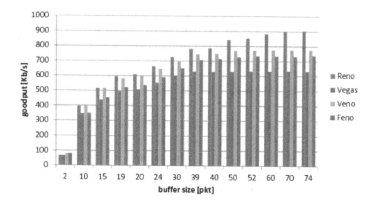

Fig. 6. Goodput values for 10 desynchronized flows

Table 4. Goodput and fairness values in the case of 10 desynchronized flows

Protocol	Reno	Vegas	Veno	Feno
Bo	74	39	52	52
Goodput [kb/s]	903.3	629.3	772.1	734.9
Fairness	0.9976	0.9995	0.9984	0.9978
Bm	74	74	74	74
Goodput [kb/s]	903.3	632.3	776.1	737.7
Fairness	0.9976	0.9996	0.9985	0.9982

In relation to the case when the flows are synchronized, higher goodput values are achieved for smaller values of Bo. It can be seen by comparing results from Table 1 and Table 4.

It can be seen that buffer sizes obtained from [1]-[5] do not match the optimal, except in the case of Reno where optimal value is the same as Bm.

Scenario with $n = 100$ TCP Flows. The simulation results show that there are no discarded flows even for the smallest buffer sizes. Mean goodput values are shown in Fig. 7. Fairness values are shown in Fig. 8. For buffer sizes from 2 to 40, goodput constantly grows, while fairness decreases. For buffer sizes greater than 40 increase of fairness occurs, while goodput is almost unchanged.

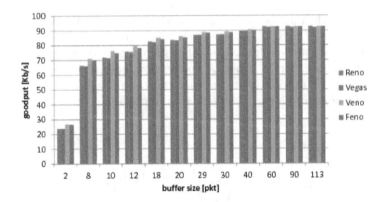

Fig. 7. Goodput values for 100 desynchronized flows

In this scenario the choice of buffer size affects mean goodput and fairness values of the considered protocols, which is presented in Fig. 7 and Fig. 8. Optimal buffer sizes are determined as given in Table 5. Compared to the case of synchronized flows, lower goodput and fairness are obtained. It can be concluded that optimal buffer sizes do not correspond to any values reported in [1]-[3], [5]. However, the optimal buffer sizes are in the range of those suggested in [4].

Scenario with $n = 40$ TCP Flows. Simulation results for this scenario are shown in Fig. 9. Optimal buffer size $Bo = 26$ is obtained by the same criteria as in the case of synchronized flows. There are no discarded flows. Values of goodput for Bo and Bm are shown in the Table 6.

As can be seen from Fig. 9, the increase of buffer size from Bo to Bm leads to changes of goodput values (especially in the case of Reno), but not to the satisfaction of TCP friendly behaviour. The optimal value does not correspond to any of the values obtained from [1]-[5].

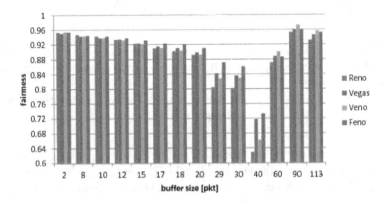

Fig. 8. Fairness values for 100 desynchronized flows

Table 5. Goodput and fairness values in the case of 100 desynchronized flows

Protocol	Reno	Vegas	Veno	Feno
Bo	17	18	15	18
Goodput [kb/s]	81.7	82.1	83	84.2
Fairness	0.9097	0.9102	0.9196	0.9205
Bm	113	113	113	113
Goodput [kb/s]	92.7	92	92.6	92.5
Fairness	0.9317	0.946	0.9564	0.9492

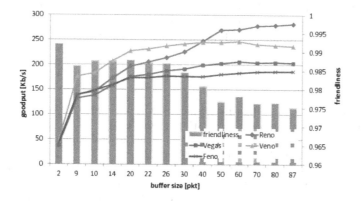

Fig. 9. Goodput and friendliness values for 40 desynchronized flows

Table 6. Goodput values in the case of 40 desynchronized flows

Protocol	Reno	Vegas	Veno	Feno
Goodput [kb/s], Bo	213.9	187.9	237.5	177
Goodput [kb/s], Bm	281.7	203.3	236.9	186.9

4   Conclusion

In this paper is analyzed influence of buffer size on TCP performance in heterogeneous environment that include wired links, last hop wireless links, and traffic in reverse direction. Synchronized and desynchronized flows are specially considered.

By simulating different conditions, with different number of flows and observing different parameters that determine the performance of protocols, results clearly shows that optimal buffer size should not be chosen only depending on goodput and number of flows, but also on other factors such as the TCP protocol that is used, fairness, friendliness and when buffer do not discard flows.

The results show different performance of TCP protocols in the case of synchronized flows than those obtained when flows are desynchronized.

Protocols which performances are observed have similar algorithms for congestion control, but in general they show different performances. In some cases the optimal buffer sizes are the same as those reported in literature, but in number of other cases it was shown that optimal buffer sizes are different from those values.

It can be concluded that selection of the optimal buffer size should not be static, but dynamically adjustable. This requires new algorithm for proper buffer dimensioning. Algorithm should consider a number of network environment parameters. In the future work we will investigate in details the influence of: different wireless channel types, ACK compression and degree of synchronization between the flows to buffer size dimensioning.

References

1. Villamizar, C., Song, C.: High Performance TCP in ANSNET. ACM SIGCOMM Computer Communications Review 24(5), 45–60 (1994)
2. Appenzeller, G., Keslassy, I., McKeown, N.: Sizing Router Buffers. In: ACM SIG-COMM, Portland, OR, USA, pp. 281–292 (2004)
3. Wischik, D., McKeown, N.: Part I: Buffer Sizes for Core Routers. ACM SIGCOMM Computer Communications Review 35(2), 75–78 (2005)
4. Enachescu, M., Ganjali, Y., Goel, A., McKeown, N., Roughgarden, T.: Part III: Routers with Very Small Buffers. ACM SIGCOMM Computer Communications Review 35(2), 83–89 (2005)
5. Hassayoun, S., Ross, D.: Loss Synchronization and Router Buffer Sizing with High-Speed Versions of TCP. In: IEEE 34th Conference on Local Computer Networks, Zurich, Switzerland, pp. 569–576 (2009)
6. Stevens, W.: TCP Slow Start, Congestion Avoidance, Fast Retransmit, and Fast Recovery Algorithms. Request for Comments (RFC) 2001 (1997)
7. Brakmo, L.S., Peterson, L.L.: TCP Vegas: End to End Congestion Avoidance on a Global Internet. IEEE Journal on Selected Areas in Communication 13(8), 1465–1480 (1995)
8. Shifeng, X., Zehua, G., Feng, G., Nan, W., Ronghua, Z.: An Enhanced TCP Veno over Wireless Local Area Networks. In: 5th Int. Conf. on Wireless Communications, Networking and Mobile Computing, WiCOM 2009, Beijing, China, pp. 4038–4041 (2009)
9. Hwang, J.-H., Yoo, S.-H., Yoo, C.: TCP Feno: Enhancement for Higher Accuracy of Loss Differentiation over Small Buffer Heterogeneous Networks. In: IEEE 34th Conference on Local Computer Networks, Zürich, Switzerland, pp. 249–252 (2009)
10. Tian, Y., Xu, K., Ansari, N.: TCP in Wireless Environments: Problems and Solutions. IEEE (Radio) Communications Magazine 43(3), S27–S32 (2005)
11. Haßlinger, G., Hohlfeld, O.: The Gilbert-Elliott Model for Packet.Loss in Real Time Services on the Internet. In: Proc. 14th GI/ITG Conference on Measurement, Modeling, and Evaluation of Computer and Communication Systems (MMB), pp. 269–283 (2008)

12. Grieco, L.A., Mascolo, S.: Performance Evaluation and Comparison of Westwood+, New Reno, and Vegas TCP Congestion Control. ACM Computer Communications Review 34(2) (2004)
13. McCanne, S., Floyd, S.: Network Simulator ns2.31,
    `http://www.isi.edu/nsnam/ns`
14. Wei, D. X., Cao, P.: A Linux TCP Implementation for NS2,
    `http://netlab.caltech.edu/projects/ns2tcplinux`
15. Hwang, J.-H., Yoo, S.-H., Yoo, C.: TCP Feno Implementation for NS2,
    `http://os.korea.ac.kr/research/research.html?title=feno`
16. Floyd, S.: Metrics for the Evaluation of Congestion Control Mechanisms. IETF Internet-draft (2006)

GetTCP+: Performance Monitoring System at Transport Layer

Aleksandr A. Sannikov, Olga I. Bogoiavlenskaia, and Iurii A. Bogoiavlenskii

Petrozavodsk State University,
Lenin St., 33, 185640, Petrozavodsk, Russia
{sannikov,olbgvl,ybgv}@cs.petrsu.ru
http://cs.petrsu.ru

Abstract. Problem of the monitoring of the network performance is important task for different classes of network applications and services. In this paper the system for monitoring of network connections at transport layer is presented. In contrast to existing analogs the monitor is able to provide details on network stack operation visible only at Linux kernel level since the monitor presented operates in both kernel and user space. The paper describes high level architecture of the system, important features of the implementation and testing results.

Keywords: Monitoring, Network Stack, TCP, Linux Kernel.

1 Introduction

Data communication performance monitoring is one of the most important and topical problems since monitoring data provide foundation for network design, development, and administration solutions. The end-to-end path performance plays the key role in this research since it essentially contributes to the end user impression of quality of service available on the path.

In this work we present monitoring system GetTCP+ which collects information on network connections at OSI [1] transport layer and derives performance metrics for the end-to-end paths which are of particular interest for users. The raw data are collected at the OS kernel level which lets monitor to have access to the data unavailable at user's space (e.g. congestion window size) and allows avoiding data distortion and/or variables interpretation problems. Several modules of GetTCP+ are based on facilities of GetTCP monitor [2].

Transport layer and namely Transmission Control Protocol (TCP) [3] is chosen for monitoring since TCP is the only instance that provides flow control solutions and is responsible for distrubuted control of connections access to the network infrastructure. This work does not consider UDP protocol and its extentions, e.g. RTP, since they do not realize flow control algorithms and do not perform data delivery control. Therefore monitoring of TCP behavior reveals essentially wider range of information about end-to-end path performance. Also, for some particular monitoring aspects the developed system captures certain data at network layer, namely Internet Protocol (IP v4 and v6) as well.

S. Balandin et al. (Eds.): NEW2AN/ruSMART 2013, LNCS 8121, pp. 236–246, 2013.

Therefore related works in the area presenting several systems of network monitoring which are widely used, e.g. IOS NetFlow [4] and its analogs, *tcpdump* [5], *iperf* [6] and others. Also, one have to mention systems for processing and analysis of monitoring data e.g. Nagios [7], Ganglia [8], *tcptrace* [9]. Distinctly general network monitoring software GetTCP+ collects monitoring data at the OS kernel level and hence gets information that either is totally unavailable at higher OS levels or could not be reliably obtained by network monitoring software of general purpose. Thus, the example demonstrating erroneous estimations of TCP segment size provided by *tcpdump* which was discovered and corrected by the developed system will be presented further.

The rest of the paper is organized as follows. Section 2 describes general system's architecture, identifies modules which use and/or modify GetTCP libraries and presents original facilities of GetTCP+ as well. Section 3 provides some details of the implementation, section 4 contains description of testbed and tests of the system performed. The conclusion offers summary and directions of future research.

2 System Architecture

The system architecture consists of two main units. These are data collection (DC) subsystem and end-to-end path performance metrics storage. The architecture is presented on Fig.1.

Data collection subsystem is based on GetTCP kernel module and *libgettcp* library [2]. The library provides tools for management of control trace points sets and OS kernel module interface for communication and data transfer from kernel space to the user space. Due to the monitoring purposes GetTCP system was significantly modified as well. A new trace point sets for processing connection events and adds segment related events. In particular flow filtering tool, dynamically controlled parameters facility and support of general segmentation offloading mechanism are implemented. Finally, several bugs were fixed as well and the system was ported for modern Linux kernels (v2.6.38-3.1.10). Using *libgettcp* interface and facilities DC subsystem extracts information about end-to-end connection state. Also, it allows processing of single segment-related events.

DC subsystem consists of two parts: kernel module collecting monitoring data and user-space interface transferring data into user space. Filtering mechanism allows extracting flows important for monitoring. When some connection or segment related event rises, trace point handler generates data entry which contains information about the event and the state of network connection. Then this entry is transferred to user-space. At present, GetTCP+ provides following list of end-to-end path metrics: source and destination hosts, maximal congestion window size reached, total volume of transmitted data, number of sent segments, loss rate, maximal segment size, also sequence of congestion window size and round trip time sequence (for each TCP segment transmitted) if required. Thus, with monitoring process organization one could get any information about transport layer connections behavior.

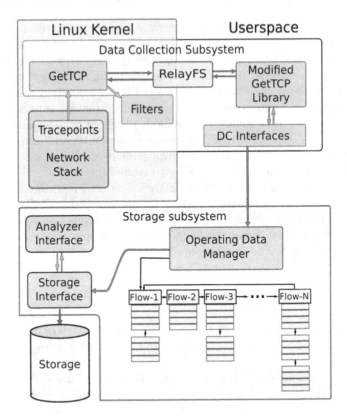

Fig. 1. System architecture

2.1 Storage Subsystem

The storage subsystem consists of three units. These are operating data manager, storage interface and analyzer interface. Operating data manager(ODM) processes running data for on-going TCP flows. A record about every transmitted segment is placed into dynamic memory buffer associated with the flow. When the flow ends, ODM processes the content of the buffer and saves a set of metrics into storage, namely the source and destination, total data sent, flow duration, segment loss rate, MSS, receiver advertised window, mean congestion window, mean RTT. These metrics has been chosen since they are needed to derive current or future TCP performance of the end-to-end path, by direct evaluation or through performance models. After processing ODM instantly removes dynamic buffer to free memory for further usage. On demand ODM can save full sequence of segments data stored in the buffer as well.

The storage interface provides inter-operation between long-time storage and internals of GetTCP+ such as DC and ODM subsystems, hence accumulating the history of end-to-end path performance demonstrated.

The specific features of monitoring data can be denoted as following: data saved never need modification, topicality of the data collected eventually expires, domination of sequential access, data are processed by big slices. Due to these reasons, the storage is based on a file system objects. The information about each sub-network is stored in the separate directory as it is shown on figure 2.

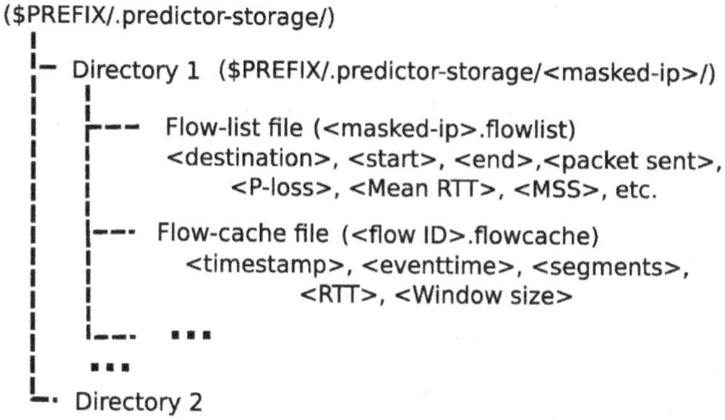

```
($PREFIX/.predictor-storage/)
|
|-- Directory 1  ($PREFIX/.predictor-storage/<masked-ip>/)
|   |
|   |--- Flow-list file (<masked-ip>.flowlist)
|   |        <destination>, <start>, <end>,<packet sent>,
|   |            <P-loss>, <Mean RTT>, <MSS>, etc.
|   |
|   |--- Flow-cache file (<flow ID>.flowcache)
|   |        <timestamp>, <eventtime>, <segments>,
|   |            <RTT>, <Window size>
|   |
|   |---.  ...
|
|--.  ...
|
L--. Directory 2
```

Fig. 2. Storage structure

This directory contains the flow-list file with common information about flows related with specified sub-network. In particular, the information contains: flow ID, host and interface information,total data sent, flow duration and some metrics: loss rate, mean round trip time, mean window size. The example of flow-list records follows:

```
Flow-ID,        Source,   Port, Dst,        Port, Start(sec), (usec), End(sec),   (usec), WMax, MSS, Sent,  Lst, RTT(msec)
1839A6F7310580, 127.0.0.1, 55256, 79.133.201.85, 35091, 1334909306, 151608, 1334909474, 692132, 166, 1424, 147061, 78,  107207.893
18DE58F7310A80, 127.0.0.1, 55512, 79.133.201.85, 35091, 1334909474, 805528, 1334909622, 350665, 166, 1424, 147168, 142, 104185.041
196F2CF7310A80, 127.0.0.1, 55768, 79.133.201.85, 35091, 1334909623, 105733, 1334909756, 216969, 166, 1424, 147020, 43,  103939.189
```

The per-flow cache can be created out of dynamic buffer data records by demand. It contains full sequence of data about segments sent. In the cache file every record contains the time stamp, sequence number, congestion window size, RTT duration etc. Thus, complete description of TCP flow behavior can be reconstructed.

The current implementation of the storage offers three types of data presentation: as a log-file, CSV-formatted file and binary representation. Each of them is available in two modes, namely verbose and standard. In a verbose mode cache files for each flow are stored. Let us notice that verbose mode significantly increases volume of stored data. For example the size of cache-file is equal to 10Mb for 200Mb of transmitted data. At the same time, the size of one record in flow-list file with general information and mean metrics is 200-300 bytes approximately.

First type of data presentation is log-file form. In this case each line for standard mode represents one flow. Such form improves visibility of collected data, but increases volume of data stored. This mode simplifies GetTCP+ deployment. Second type is CSV-formatted file. In this case volume of stored data and its visibility decreases insignificantly, although this mode simplifies automatic processing of monitoring results. Finally, the third form is raw binary data storage. This form is equal to CSV-form by structure and significantly increases performance of the system and reduces volume of data stored. Also some special approaches in data storage can be applicable in this form, e.g. flow indexing.

The interface of analysis subsystem provides access to collected data to any analytical application in necessary form. This subsystem can be implemented as set of plug-in modules according to the requirements of analytical application.

2.2   Tracepoints

Let us consider the set of the trace points implemented more in detail. The set is used to get information about TCP flows and separate segments. The tracepoints define breakpoints inside kernel image and associate handlers with them. When control flow reaches a trace point the correspondent handler invokes. GetTCP+ implements a set of trace point handlers located in the network stack of Linux kernel. For the aims of monitoring four events are essential and henceforth, four tracepoints are defined.

The first two of them are *flow_start_event* and *flow_end_event* which are connected with the start and termination of a TCP flow. The events provide common flow information and their handlers are associated to TCP state machine as it is shown on Figure 3.

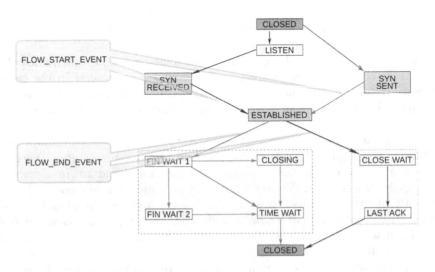

Fig. 3. Flow state-related tracepoints

When the machine changes the state of a connection to "established", *flow_start_event* event is been raised. The event handler performs filtering of the end-to-end path an if it is under monitoring the handler generates record containing information about network device and destination host. When TCP state machine leaves established state *flow_stop_event* is rising. In this case handler provides information about flow duration, maximal congestion window size reached and maximum segment size of the connection.

For per-segment monitoring two other events are used. This events are associated with reached list of unacknowledged segments from retransmission mechanism of TCP, as shown on Figure 4. Each segment sent is added to this list. If the segment is unacknowledged during RTO or treated lost due to SACK information or triple acknowledgment is considered lost and then it is retransmitted by TCP implementation. Thus we can monitor losses. If the segment is acknowledged is removed from the retransmission list. Hence both the segment sent and it's acknowledgment are avaliable the same time. So DC subsystem can obtain full information about the segment.

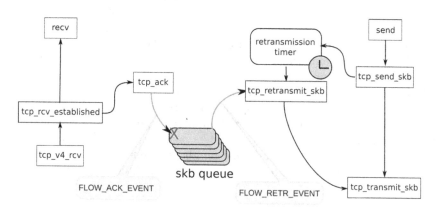

Fig. 4. Segment-related tracepoints

Every unacknowledged segments are considered as lost. The information about connection state is generated by *flow_retr_event* event handler for all such segments.

3   Implementation Special Features

To monitor the flow a special mechanism is required for clustering flows by destination hosts. It should provide the ability of host identification and binding of network flows to such hosts. IP protocol is used for this aims. IP-address can identify both single hosts and sub-networks.

During GetTCP+ development IPv4 and IPv6 facilities were added. It allows collection, storing and processing information about hosts or sub-networks for any IP-based network.

3.1   Segmentation Offloading

The Linux kernel since v2.6.13 uses the set of extensions and improvements of network stack performance. For example, TCP CUBIC flow control algorithm differs (as shown in [10]) from original version proposed in [11]. Some of these changes can affect monitoring tools distorting final results of monitoring. In particular, TSO – TCP segmentation offloading mechanism, may have influenced collected data about TCP flows.

TSO delegates to network interface card the task of splitting big data frames into segments of the size acceptable by data communication network. By this way network stack becomes able to process segments of the sizes several times bigger than MTU. As a result, CPU workload reduces. This technology is especially appropriate in high-performance networks such as 1000BASE-T. As it was shown in [12], TSO significantly increases performance of the network stack. TSO usage looks like transmission of big-size segments at higher layers, because of splitting segments at the network device. So the user space application is not able to estimate the real segment size. Thus, widely used *tcpdump* utility provides erroneous information about segment sizes in high-performance network with the throughput about 50 Mb/sec which enables TSO option. One can observe on Fig. 5 dynamics of segment sizes transmitted during a connection visible by *tcpdump*. At the same time, the real segment size always was equal to 1424 bytes. Meanwhile, according to the data provided by *tcpdump*, the segment size reached 22784 bytes. This is equal at least to 16 transmitted segments.

GetTCP+ prototype is able to operate at kernel's level tracks segmentation offloading and provides correct information about segment size and count. DC subsystem tracks TSO state and parameters passed to it during transmission. One is able to provide correct information about segments characteristics immediately after splitting data to transmit by network interface card under such approach.

3.2   Kernel Module Configuration and Flow Filtering

The run-time configuration interface allows passing specific options to kernel module without reloading, in contrast to the method of passing options as arguments to kernel module. This approach lets an end user change parameters of monitoring dynamically which simplifies the system deployment and exploiting.

At the user space the configuration interfaces are presented through *gettcp_conf_setup(name, value)* function added to *libgettcp* library. This function gets parameter's name and value as a string. Kernel module provides the set of interface functions for secure access to parameter list from handler functions.

The filtering mechanism allows extracting flows important for monitoring. Thus, end user can denote flows related to particular host, sub-network or network interface only. The implementation of filtering in kernel space reduces data flow transmitted to user-space. The filtering is based on two lists: the list of devices and the list of sub-networks. For their management several functions were introduced in the user-space:

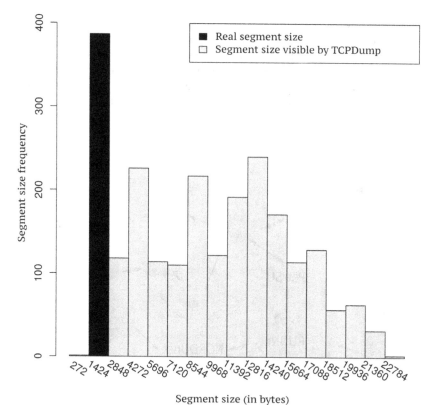

Segment size (in bytes)

Fig. 5. Histogram of segment sizes provided by tcpdump

- *gettcp_conf_adddev(dname)* – adds device with a specified name to the monitoring list.
  The monitoring is performed only for devices included in this list. Behavior of this filter can be inverted with *DEV_FLTR_EXCLUDED* option.
- *gettcp_conf_addadr(adr, mask)* - adds sub-network or a single host to monitoring list. The filtering is performed by address and network mask. Behavior of this filter is similar to the device filter and can be controlled by *ADDR_FLTR_EXCL*.

The filtering trace point handler is invoked at flow-start. The value of *probed_sock* field from *tcp_sock* structure makes the monitoring for the flow necessary. Further, at the trace points involving processing time, the monitor behavior for the flow is defined by *probed_sock* value. If flow is filtered, all related events are ignored and no data is produced.

4   Testing GetTCP+

GetTCP+ was tested for several network fragments with different characteristics and structure. In particular, the fragments with high performance and the fragments with low throughput, the high round trip time and the loss rate were tested. The four series of experiments were performed from testbed presented on Fig. 6. The monitor GetTCP+ run on the host A which is Intel Celeron 2.20 GHz with RAM 1G and the network connections to Ethernet LAN 1000Base-T and 3G-network on 256Kb/s. The first series of experiments were conducted for the network path with high throughput and low round trip time. This fragment consists of two host connected via router by gigabit Ethernet (A⇒B route on Fig. 6). In this case GetTCP+ was tested under high load.

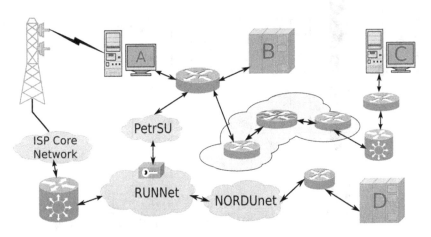

Fig. 6. Testbed

The second and third series of experiments were performed for routes with different throughput and round trip time (i. e. A⇒C and A⇒PetrSU⇒D routes). The fourth series of experiments were performed on the fragment with high loss rate and round trip time (A⇒ISP Core Network⇒D route). During every experiment TCP flow of 200 Mbyte was generated by *iperf* tool at the source host. At the same time these flows were under GetTCP+ monitoring. General information about flows i.e. connection duration, total data sent and average throughput were compared to those provided by *iperf* reports and they correspond completely. The sequences of congestion window sizes provided by cache file follow current TCP NewReno standard and CUBIC implementation for Linux version 2.6.38 completely as well. The round trip time estimations were tested against those provided by *ping* utility and they demonstrated accordance as well.

Also, the delays were estimated. The delays are brought into Linux network stack performance by developed trace point handlers. Linux network stack performance. The execution time of particular functions was measured by *Ftrace* tool. This is internal function's tracer included in the main line Linux kernel since v2.6.27 and it could be used for estimation of the particular function execution time [13]. The average delays measured are following:

Notice, that handlers *tcp_end_event()* and *tcp_start_event()* are invoked once per flow, handler *tcp_retr_event()* is invoked only for lost segments and the loss

Event	Handler	Mean delay
flow_start_event	tcp_start_event()	16.904 usec
flow_ack_event	tcp_ack_event()	0.628 usec
flow_retr_event	tcp_retr_event()	1.172 usec
flow_end_event	tcp_end_event()	2.480 usec

rate in the experiments conducted did not exceed 5%. Meanwhile, average processing time is 8.406 usec for standard *tcp_transmit_skb()* function which invokes *tcp_ack_event()* handler, and for standard function *tcp_retransmit_skb()* it is 17.8 usec. Henceforth, GetTCP+ kernel space modules does not bring significant delays into performance of the Linux network stack. Thus, GetTCP+ was tested in different networking environments, and it has shown high stability and performance.

5   Conclusion

The monitoring system GetTCP+ for observation on the end-to-end paths performance at transport layer was developed. This system produces general and/or detailed data of TCP flows performance for the source-destination pairs. The system provides filtering flows, data storage tools, dynamic settings control, using trace points handlers.

In contrast to other existing systems, the certain modules of GetTCP+ operate at OS Linux kernel level. Thus, the system provides accurate and complete information about connection's behavior. It provides correct data which otherwise could be distorted by the monitoring tools operating at user space, e.g. *tcpdump*. The system processes specific important features of network stack, such as TCP segmentation offloading. The interfaces provided by the system allow its integration to the network analysis tools.

For future development we plan to implement an analytical component into the system. Implementation of external interface for storage system will expand the area of GetTCP+ applications as well.

References

1. International Standard ISO/IEC 7498-1, p. 68 (1996)
2. Ponomarev, V.A., Bogoyavlenskaya, O.Y., Bogoyavlenskiy, Y.A.: Configurable Kernel-Level Monitoring System of the TCP Behavior. In: Information Technologies 2010, vol. 1, pp. 54–56 (2010)
3. Allman, M., Paxson, V., Blanton, E.: RFC 5681: TCP Congestion Control (2009), http://datatracker.ietf.org/doc/rfc5681/
4. Cisco IOS NetFlow, http://www.cisco.com/en/US/products/ps6601/products_ios_protocol_group_home.html
5. tcpdump/Libpcap public repository, http://www.tcpdump.org/
6. Iperf - TCP/UDP Bandwidth Measurement tool, http://iperf.fr/

7. Josephsen, D.: Building a Monitoring Infrastructure with Nagios, 1st edn. (2007) ISBN 0-13-223693-1
8. Massie, M.L., Chun, B.N., Culler, D.E.: The Ganglia distributed monitoring system: design, implementation, and experience. Parallel Computing 30 (2004)
9. Tcptrace Homepage, http://www.tcptrace.org/
10. Leith, D.J., Shorten, R.N., McCullagh, G.: Experimental evaluation of Cubic-TCP. In: Proceedings of the 6th International Workshop on Protocols for Fast Long-Distance Networks, March 5-7 (2008)
11. Ha, S., Rhee, I., Xu, L.: CUBIC: A New TCP-Friendly Hight-Speed TCP Variant. ACM SIGOPS Operating Systems Review - Research and Developments in the Linux Kernel 42(5), 64–74 (2008)
12. Linux: TCP Segmentation Offload (TSO), http://kerneltrap.org/node/397
13. Bird, T.: Measuring Function Duration with Ftrace. In: Proceedings of the Linux Symposium, pp. 47–54 (July 2009)

A Game Theoretical Perspective
on Small-Cell Open Capacity Sharing
in Cognitive Radio Environments

Ligia C. Cremene, Noémi Gaskó, Marcel Cremene, and Dumitru Dumitrescu

Babes-Bolyai University of Cluj-Napoca, Romania
Technical University of Cluj-Napoca, Romania

Abstract. Small-cell, open capacity sharing scenarios in Cognitive Radio (CR) environments are studied from a game theoretical (GT) perspective. Simultaneous capacity requests in small-cell scenarios are modelelled as strategic interactions between CRs and analysed as resource access games. CR capacity access competition is modelled based on discrete reformulations of the Bertrand GT model. Detected equilibria describe stable game situations. Numerical simulations identify situations where Nash equilibrium (NE) is both fair and Pareto efficient or where there are multiple NE solutions to choose from, indicating a flexible range for CR strategies. Adding to the analysis are the joint Nash-Pareto solutions (intermediate between Nash and Pareto) capturing heterogeneous behaviour of players. Stable and equitable states are detected even when players have different biases.

Keywords: cognitive radio, computational game theory, equilibrium, resource access, small-cells, uncoordinated deployment.

1 Introduction

Emerging communication systems are flexible, dynamic, context-aware, and spectrum-aware [1], [2]. The underlying technologies of future communication systems are Cognitive Radio Technologies (CRT) [3]. CRT are seen as enablers of efficient usage of spectrum, heterogeneous operation, dynamic capacity flow, and green communications. Cognitive Radio (CR) [4] is also seen as the solution for managing the ever rising complexity of nowadays telecommunication systems [5].

We witness the passing from regulation to self-organization: multi hop, peer-to-peer, or grid configurations, macro cells complemented with layers of small cells. New network visions are being proposed by telecommunications companies (e.g. Liquid Radio [6], Liquid Net, Light Radio [7]) having flexibility as their middle name. Users enjoy high data rates on the move and value wide-area availability of quality broadband connectivity [6]. Moreover, the aim is to flow spare capacity where it is needed, when it is needed.

There is increased interest in the deployment of small cells to improve coverage and overall capacity. Femtocells are actually becoming the new paradigm in

S. Balandin et al. (Eds.): NEW2AN/ruSMART 2013, LNCS 8121, pp. 247–259, 2013.

delivering high data rate communications. Massive and uncoordinated deployment of femtocells poses significant challenges to efficient radio resource sharing [8], [9], [10].

We have chosen to analyze capacity sharing in high-density, uncoordinated deployment, small-cell environment, from a game theoretical perspective.

Simultaneous capacity-requests in small-cell scenarios are modelled as strategic interactions between CRs and analyzed as resource access games.

Computational Game Theory (CGT) offers a fertile framework for the analysis of CR interactions in their pursuit for achieving the desired individual payoffs [3], [11], [12], [13], [14], [15], [16], [8], [9], [17]. Cognitive radio interactions are strategic interactions: each CR's payoff depends on the other CR actions. CGT analysis proves useful in devising local interaction models as well as rules of behaviour in CR environments [18], [5].

A reformulation of the Bertrand oligopoly[19], [20] for open capacity sharing modelling is considered. Although it has been intensively used for spectrum trading and pricing, the Bertrand model may also be reformulated in terms of radio resource access, capturing very general and intuitive scenarios. Several types of equilibria are detected and their significance is discussed. An evolutionary equilibrium detection method is used [21], [22].

The paper is structured as follows: Section 2 introduces some theoretical aspects of game equilibria and provides basic insights to their definition and significance. Section 3 describes the reformulation of the Bertrand game theoretic model for simultaneous capacity requests in a high-density, small-cell, CR environment. Section 4 discusses simulation results obtained for various instances of the game, analysing three different open capacity sharing scenarios. Conclusions are presented in Section 5.

2   Game Theoretical Insights

A strategic-form game model has three main components [23]: a finite set of players, a set of actions, and a payoff or utility function which measures the outcome for each player determined by the actions of all players.

Formally a game can be described as a system $G = ((N, S_i, u_i), i = 1, ..., n)$, where:

- N represents a set of players, and n is the number of players;
- for each player $i \in N$, S_i is the set of available actions,

$$S = S_1 \times S_2 \times ... \times S_n$$

  is the set of all possible situations of the game and $s \in S$ is a strategy (or strategy profile) of the game;
- for each player $i \in N$, $u_i : S \to R$ represents the payoff function (utility) of the player i.

2.1   Nash Equilibrium

In non-cooperative GT the most used equilibrium concept is the Nash equilibrium (NE) [24]. Playing in the Nash sense means that no player can improve her payoff by unilaterally changing her strategy. In other words, no player has any incentive to unilaterally deviate from a NE (this would result in a reduced individual payoff).

Let us denote by (s_i, s^*_{-i}) the strategy profile obtained from s^* we say that by replacing the strategy of player i, s^*_i, with s_i :

$$(s_i, s^*_{-i}) = (s^*_1, ..., s_i, ..., s^*_n).$$

A strategy profile $s^* \in S$ is a Nash equilibrium if the inequality

$$u_i(s_i, s^*_{-i}) \le u_i(s^*),$$

holds $\forall i = 1, ..., n, \forall s_i \in S_i, s_i \ne s^*_i$.

2.2   Pareto Equilibrium

Considering two strategies s and s^* from S, strategy s Pareto dominates strategy s^* if the payoff of each player using strategy s is greater than or equal to the payoff of the player using strategy s^*, and at least one payoff is strictly greater.

A strategy profile $s^* \in S$ is Pareto efficient, when it does not exist a strategy $s \in S$, such that

$$u_i(s) \ge u_i(s^*), i \in N,$$

with at least one strict inequality.

Informally, the Pareto optimality (or Pareto efficiency) is a strategy profile where no strategy can increase one player's payoff without decreasing any other player's payoff [23].

2.3   Joint Nash-Pareto Equilibrium

The recently introduced Nash-Pareto and Pareto-Nash equilibrium concepts [25] capture game situations where players are biased towards different types of rationality - Nash or Pareto.

In an n-player game, let us consider that each player acts based on a certain type of rationality. Let us denote by r_i the rationality type of player i, $i = 1, ..., n$..

For opportunistic or open access on a CR scene, a two-player game, where r_1 = Nash and r_2 = Pareto, may be considered. The first player is biased towards the Nash equilibrium and the other one is Pareto-biased. In this case, a new type of equilibrium, called the joint Nash-Pareto equilibrium, arises and may be considered in CR interaction analysis [25].

The two types of rationality may be associated to different CR behaviours: either oriented towards maintaining a certain payoff, not deviating unilaterally from the current strategy (Nash-biased) or oriented towards getting maximum payoff (Pareto-biased).

3   Small-Cell, Open Capacity Sharing Scenario

Massive and uncoordinated deployment of femtocells poses significant challenges to efficient radio resource sharing [8], [9]. Voice traffic is still considered a high-priority application both by users and operators, so low-rate channels may still be exploited in both licensed and unlicensed bands, under the new umbrella of open and opportunistic spectrum sharing.

We have chosen to analyse, from a game theoretical perspective, opportunistic capacity sharing scenarios that may arise in high-density, uncoordinated deployment CRs. Simultaneous capacity requests are modelled and analysed as a resource access games with different parameters.

In a general scenario a number of n ($n \geq 2$) independent cognitive radios coexist in the same area. Local interactions take place in their pursuit for capacity access. The CRs are sharing a same bandwidth B. This bandwidth is divided into N orthogonal sub-channels and may provide a total capacity amount of W. Each CR needs to transfer a certain amount of data in a given time. This means that each CR needs to implement a certain number of sub-channels of B in order to transfer its data. The channels may provide equal or different data rates.

We consider situations where each CR requests a bit rate of at least W/n, thus indicating a total capacity W that is insufficient for all CRs. Another assumption is that the CRs are offering support for some real time applications so they cannot accept a bit-rate below what they requested. However, they can retry later and demand a different bit-rate.

The aim of the present research is to investigate the various equilibria that appear in scarce-resource, simultaneous access scenarios. The approach is based on one-shot, non-cooperative game model, continuous and discrete instances. Non-cooperative in terms of GT model does not mean non-collaborative in terms of technological aspects. CRs do not know the other CR requests and cannot anticipate the demand of another CR (its future action).

3.1   Bertrand Model Reformulation

A well known game theoretical model - Bertrand oligopoly [19] - is chosen as support for simulation, due to its simple and intuitive form and suitability for resource access modelling.

In the Bertrand economic competition producers compete by varying the product price and thus adjusting the demand. Their strategy is the price. Players decide their actions independently and simultaneously. The model was extensively used for pricing problems, including spectrum trading (e.g. [16]). We propose a reformulation in terms of resource access, namely opportunistic capacity access.

The Bertrand competition for capacity access may be reformulated as captured by Table 1. Constant, equal power levels are assumed.

We consider n cognitive radios competing for an available capacity W. The objective of each CR is to activate a subset of channels c_i in order to satisfy its current capacity demand level D_i (e.g. target throughput).

Table 1. Capacity sharing game model

Players	The CRs opportunistically sharing an available capacity W (attempting to simultaneously access a set of available channels in order to transmit a target number of symbols in a given time).
Actions	The set of available actions is a vector $s = (s_1, s_2, ..., s_n)$ indicating the individual capacity requests of the CRs (target number of non-interfered symbols per channel).
Payoffs	The capacity amount accessed by each CR from the available capacity W.

The strategy of each CR i is actually a target number of non-interfered symbols per channel, s_i.

The demand function of accessing a channel is a decreasing function D_i of s_i, the demand for additional channels decreases if the currently accessed channels are high-throughput ones. (The bigger the chunk a CR requests and gets, the smaller the need for additional capacity).

Let us assume a linear demand function defined as:

$$D_i(s_i) = \begin{cases} W - s_i, & \text{if } s_i \leq W; \\ 0, & \text{if } s_i > W. \end{cases} \quad (1)$$

The demand D_i becomes zero when the requested throughput s overflows the available capacity W.

The payoff of CR i may be written as a difference between a goodput $s_i D_i(s_i)$ and a cost $C_i(D_i(s_i))$ of accessing the requested capacity:

$$u_i(s) = s_i \frac{D_i(s_i)}{m} - C_i(\frac{D_i(s_i)}{m}), i = 1, ..., n, \quad (2)$$

where C_i is a cost function and m is the number of CRs with the lowest capacity request. We consider $m = 1$. We also consider a linear expression of the cost function C_i:

$$C_i(D_i(s_i))) = KD_i(s_i)), i = 1, ..., n, \quad (3)$$

where $K > 0$ is a constant.

The payoff of CR i may be further written as:

$$u_i(s) = (s_i - K)D_i(s_i), i = 1, ..., n. \quad (4)$$

Consider a 2-player game. According to the Bertrand standard model, from (1) and (4) the payoff function of CR i may be written as:

$$u_i(s_1, s_2) = \begin{cases} (W - s_i)(s_i - K), & \text{if } s_i < s_j; \\ \frac{1}{2}(W - s_i)(s_i - K), & \text{if } s_i = s_j; \\ 0, & \text{if } s_i > s_j. \end{cases} \quad (5)$$

where s_i and s_j are the individual strategies of CR i and CR j.

The imagined scenario proposes that the player requesting a higher through-put, in a scarce-resource situation, will not get it and the one asking for less is served. Here $(W - s_i)$ is an adimensional measure of resource saving which indicates the degree of acceptability of CR i's capacity request. The payoff of each CR is the accessed capacity amount from the available capacity W (number of useful symbols multiplied by an adimensional measure of resource saving).

3.2   Equilibrium Interpretation

Three types of equilibria are analysed in this paper: Nash, Pareto and the joint Nash-Pareto equilibrium. Each equilibrium type is related to a certain CR be-haviour. Nash equilibrium corresponds to a situation where CRs are purely self-ish: a CR does not care about another CR's payoff. Nash equilibrium frequently describes sub-optimal solutions.

Nash equilibrium is important because it captures a situation from which no player has any incentive to deviate unilaterally. Achieving NE in a resource-access game would equivalate the existence of self-enforcing rules in the CR environment [18],[17]. The challenge is to design resource access rules that lead to a NE that is both fair and efficient.

Pareto equilibrium indicates a set of optimal solutions (which are not nec-essarily equitable). It may correspond to a situation where the CRs exchange messages between them and adhere to convention induced by the Pareto set: no CR will improve its payoff if other CR payoffs are degraded.

The joint Nash-Pareto equilibria correspond to situations where some CRs are playing selfish (Nash) while others respect the convention stated before (Pareto). This kind of equilibrium is particularly interesting in showing which type of player: Nash- or Pareto-biased, is favoured in a situation where both types of players coexist.

4   Capacity Sharing Scenarios. Numerical Simulations and Discussion

Various instances of the general scenario described in the previous Section are considered: a) equal-rate channels, with equal costs, b) equal rate channels with different costs, and c) different rate channels.

The continuous form of the game is presented as a reference point, as most literature considers it. Challenging discrete instances of the game are considered in our simulations. The discrete form of the game is more realistic as it captures the discrete nature of choices made by CRs in their pursuit for capacity.

For the sake of accuracy and simplicity in illustrating the detected game equilibria, we have chosen to represent the two dimensional cases. Two CR si-multaneous capacity request scenarios are therefore considered in the following.

Equilibria are detected using evolutionary computing methods described in [21], [25].

4.1 Capacity Sharing Scenario No. 1: Equal-Rate Channels, Equal Costs

The standard 2-player Bertrand game is considered. Continuous and discrete instances of the game are simulated. The simulation parameters are: W = 12 which represents the available capacity, e.g. in Mbps) and K = 3 which stands for cost of accessing the rate available on one channel.

The individual payoff of each interacting CR is:

$$u_i(s_1, s_2) = \begin{cases} (W - s_i)(s_i - K), & \text{if } s_i < s_j; \\ \frac{1}{2}(W - s_i)(s_i - K), & \text{if } s_i = s_j; \\ 0, & \text{if } s_i > s_j. \end{cases} \tag{6}$$

Fig. 1 qualitatively illustrates the strategies and outcomes of two CRs simultaneously trying to access a limited capacity W.

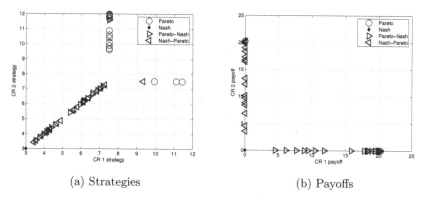

(a) Strategies   (b) Payoffs

Fig. 1. Scenario 1.1. Equal-rate, equal-cost channels - *continuous* Bertrand game reformulation - two CR simultaneous access ($W = 12$, $K = 3$). Evolutionary detected equilibrium (a) strategies: Nash (3;3), Pareto, Nash-Pareto, Pareto-Nash, and (b) payoffs: Nash (0;0), Pareto (0;20.25), (20.25;0) Nash-Pareto, Pareto-Nash (Scenario 1).

The NE in the continuous modelling of the equal-rate, equal-cost channels means zero payoff for each CR (Fig. 1 b). This is also known as the Bertrand paradox (zero-payoff outcome). For this particular scenario the Pareto strategy ensures the maximum possible payoff for one CR at a time: (0,20.25), (20.25,0) (Fig. 1 b). This illustrates a win-lose situation which is very probable in this scenario. This may also indicate that, for a high-density scene, some sort of scheduling or sequential access scheme (centralized approach) may be needed.

Fig. 2 captures Nash, Pareto, Nash-Pareto, and Pareto-Nash equilibria achieved in a 2-CR capacity sharing situation (discrete strategies).

The discrete instance of the game (Fig. 2 a) reveals an additional NE strategy besides (3,3), which is (4,4) and also non-zero NE payoffs: (4,4),(0,0) (Fig. 2 b).

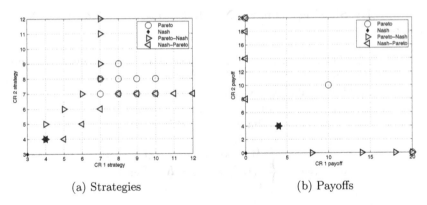

(a) Strategies   (b) Payoffs

Fig. 2. Scenario 1.2. Equal-rate, equal-cost channels - *discrete* Bertrand game reformulation - two CR simultaneous access ($W = 12$, $K = 3$). Evolutionary detected
equilibrium (a) strategies: Nash (3;3), (4;4), Pareto, Nash-Pareto, Pareto-Nash, and
(b) payoffs: Nash (0;0), (4;4), Pareto (20;0),(0;20),(10;10), Nash-Pareto, Pareto-Nash.

The Pareto strategy set is relatively small (only 9 strategies) and only three efficient outcomes (Fig 2 b): two unbalanced ones (winner-takes-all): (20,0), (0,20)
and an equitable one: (10, 10).

In this scenario Pareto strategies ensure higher payoffs than NE (Fig. 2 b). We
may also notice that heterogeneity of players introduces unbalanced outcomes
(PN and NP solutions are on the axes). The Pareto player gets the non-zero
outcome and the Nash player gets the zero-outcome.

Joint equilibria (N-P and P-N) illustrate intermediate situations between NE
and Pareto solutions which in this case are unbalanced solutions (Fig. 2 b). One
NP and one PN solutions overlap one of the NE solutions: (4,4), indicating that
a stable and, moreover, equitable state exists even when players have different
biases.

4.2 Capacity Sharing Scenario No. 2: Equal-Rate Channels, Different Costs

This scenario considers 2 CRs simultaneously trying to access an available capacity W by accessing equal-rate channels but with different individual costs,
K_1 and K_2.

In the case of the Bertrand game with different costs, the individual CR payoff
is given by ($i = 1, 2$):

$$u_i(s_1, s_2) = \begin{cases} (W - s_i)(s_i - K_i), & \text{if } s_i < s_j; \\ \frac{1}{2}(W - s_i)(s_i - K_i), & \text{if } s_i = s_j; \\ 0, & \text{if } s_i > s_j. \end{cases} \tag{7}$$

For numerical simulations $W = 12$, $K_1 = 1$, and $K_2 = 3$ are considered.

The asymmetry introduced by the different costs is captured in Fig. 3. At NE
only CR_1 (the one with the lowest cost) gets non-zero payoffs: (10,0) and (18,0).

CR_2 (having a higher access cost than CR_1) manages to get non-zero payoffs only in Pareto-Nash situations (where CR_1 plays Pareto, is other-regarding, and CR_2 plays Nash, is self-regarding). We may also notice that the maximum payoff for CR_1 is higher than for CR_2 (Fig. 3 b).

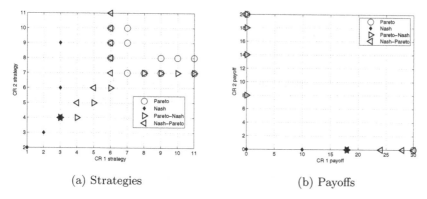

(a) Strategies   (b) Payoffs

Fig. 3. Scenario 2. Equal-rate, different cost channels - asymmetric, discrete Bertrand game reformulation - two CRs ($W = 12$, $K_1 = 1$, $K_2 = 3$). Evolutionary detected equilibrium (a) strategies: Nash (1;2), (2;3), (3;6), (3;9), (3;4), Pareto, Nash-Pareto, Pareto-Nash and (b) payoffs: Nash (0;0),(10;0),(18;0), Pareto ((30;0),(0;20)), Nash-Pareto, Pareto-Nash.

The NE outcome is a set of unbalanced solutions - 3 NE on the CR1 axis (Fig. 3 b): (0,0), (10,0), (18,0) and the Bertrand paradox is present again (NE payoff: (0,0)), but is not the only possible outcome.

The Pareto set is again relatively small - 12 strategies yielding only 2 Pareto efficient outcomes (30,0); (0,20). This is again a win-or-lose situation (one of the interacting CRs gets all).

In heterogeneous situations the Nash player gets all and the Pareto player gets zero (Fig. 3, b)).

4.3   Capacity Sharing Scenario No. 3: Different-Rate Channels, Additional Costs

A reformulation of the Bertrand model with differentiated products [26] captures a high-density capacity sharing situation where the CRs simultaneously access different-rate channels in their attempt to satisfy a certain capacity request. The demand function is:

$$D_i(s_1, s_2) = \begin{cases} 1, & \text{if } s_i \leq s_j - t; \\ \frac{s_j - s_i + t}{2t}, & \text{if } s_i \in (s_j - t, s_j + t); \\ 0, & \text{if } s_i \geq s_j + t. \end{cases} \tag{8}$$

where s is the capacity request of CR i and t is an additional cost accounting for the heterogeneity of CR technological means; it represents the differentiation degree between the channels.

Considering (8), the utility function for the CRs accessing different-rate channels is:

$$u_i(s_1, s_2) = (s_i - c)D_i(s_1, s_2), i = 1, 2. \qquad (9)$$

Simulation parameters are: $W = 12$ (available capacity, e.g. in Mbps) and $K = 3$ (cost of accessing the rate available on one channel), and $t = 9$ (additional cost).

Fig. 4 captures the equilibrium strategies and payoffs for the simultaneous capacity requests of two CRs, considering different-rate channels and additional costs of accessing those channels.

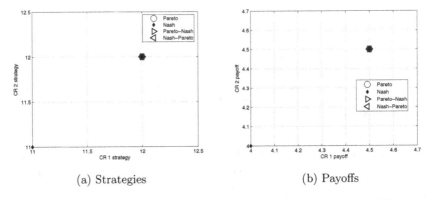

(a) Strategies   (b) Payoffs

Fig. 4. Scenario 3. Different-rate, additional-cost channels - discrete, differentiated Bertrand game reformulation - two CRs ($W = 12$, $K = 3$, $t = 9$). Evolutionary detected equilibrium (a) strategies: Nash (11;11), (12;12), Pareto (12;12), Nash-Pareto (12;12), Pareto-Nash (12;12) and (b) payoffs: Nash (4;4),(4.5;4.5), Pareto (4.5;4.5), Nash-Pareto (4.5;4.5), Pareto-Nash (4.5;4.5).

Two NE strategies are detected: (11,11) and (12,12), among them one is Pareto efficient: (12, 12). What is also interesting and may prove to be an advantage is that the set of Pareto efficient solutions includes only one solution: (12, 12) / (4.5, 4.5) and is not a large, hard-to-process set. Also, the joint strategies NP and PN (where the players are biased towards different equilibria) are Pareto efficient and overlap one NE. NE is the most stable game situation as it is maintained even for heterogeneous strategies (NP and PN overlap NE).

The most significant finding is that, in this scenario, one of the NE - (12, 12) - is both Pareto efficient and equitable and all equilibria in this scenario are equitable.

5 Conclusions

Bertrand game is reformulated in terms of radio resource access for small-cell, open capacity sharing in CR environments. CR interaction analysis reveals various equilibrium states among which some are efficient and equitable. These are the most advantageous states (the ones that GT analysis usually seeks).

Moreover, systematic exploration of discrete game models points to a new approach in using GT tools and concepts for CR interaction analysis. Evolutionary equilibrium detection techniques are the underlying enablers of this more realistic approach.

The Bertrand model proves valuable in estimating the chances in a crowded spectrum, high-density scenario, where the resources become scarce. The observations may be especially relevant for designing new rules of behaviour for heterogeneous cognitive radio environments. Three simultaneous capacity-request scenarios, in a scarce-resource context, are analysed.

An interesting particularity of the Bertrand game, in its different versions, is that equilibria are not robust to changes in parameters (like for Cournot game for instance [27], [28], [29]).

In the proposed approach we identify situations where NE is both fair and Pareto efficient or where there are sometimes multiple NEa to choose from. Adding to the flexibility are the joint Nash-Pareto solutions (intermediate between Nash an Pareto) capturing heterogeneous behaviour of players. Stable and, moreover, equitable states exist even when players have different biases.

Future work may consider variations of these general scenarios, for instance some players may adequate their strategies to the total available resources, some may not.

Acknowledgements. This work was supported by CNCS-UEFISCDI, Romania, project TE 252/2010. The authors contributed equally to this work.

References

1. Akyildiz, I.F., Lee, W.-Y., Vuran, M.C., Mohanty, S.: Next generation/dynamic spectrum access/cognitive radio wireless networks: a survey. Comput. Netw. 50(13), 2127–2159 (2006)
2. DiBenedetto, M.-G., DiBenedetto, M.D., Guerino, G., DeSantis, E.: Analysis of cognitive radio dynamics. In: Hossain, E., Bhargava, V. (eds.) Cognitive Wireless Communication Networks, pp. 425–438. Springer US (2007)
3. Neel, J., Buehrer, R.M., Reed, B.H., Gilles, R.P.: Game theoretic analysis of a network of cognitive radios. In: The 2002 45th Midwest Symposium on Circuits and Systems, MWSCAS 2002, vol. 3, pp. 409–412 (2002)
4. MItola, J.: Cognitive radio: An integrated agent architecture for software defined radio. PhD thesis, Royal Inst. Technol (KTH), Stockholm (2000)

5. Doyle, L.E.: Essentials of Cognitive Radio. Cambridge U.P (2009)
6. Nokia Siemens Networks White paper, Liquid radio. Let traffic waves flow most efficiently (2011)
7. Alcatel Lucent Technology White paper, Light Radio (2011)
8. da Costa, G.W.O., Cattoni, A.F., Kovacs, I.Z., Mogensen, P.E.: A fully distributed method for dynamic spectrum sharing in femtocells. In: 2012 IEEE Wireless Communications and Networking Conference Workshops (WCNCW), pp. 87–92 (2012)
9. da Costa, G.O.W., Cattoni, A.F., Kovacs, I.Z., Mogensen, P.E.: Scalable spectrum sharing mechanism for local area networks deployment. IEEE Trans. on Veh. Technol. 59(4), 1630–1645 (2010)
10. Sodagari, S., Bilen, S.G.: On cost-sharing mechanisms in cognitive radionetworks. Eur. Trans. Telecomm. 22, 515–521 (2011)
11. Neel, J.: Analysis and Design of Cognitive Radio Networks and Distributed Radio Resource Management Algorithms. PhD thesis, Faculty of the Virginia Polytechnic Institute and State University (2006)
12. Xiao, Y., Bi, G., Niyato, D., da Silva, L.A.: A hierarchical game theoretic framework for cognitive radio networks. IEEE J. Sel. Area. Comm. 30(10), 2053–2069 (2012)
13. MacKenzie, A.B., Wicker, S.B.: Game theory in communications: motivation, explanation, and application to power control. In: Global Telecommunications Conference, GLOBECOM 2001, vol. 2, pp. 821–826. IEEE (2001)
14. Maskery, M., Krishnamurthy, V., Qing, Z.: Game theoretic learning and pricing for dynamic spectrum access in cognitive radio. In: Hossain, E., Bhargava, V. (eds.) Cognitive Wireless Communication Networks, pp. 303–325. Springer US (2007)
15. Huang, J.W., Krishnamurthy, V.: Game theoretic issues in cognitive radio systems. J. of Comm. 4, 790–802 (2009)
16. Niyato, D., Ekram, H.: Microeconomic models for dynamic spectrum management in cognitive radio networks. In: Hossain, E., Bhargava, V. (eds.) Cognitive Wireless Communication Networks, pp. 391–423. Springer US (2007)
17. Wang, B., Yongle, W., Liu, K.J.R.: Game theory for cognitive radio networks: An overview. Comput. Netw. 54(14), 2537–2561 (2010)
18. Etkin, R., Parekh, A., Tse, D.: Spectrum sharing for unlicensed bands. IEEE J. Select. Areas Comm., 517–528 (2007)
19. Bertrand, J.: Book review of theorie mathematique de la richesse sociale and of recherches sur les principles mathematiques de la theorie des richesses. J. de Savants 67, 499–508 (1883)
20. Osborne, M.J.: An Introduction to Game Theory. Oxford University Press (2009)
21. Lung, R.I., Dumitrescu, D.: Computing Nash equilibria by means of evolutionary computation. Int. J. of Comput., Commun. & Control 3, 364–368 (2008)
22. Dumitrescu, D., Lung, R.I., Mihoc, T.D.: Generative relations for evolutionary equilibria detection. In: 11th Annual Conference on Genetic and Evolutionary Computation, pp. 1507–1512 (2009)
23. Fudenberg, D., Tirole, J.: Multiple Nash Equilibria, Focal Points, and Pareto Optimality, Game Theory. MIT Press (1983)
24. Nash, J.F.: Non-cooperative games. Annals of Mathematics 54, 286–295 (1951)
25. Dumitrescu, D., Lung, R.I., Mihoc, T.D.: Evolutionary Equilibria Detection in Non-cooperative Games. In: Giacobini, M., et al. (eds.) EvoWorkshops 2009. LNCS, vol. 5484, pp. 253–262. Springer, Heidelberg (2009)

26. Hotelling, H.: Stability in competition. The Economic Journal 39, 41–57 (1929)
27. Cremene, L.C., Dumitrescu, D.: Analysis of cognitive radio scenes based on non-cooperative game theoretical modelling. IET Communications 6(13), 1876–1883 (2012)
28. Cremene, L.C., Dumitrescu, D., Nagy, R., Cremene, M.: Game theoretical modelling for dynamic spectrum access in TV whitespace. In: 6th International ICST Conference on Cognitive Radio Oriented Wireless Networks and Communications, CROWNCOM, pp. 336–340 (2011)
29. Cremene, L.C., Dumitrescu, D., Nagy, R., Gaskó, N.: Cognitive radio simultaneous spectrum access/ one-shot game modelling. In: IEEE, IET International Symposium on Communication Systems, Networks and Digital Signal Processing, CSNDSP 2012, pp. 1–6 (2012)

Optimization of a Decentralized Medium Access Control Scheme for Single Radio Cognitive Networks

Miguel Luís[1,2], Rodolfo Oliveira[1,2], Rui Dinis[1,2], and Luis Bernardo[1]

[1] CTS, Uninova, Depto. de Eng. Electrotécnica, Faculdade de Ciências e Tecnologia, FCT, Universidade Nova de Lisboa, 2829-516 Caparica, Portugal
nmal@campus.fct.unl.pt
[2] Instituto de Telecomunicações, Lisboa, Portugal

Abstract. The medium access control (MAC) of decentralized cognitive radio networks has been a topic of interest in the last years due to the lack of a central coordinator and the necessity of self-organizing procedures that effectively lead the nodes to act autonomously but efficiently. This work addresses a scenario where multiple non-licensed cognitive radios communicate with an access point when licensed users do not use the spectrum. We consider a MAC protocol for the non-licensed users which uses a double stage to schedule each node's transmission. Non-licensed users perform spectrum sensing in a synchronous way and the proposed MAC is opportunistically employed when the channel is sensed idle. In the first stage the number of competing non-licensed nodes is decreased to reduce the number of collisions. In the second stage we adopt a reservation procedure to schedule the non-licensed users competing for the medium. Adopting a traditional energy-based sensing, we characterize the performance of the considered protocol by capturing the influence of the sensing in the MAC's performance. Finally we present several results to evaluate the performance of the proposed scheme achieved in optimal conditions, which are compared to a cognitive "slotted-aloha"-like protocol. The obtained results demonstrate the effectiveness of the proposed solution.

Keywords: Cognitive Radio, Distributed Medium Access Control, Performance Analysis.

1 Introduction

Cognitive radio networks (CRNs) are an effective solution to alleviate the increasing demand for radio spectrum [1]. In these networks non-licensed users, usually denominated secondary users (SUs), must be aware of the activity of the licensed users, denominated primary users (PUs), in order to dynamically access the spectrum without causing them harmful interference. In decentralized cognitive radio networks (dCRNs) the SUs are not managed by a central coordinator, meaning that SUs must adopt distributed policies able to manage the network in an efficient way.

S. Balandin et al. (Eds.): NEW2AN/ruSMART 2013, LNCS 8121, pp. 260–271, 2013.

This work considers a single-radio dCRN, indicating that SUs are only equipped with a single transceiver and can not sense the spectrum and simultaneously transmit the waiting packets. The work addresses a scenario where multiple SUs wish to communicate with an access point when PUs do not use the spectrum. We assume a cognitive radio system where SUs synchronously and periodically sense the channel in order to determine the level of PU's activity. SUs sense the channel activity using an energy-based sensing scheme, but the proposed scheme can be replaced by other spectrum sensing techniques, such as the ones described in [2].

The operation of the SUs is organized in time frames. Each time frame is organized in two periods. The spectrum sensing is performed in the first period in order to evaluate if a SU is allowed to access the channel in the second period. Each SU employs a double stage random MAC to access the channel. The first stage of the MAC protocol aims at decreasing the number of collisions between SUs by reducing the number of competing SUs. The second stage of the protocol starts with a reservation phase where SUs schedule their transmissions and is followed by a transmission phase where SUs effectively access the channel. Since all SUs must be synchronized in order to start the sensing period simultaneously, we consider that the access point can transmit a tone in a narrow band, which is used by SUs to resynchronize and identify the first and the second MAC stages.

We start to detail the considered MAC protocol and the analytical steps that characterize the aggregate throughput achieved by the SUs. By modeling the influence of the sensing duration and its accuracy in MAC's performance, we compute the optimal protocol's parameterization that maximizes SU's throughput constrained to the sensing duration. Finally we present several results to evaluate the performance of the proposed scheme, which are compared to a cognitive "slotted-aloha"-like protocol. The obtained results demonstrate the effectiveness of the proposed solution, making it a good candidate for future decentralized single radio cognitive networks.

The rest of this paper is organized as follows. In the next section we review a few works related to ours. Section 3 describes the adopted MAC scheme and presents a simple model to characterize its throughput. Performance results are discussed in Section 4. Finally, some concluding remarks are given in Section 5.

2   Related Work

The design of efficient MAC protocols for dCRNs is a challenging task due to necessity of efficient self-organizing procedures that effectively guide the SUs to act autonomously. Recently, several MAC schemes have been proposed for dCRNs. [1] describes a new MAC protocol which considers a time frame organization where each node starts to sense the radio spectrum before trying to transmit data. If a node does not detect PUs' activity during the sensing period, it applies a random backoff during the transmission period and the channel is granted to the first competing node accessing the channel. A collision can occur if the first access during the transmission period is attempted by multiple nodes.

[3] proposes a carrier sense multiple access/collision avoidance (CSMA/CA) protocol that exploits statistics of channel usage by PUs for decision making on SUs' channel access. Basically, the protocol uses a common control channel to negotiate the transmission parameters with the receiver. A successful transmission rate is then defined from the exchanged information, which is used to adapt the level of contention used by SUs to access the medium. The work in [4] proposes a MAC for CSMA-based PU systems. A CSMA scheme is also used to regulate the access of SUs and the main feature of this protocol is the dynamic adaptation of SUs' physical layer parameters (e.g. transmission power) to allow simultaneous SUs and PUs transmission when interference to and from the PUs are acceptable. More recently, [5] proposed a two-level opportunistic spectrum access strategy to optimize the system performance of the secondary network. At the first level, SUs maintain a sufficient detection probability to avoid interference with PUs. At the second level, two contention-based MAC protocols called slotted cognitive radio Aloha and cognitive-radio-based carrier-sensing multiple access were proposed to deal with the packet scheduling of the secondary network. The next section introduces the principle of operation of the scheme proposed in this work.

3   System Description

In this work we consider that n SUs may transmit data to an access point in an opportunistic way, when the channel is not used by PUs. SUs are equipped with a single radio transceiver. However, because SUs are unable to distinguish SUs and PUs transmissions, each SU will have to divide its operation cycle (time frame structure) into spectrum sensing and spectrum access periods, with durations T_S^{SU} and T_D^{SU} respectively, as illustrated in Figure 1. The time frame lasts for $T_F^{SU} = T_S^{SU} + T_D^{SU}$ and is divided into N_T slots, where each slot duration is given by the channel sampling period [6]. The first N_S slots are allocated to the spectrum sensing (channel sampling) and the remaining ones ($N_S + 1$ to N_T) are used to access the channel. In this paper we consider that SUs always have a packet to transmit. We assume that SUs are synchronized in the first slot of the sensing period and the access point transmits a tone to guarantee resynchronization.

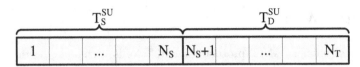

Fig. 1. SU's time frame structure

3.1   Spectrum Sensing Period - T_S^{SU}

SUs sense the channel during the period T_S^{SU} using an energy-based sensing (EBS) technique [7]. To distinguish between occupied and vacant spectrum bands, SUs sample the channel during the sensing period, and for each sample k two hypotheses can be distinguished

$$
\begin{aligned}
\mathcal{H}_0 &: x(k) = w(k) & k &= 1, 2, ..., N_S \\
\mathcal{H}_1 &: x(k) = w(k) + s(k) & k &= 1, 2, ..., N_S,
\end{aligned}
\tag{1}
$$

where $s(k)$ denotes the transmitted signals by PUs, modeled as a Gaussian variable with mean μ_s and variance σ_s^2, i.e., $s(k) = \mathcal{N}(\mu_s, \sigma_s^2)$. $w(k)$ is assumed to be a zero-mean additive white Gaussian noise (AWGN) with unit variance, i.e., $w(k) = \mathcal{N}(0, 1)$. The condition $\mathcal{H}_0$ represents the case when PUs are absent. $\mathcal{H}_1$ indicates that there exists a signal transmitted by a PU.

EBS relies on the classical energy detector [7]. In the detection stage, each SU calculates the amount of energy received in N_S samples, given by

$$
Y = \sum_{i=1}^{N_S} |x(i)|^2,
\tag{2}
$$

and compares it with the energy threshold γ to decide whether a PU is present or absent. Under the hypothesis $\mathcal{H}_0$, the decision variable Y follows a central chi-square distribution with $2N_S$ degrees of freedom. Under the hypothesis $\mathcal{H}_1$, Y follows a non-central chi-square distribution also with $2N_S$ degrees of freedom, and a non centrality parameter λ, denoting the signal-to-noise ratio (SNR) [8]. However, if the number of samples N_S is large enough[1], it is possible to use the Central Limit Theorem (CLT) to approximate the chi-square distribution to a Gaussian distribution [9]:

$$
Y \sim \begin{cases} \mathcal{N}(N_S, 2N_S), & \mathcal{H}_0 \\ \mathcal{N}(N_S + \lambda, 2(N_S + 2\lambda)), & \mathcal{H}_1. \end{cases}
\tag{3}
$$

Therefore, for a single SU the probability of detection (P_D) and probability of false alarm (P_{FA}) are represented by

$$
P_D = Pr(y > \gamma | \mathcal{H}_1) = Q\left(\frac{\gamma - (N_S + \lambda)}{\sqrt{2(N_S + 2\lambda)}} \right),
\tag{4}
$$

$$
P_{FA} = Pr(y > \gamma | \mathcal{H}_0) = Q\left(\frac{\gamma - N_S}{\sqrt{2N_S}} \right),
\tag{5}
$$

where $Q(.)$ is the complementary distribution function of the standard normal distribution.

3.2   Medium Access Control

Figure 2 introduces the MAC protocol. SUs may declare an idle or busy transmitting period with duration $T_D^{SU} = T_F^{SU} - T_S^{SU}$, if the EBS applied during the

---

[1] The sampling rate must satisfy the Nyquist sampling theorem.

N_S slots indicates absence or presence of the PU. We adopt the terminology idle frames to denote the frames detected idle by the EBS system and busy frames to denote the opposite. SUs start to decide their medium access from the moment when the present frame is declared idle or busy, which occurs after the N_S-th channel sampling slot.

If a frame is declared busy by the EBS, as is the case for the second and fourth frames depicted in Figure 2, the nodes do not perform any operation until the sensing decision of the next frame.

If a frame is declared idle, as is the case of the first and third frames shown in Figure 2, then a SU may access the medium during the time interval equivalent to $T_F^{SU} - T_S^{SU}$, depending on a random decision. This means that SUs are only granted to randomly access the medium after sensing an idle frame, which occurs with probability

$$P_{idle} = P_{OFF}^{PU}(1 - P_{FA}) + P_{ON}^{PU}(1 - P_D). \tag{6}$$

P_{idle} accounts with the sensing accuracy by including the EBS' probabilities of misdetection $(1 - P_D)$ and correct rejection $(1 - P_{FA})$.

SUs may access the medium in the beginning of the slot $N_S + \varphi$ of each idle frame. φ represents the duration of the synchronization tone sent by the access point in each idle frame. To avoid collisions between SUs, their medium access decision is randomized. One solution to randomize SUs' medium access is to use the traditional "slotted-aloha" protocol, where SUs randomly choose an idle frame to transmit with probability p_{aloha}. This behavior is used in this work for comparison purposes.

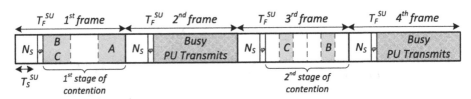

Fig. 2. MAC protocol - reservation's schedulling phase (e.g. $cw_1 = 3$ and $cw_2 = 6$)

It is well known that slotted-aloha achieves low throughput. Motivated by this fact we consider a different scheme, where the nodes adopt two stages of contention. The first stage of the scheme is applied during the first frame detected idle (idle frame). Once again this occurs with probability P_{idle} described in Eq. (6). The transmission period $T_D^{SU} - \varphi$ of this frame is divided into cw_1 mini-slots[2]. The nodes can randomly transmit at most one mini-packet in one of the mini-slots with probability $\tau_1 = 1/cw_1$, which serves to announce its intention to access the medium. The first stage of this scheme finishes at the end of the first timing frame detected idle. When this occurs, we have defined a method to

---

[2] Note that we use the term mini-slots for MAC purposes and the term slot, adopted in Section 2, is only used for channel sensing purposes.

identify the SUs that will compete in the second stage. We have adopted a simple criterion: if a node listens an idle mini-slot before transmitting its mini-packet, it knows that it should compete in the second stage. In Figure 2, SUs B and C transmit a mini-packet in the first mini-slot while SU A transmits the mini-packet in the third mini-slot. This means that only SUs B and C will compete in the second contention stage. Assuming that we have n SUs competing for the medium in the first stage, then the expected number of nodes selected to compete in the second contention stage is given by

$$n_2 = max\left(1, \sum_{k=1}^{n} k\binom{n}{k}\tau_1^k(1-\tau_1)^{(n-k)}\right). \tag{7}$$

Nodes selected in the first stage do not start accessing the medium in the first idle frame of the second stage. Instead, the transmission period $T_D^{SU} - \varphi$ of the first idle frame in the second stage is divided into cw_2 mini-slots, which are used to reserve future idle frames for channel access. This is called the reservation phase of the second stage. Admitting that we have n_2 nodes competing for the medium in the second stage, they can access the medium by selecting a single frame in the interval $\{1, 2, ..., cw_2\}$ of future idle frames, meaning that a frame is chosen with probability $\tau_2 = 1/cw_2$. Some of the cw_2 idle frames may not be selected for transmission, and consequently they represent a situation of underutilization of idle frames from SUs. This fact motivates a reservation scheme introduced in the second contention scheme.

To exemplify the reservation scheme we consider the example depicted in Figure 2. During the first stage 2 SUs B and C, are selected for the second stage. During the second stage the SU C transmits a mini-packet in the second mini-slot and the SU B transmits its mini-packet in the fifth mini-slot. After elapsing cw_2 mini-slots, the SUs know how many mini-slots were occupied by one or more mini-packets, because they sense the channel activity at each mini-slot. The information about the mini-slots found busy is then used to reserve the future idle frames to the nodes that have manifested their intention on accessing the channel during the cw_2 mini-slots. In the previous example, all SUs have sensed transmissions in the second and in the fifth mini-slots. From this information the SUs reserve an equal number of idle frames (two) in the next transmission phase, and thus the second stage lasts for a number of idle frames equal to the number of busy mini-slots observed in the reservation stage plus the idle frame where the cw_2 mini-slots were defined. Figure 3 illustrates an hypothetic scenario for the transmission phase after the reservation phase occurred in the second stage in Figure 2, where SUs C and B transmit.

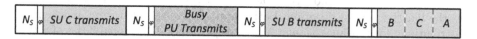

Fig. 3. MAC protocol - transmission's phase (e.g. $cw_1 = 3$ and $cw_2 = 6$)

To determine a bound for the maximum achievable throughput of this scheme, we start to define the expected number of idle mini-slots in the second stage. Since n_2 nodes compete in the second stage (computed from (7)), the expected number of mini-slots found idle in the second stage is given by

$$\Gamma_{idle} = cw_2(1 - \tau_2)^{n_2}, \tag{8}$$

where $(1 - \tau_2)^{n_2}$ is the probability of finding an idle mini-slot and $\tau_2 = 1/cw_2$ is the probability of a SU transmitting the mini-packet in a given mini-slot. The expected number of idle frames reserved for SUs access (transmission or collision) during the cw_2 mini-slots is given by

$$\chi_{res} = cw_2 - \Gamma_{idle}, \tag{9}$$

which shows that at most cw_2 future idle frames are reserved for the transmission phase when none of the cw_2 mini-slots is sensed idle ($\Gamma_{idle} = 0$). Note that the first and the second contention stages last for $2 + \chi_{res}$ idle frames, since 2 idle frames are used to implement the cw_1 and cw_2 mini-slots and the remaining ones are reserved for SUs transmissions. Given that the maximum number of reserved frames is cw_2, we define the performance weight $\alpha_C = cw_2/(2 + \chi_{res})$, which expresses the relative gain ($\alpha_C > 1$) due to the suppression of unused idle frames in the second stage and also accounts with the two idle frames used for implementing cw_1 and cw_2 mini-slots. The throughput of this scheme can be approximated by

$$S_{DS} = n_2\tau_2(1 - \tau_2)^{n_2-1}P_{OFF}^{PU}(1 - P_{FA})\alpha_s\alpha_C, \tag{10}$$

which is a bound for the maximum achievable throughput for the case when all SUs have an homogeneous detection of the spectrum sensing occupancy. $\alpha_s = (T_D^{SU} - \varphi)/(T_S^{SU} + T_D^{SU})$ represents the loss of throughput due to the period when SUs sense the channel and receive the synchronization tone. Note that the synchronization tone sent by the access point also indicates what is the idle frame where SUs should run the first contention stage and the reservation phase of the second stage. Nodes that have detected busy channel during the sensing period followed by the reception of a tone indicating the beginning of the first or the second contention stages postpone their transmission because they are not coherent with the spectrum sensing performed by the access point. The SUs postpone their transmission until receiving a new tone indicating the first contention stage after having detected an idle frame.

3.3   PU Activity Model

Each PU has an ON-OFF behavior, meaning that it is active during T_{ON}^{PU}, and inactive during the T_{OFF}^{PU} interval. The PU's activity is modeled by two random processes. The first one models the active period duration and the second models the inactive period duration. Both durations are sampled from exponential distributions with mean λ_{ON} and λ_{OFF}, respectively. The probabilities of a PU staying OFF and ON are respectively given by $P_{OFF}^{PU} = \lambda_{OFF}/(\lambda_{OFF} + \lambda_{ON})$ and $P_{ON}^{PU} = \lambda_{ON}/(\lambda_{OFF} + \lambda_{ON})$.

4   Protocol's Optimization and Performance Analysis

This section evaluates the performance of the previously described decentralized MAC scheme. We have considered a scenario formed by one PU transmitter-receiver pair and multiple SUs transmitting to a single SU Access Point. The operation mode of SUs and PUs is as described in Sections 3.2 and 3.3. The mean duration of PU's active period (λ_{ON}) was set to 140ms, while the mean duration of the inactive period (λ_{OFF}) was set to 327ms in order to obtain a probability of a PU staying ON of approximately 30% (P_{ON}^{PU}=0.3).

The adopted parameters for the PU's transmitting signal and for the energy detector implemented in the SUs are described in Table 1. The energy detector threshold (γ) was defined to 38.3 Joules following the parametrization criterion C_4 presented in [10] and considering that PUs' SNR is equal to 5dB.

Table 1. Parameters used in the energy detection

Sensed band	10 kHz	Channel Sampling Period	50 μs
$T_S^{SU} + T_D^{SU}$	20.0 ms	N_S^{min}	20
μ_s	3.16 (5dB)	σ_s^2	3
μ_w	1 (0dB)	σ_w^2	1
λ (SNR)	5 dB	γ	38.3 J

4.1   Slotted Aloha Benchmark Protocol

In order to have a benchmark for the performance of our proposal, we have implemented a "slotted-aloha"-like medium access protocol which adopts the same basis regarding the channel sensing as the previous proposal: the SUs declare an idle or busy transmitting period with duration $T_F^{SU} - T_S^{SU}$, if the EBS applied during the N_S slots indicates absence or presence of the PU. If a frame is declared idle, then a SU may access the medium during the time interval equivalent to $N_T - N_S$, depending on a randomly decision. This means that SUs are only granted to randomly access the medium after sensing an idle frame, which occurs with probability P_{idle} given by (6). To avoid collisions between SUs, their medium access is randomized: when a SU has a new packet to transmit, it randomly chooses an idle frame to transmit. This frame is chosen from the interval 1, 2, ..., cw of the next cw idle frames, which means that a frame is chosen with probability $\tau = 1/cw$. A SU only has a chance to successfully transmit in the chosen frame if the frame is declared idle due to a correct rejection, $P_{OFF}^{PU}(1 - P_{FA})$, and if the SU is the unique node granted to access the medium on that frame, $\tau(1 - \tau)^{(n-1)}$. Consequently, the aggregate throughput achieved by n SUs is given by

$$S_{SA} = n\tau(1 - \tau)^{(n-1)} P_{OFF}^{PU}(1 - P_{FA})\alpha_s. \tag{11}$$

4.2   Protocol's Optimization

Our MAC proposal is optimized to achieve the maximum aggregated throughput in the secondary network guaranteing a given level of protection to PUs. For that, both MAC schemes were optimized for different channel conditions (SNR) and different number of SUs, by finding the optimal number of sensing samples N_S and mini-slots (cw, cw_1 and cw_2). This optimization process introduces the concept of cross-layer between the PHY and the MAC layers because the optimal number of samples N_S is highly dependent on the performance of the spectrum sensing task, while the optimal number of mini-slots is related with the number of SUs.

In both schemes, the EBS was parameterized according to the Table 1. In order to guarantee that all the SUs have the same probability of false alarm (Eq. (5)), we fixed the energy detector threshold (γ) and maximized the throughput achieved by the "slotted-aloha"-like MAC protocol as follows

$$\max_{cw,N_S} \quad S_{SA},$$

$$\text{s.t.} \quad P_D \geq P_D^{min}$$

$$N_S^{min} \leq N_S \leq N_T$$

$$1 \leq cw \leq cw^{max}.$$

Basically the optimal throughput is computed taking into account several constraints: $P_D \geq P_D^{min}$ guarantees a minimum level of protection to PUs; $N_S \geq N_S^{min}$ assures the minimum number of sensing samples needed to apply the approximation of Eq. (3); and the slotted-aloha medium access probability ($1/cw$) is bounded. The optimization process identifies the medium access probability and the sensing duration that leads to the maximum throughput.

Similarly, our proposal was optimized as follows

$$\max_{cw_1,cw_2,N_S} \quad S_{DS},$$

$$\text{s.t.} \quad P_D \geq P_D^{min}$$

$$N_S^{min} \leq N_S \leq N_T$$

$$1 \leq cw_1 \leq cw_1^{max}$$

$$1 \leq cw_2 \leq cw_2^{max}.$$

In order to protect the PUs, P_D^{min} was set to 95%. cw_1 and cw_2 mini-slots were bounded to cw_1^{max} and cw_2^{max} to assure that a mini-packet can be successfully sent within the duration of cw_1 and cw_2. N_T represents the total number of slots in a SU's frame.

4.3   Performance Evaluation

We have implemented both schemes in a simulator. Figure 4 illustrates the throughput obtained for different values of SNR, where 25 SUs are competing for the medium. Due to lack of space it is not possible to reproduce the results for different number of SUs, but a similar behavior was obtained. 10^6 SUs timing frames were simulated in both schemes. The confidence interval of the simulations at 95% of confidence level was computed but, because it is too small, we decided to not plot it in the figure. The numerical results from theoretical approximations, which correspond to the optimization of S_{DS} (Eq. (10)) and S_{SA} (Eq. (11)) for each value of SNR, are also plotted.

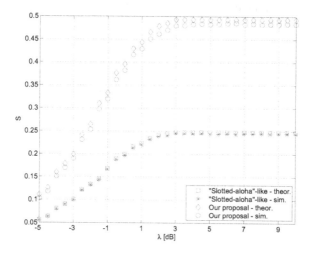

Fig. 4. Aggregated normalized throughput for $P_{ON}^{PU} = 0.3$ and $n = 25$

Observing the simulation results, which were obtained using the optimal values of cw_1, cw_2, cw and N_S for each SNR, we conclude that our proposal achieves a better throughput when compared with the "slotted-aloha"-like scheme, mainly for higher SNR values. For small values of SNR, the EBS (which has always a fixed threshold - γ) has more difficulties to distinguish between PUs and noise. For lower SNR values, the optimization process increases the number of samples N_S to increase the probability of detection, thereby decreasing the aggregated throughput. Even so, for an SNR of -5dB, which is a channel condition relatively far from the one used to parameterize the EBS, our proposal can achieve almost 0.15 of aggregated throughput outperforming the "slotted-aloha"-like protocol.

Regarding the "slotted-aloha"-like scheme it is well known that these schemes can only achieve $1/e \approx 0.38$ of throughput when $cw = 1/n$, which was the case. However, this throughput level is for a traditional MAC protocol where a node can access the spectrum in each time frame. In our case the SUs do not access in busy frames, which decreases the maximum aggregated throughput to 0.25 (for the adopted parametrization). In the proposed scheme, the double contention scheme improves the aggregated throughput mainly because it reduces the number of idle frames without SUs' transmissions, and the number of collisions between SUs is also decreased. In fact, our MAC scheme introduces properties normally presented in cross-layer schemes because the optimal number of sensing samples is mainly dependent on the performance of the spectrum sensing task while the optimal number of mini-slots is directly related with the number of number of SUs competing for the spectrum.

The theoretical approximations presented for the throughput follow the trend observed in the simulations. For the "slotted-aloha"-like scheme the theoretical approximation is quite accurate. The theoretical approximation for the throughput of our proposal can be considered a bound for the best-case scenario because we do not consider the impact of heterogenous misdetection that can occur during the cw_1 and/or cw_2 mini-slots.

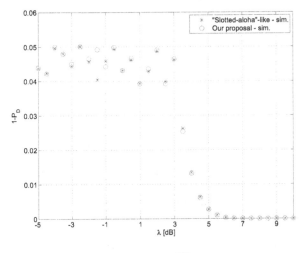

Fig. 5. Misdetections for $P_{ON}^{PU} = 0.3$ and $n = 25$

Figure 5 illustrates the number of misdetections $(1 - P_D)$ obtained in the simulations. For small values of SNR, the misdetection is close to 0.05, which corresponds to the maximum allowed for the optimization problem ($P_D^{min} = 0.95$). However, for higher values of SNR, mainly for $SNR > 5dB$, the number of misdetections become negligible due to the constraint regarding the minimum number of sensing samples (N_S^{min}), and also due to the fixed energy threshold used in the EBS.

5   Conclusions

This work proposed a decentralized MAC protocol for a Single Radio Cognitive Network, considering a scenario where multiple non-licensed cognitive radios communicate with an access point when licensed users do not use the spectrum. The proposed protocol uses a double stage to schedule each SU's transmission. In the first stage the number of competing SUs is decreased to reduce the number of collisions. In the second stage we adopt a reservation procedure to schedule the SUs competing for the medium. Adopting a traditional energy-based sensing, we characterize the performance of the considered protocol by capturing the influence of the sensing in the MAC's performance. Finally a protocol optimization was proposed to select the best contention probabilities applied in the first and in the second stages, as well as the duration of the sensing period. We have compared the performance of the proposed protocol with a simple "slotted-aloha"-like protocol and the obtained results demonstrate the effectiveness of the proposed solution.

Acknowledgments. This work was partially supported by COST Action IC0902 "Cognitive Radio and Networking for Cooperative Coexistence of Heterogeneous Wireless Networks", funded by the European Science Foundation, and by the FCT/MEC (projects PEst-OE/EEI/UI0066/2011, PEst-OE/EEI/LA0008/2013, PTDC/EEA-TEL/099074/2008, PTDC/EEA-TEL/115981/2009, PTDC/EEA-TEL/099973/2008 and PTDC/EEA-TEL/120666/2010, as well as grant SFRH/ BD/68367/2010).

References

1. Zhao, Q., Tong, L., Swami, A., Chen, Y.: Decentralized cognitive mac for opportunistic spectrum access in ad hoc networks: A pomdp framework. IEEE Journal on Selected Areas in Communications 25(3), 589–600 (2007)
2. Yucek, T., Arslan, H.: A survey of spectrum sensing algorithms for cognitive radio applications. IEEE Communications Surveys Tutorials 11, 116–130 (2009)
3. Hsu, A.-C., Wei, D.S.L., Kuo, C.-C.: A cognitive mac protocol using statistical channel allocation for wireless ad-hoc networks. In: Wireless Communications and Networking Conference, WCNC 2007, pp. 105–110. IEEE (2007)
4. Lien, S.-Y., Tseng, C.-C., Chen, K.-C.: Carrier sensing based multiple access protocols for cognitive radio networks. In: IEEE International Conference on Communications, ICC 2008, pp. 3208–3214 (2008)
5. Chen, Q., Liang, Y.-C., Motani, M., Wong, W.-C.: A two-level mac protocol strategy for opportunistic spectrum access in cognitive radio networks. IEEE Transactions on Vehicular Technology 60(5), 2164–2180 (2011)
6. Luis, M., Furtado, A., Oliveira, R., Dinis, R., Bernardo, L.: Towards a realistic primary users' behavior in single transceiver cognitive networks. IEEE Communications Letters 17(2), 309–312 (2013)
7. Urkowitz, H.: Energy Detection of Unknown Deterministic Signals. Proceedings of the IEEE 55, 523–531 (1967)
8. Digham, F., Alouini, M.-S., Simon, M.: On the energy detection of unknown signals over fading channels. In: Proc. IEEE ICC 2003, pp. 3575–3579 (May 2003)
9. Tang, H.: Some physical layer issues of wide-band cognitive radio systems. In: Proc. IEEE DySPAN 2005, pp. 151–159 (November 2005)
10. Luis, M., Furtado, A., Oliveira, R., Dinis, R., Bernardo, L.: Energy sensing parameterization criteria for cognitive radios. In: 2012 International Symposium on Wireless Communication Systems (ISWCS), pp. 61–65 (2012)

Coalitional Games with Incomplete Information among Secondary Users in Cognitive Radio Networks

Jerzy Martyna

Jagiellonian University, Faculty of Mathematics and Computer Science
Institute of Computer Science, ul. Prof. S. Lojasiewicza 6, 30-348 Cracow, Poland

Abstract. In this paper, we propose a model for coalition formation among Secondary Users (SUs) with incomplete information in Cognitive Radio (CR) networks based on the Bayesian equilibrium. This model allows us to study coalition formation among SUs with respect to the stations' information endowments. By using the proposed method, SUs can self-organize into disjoint independent coalitions. We are the able identify the cost of incomplete information on the Bayesian equilibrium. As a result, we can propose an algorithm for coalition formation among SBSs with incomplete information based on the Bayesian equilibrium. To evaluate our approach, we developed a realistic model of cognitive radio networks, and used them to make simulation experiments. The results demonstrate the practicality of our algorithm.

Keywords: cognitive radio networks, game theory, Bayesian equilibrium.

1 Introduction

Recent advances in technology have led to the development of distributed and self-configuring wireless network architectures. This is seen especially in the case of usage of the radio spectrum, which refers to the frequency segments that have been licensed to a particular primary service, but are completely unused or partly utilized at a given location or a given time. The Federal Communications Commission (FCC) reported [5] vast temporal and geographic variations in the usage of the allocated spectrum with use ranging from 15% to 85% in the bands below 3 GHz that are favoured in non-line-of-sight radio propagation. On the other hand, a large portion of the assigned spectrum is used sporadically, leading to an under utilisation of a significant amount of spectrum. Some studies have shown up to 90% of the radio spectrum remains idle in any one geographical location.

Cognitive radio (CR) [1, 9, 15] have been extensively researched in recent years as a promising technology to improve spectrum utilization. Cognitive radio networks and spectrum-sensing techniques are natural way to allow these new technologies to be deployed. These spectrum-sensing techniques and the ability to switch between radio access technologies are the fundamental requirements for transmitters to adapt to varying radio channel qualities, network congestion, interference, and quality of service (QoS) requirements.

The CR networks have three different access types as follows:

S. Balandin et al. (Eds.): NEW2AN/ruSMART 2013, LNCS 8121, pp. 272–283, 2013.

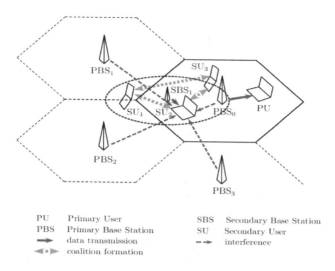

PU	Primary User	SBS	Secondary Base Station
PBS	Primary Base Station	SU	Secondary User
→	data transmission	⇢	interference
◄•►	coalition formation		

Fig. 1. Downlink/uplink CR network

- CR users can access their own base station on licensed and unlicensed bands.
- CR users can communicate with each other through ad-hoc communication on licensed and unlicensed bands.
- CR users can also access the primary base station through a licensed band.

The CR users can operate in both licensed and unlicensed bands. Therefore, we may categorize the CR applications of spectrum into three possible scenarios: (i) CR network on a licensed band, (ii) CR network on a unlicensed band, (iii) CR network on a licensed band and unlicensed band. We use Fig. 1 to illustrate the third scenario. Thus, from the users' perspective, the CR network coexists with the primary system in the same geographic location. A primary system operated in the licensed band has the highest priority to use that frequency band (e.g. 2G/3G cellular, digital TV broadcasts, etc.). Other unlicensed users and/or systems can neither interfere with the primary system in an obtrusive way nor occupy the licensed band. By using the pricing scheme, each of the primary service providers (operators) maximises its profit under the QoS constraint for primary users (PUs). All of the unlicensed secondary users (SUs) are equipped with cognitive radio technologies, usually static or mobile. The PUs are responsible for throwing unused frequencies to the SUs for a fee. While the existing literature has focused on the communications needed for CR system control, this paper assumes a network of SUs. Every SU can only have information on a small number of secondary base station (SBSs) or channels. It causes interference to SBSs and SUs.

The key to the next revolution in communications delivery is the application of artificial intelligence (i.e. cognition) to the communications devices. This will enable intelligent local decisions to made on routing, dynamic spectrum access and resources usage, etc. Such decisions can take into account mixed systems and applications, and even devices that break the rules. The study of artificial intelligence, learning, and reasoning has been around for a number of years, but it is only now that concepts such as reinforcement-based learning, game theory and neural networks are being actively applied to CR networks. Among the game-theoretic approaches to addressing resource management in these systems, the paper [12] proposed the use of Shapley value to power allocation games in CR networks.

In the literature, a coalition formation is broadly studied for cognitive radio networks. In the paper by Ghasemi *et al.* [7] amongst others, it has been shown that through collaboration among SUs, the effects of the hidden terminal problem can be reduced and the probability of detecting the PU can be improved. A collaborative detection of TV transmission in support of dynamic spectrum sensing was presented by Visotsky *et al.* [20]. The authors of the paper [21] have proposed spatial diversity techniques for improving the performance of collaborative spectrum sensing by combating the error probability due to fading on the reporting channel between the SUs and the central fusion center. Recently, in collaborative spectrum sensing among SUs in CR networks, it was show in [17] that the performance of these networks is a significant improvement. A distributed coalition formation algorithm was presented by Saad *et al.* [18] that allows the SUs to self-organize into disjointed coalitions while accounting for the tradeoff of collaborative spectrum sensing. A survey of cooperative spectrum sensing was presented by Akyildiz *et al.* [2]. Nevertheless, none of the above mentioned papers provided coalition formation among the SUs with incomplete information.

The motivation of this paper is exactly to study how game theory [6] can be implemented in the situations where CR devices have limited information. Recently, games of complete information have been used in various types of communication networks, multiple input and multiple output channels [11], interference channels [3], and combination of them. Games with complete information have been studied in the distributed collaborative spectrum sensing [17], for the solve of a dynamic spectrum sharing [16], interference minimalization in the CR networks [14], etc.

In this paper, we introduce the means of coalition formation among SUs with incomplete information in CR networks. A Bayesian equilibrium allowing us to formulate the coalition among the SUs with incomplete information is proposed. Moreover, an algorithm for building a coalition is presented. It is also shown that each SUS decides to enter or leave the coalition based on maximizing its utility function. The presented algorithm can also dynamically produce reconfigured coalitions of SUs, maintaining structural integrity. Finally, the presented system model is verified through simulation and compared with other solutions of coalition forms of games for wireless networks.

The rest of this paper is as follows. In section 2, we present the system model. Section 3 provides the proposed coalition game among the SUs with incomplete information. In section 4 we propose a distributed algorithm for coalition formation. Some simulation results are provided in section 5. Finally, conclusions are given section 6.

2   System Model

In this section, we present the model of the CRN system consisting of SUs and PUs.

Let $\Omega = \{1, \ldots, N\}$ be the transmit-receive pairs of SUs and a single PU in the ad hoc network. Each of the N SUs can continuously sense the spectrum to detect the presence of the PU. To detecting the presence of the PU each of the N SUs can use the energy detectors used in the CR networks [7, 20]. It is important that these non-cooperative SUs in the Rayleigh fading environment can detect the PU and the false alarm probability of an SU i, respectively, given by [4, 7]:

$$
\begin{aligned}
P_{det,i}^k = e^{-\frac{\lambda_{i,k}}{2}} \sum_{n=0}^{m-2} \frac{1}{n!} \left(\frac{\lambda_{i,k}}{2}\right)^n + \left(\frac{1+\overline{\gamma}_{k,i}}{\overline{\gamma}_{k,i}}\right)^{m-1} \\
\times \left[e^{-\frac{\lambda_{i,k}}{2(1+\overline{\gamma}_{ki})}} - e^{-\frac{\lambda_{i,k}}{2}} \sum_{n=0}^{m-2} \frac{1}{n!} \left(\frac{\lambda_{i,k}\overline{\gamma}_{ki}}{2(1+\overline{\gamma}_{ki})}\right)^n \right]
\end{aligned}
\tag{1}
$$

where $\lambda_{i,k}$ is the energy detection threshold selected by the i-th SU for sensing the k-th channel, m is the time bandwidth product. $\overline{\gamma}_{ki}$ is the average SNR of the received signal from the k-th SBS, where P_k is the transmit power of the k-th SBS, $g_{ki} = \frac{1}{d_{ki}^\mu}$ is the path loss between the k-th SBS and the i-th SU, d_{ki} is the distance between the k-th SBS and the i-th SU, σ^2 is the Gaussian noise variance.

Thus, as was shown in [7] the false alarm probability perceived by the i-th SU $i \in N$ over the radio channel combining SBS, is given by

$$
P_{fal,i}^k = P_{fal} = \frac{\Gamma(m, \frac{\lambda_{i,k}}{2})}{\Gamma(m)}
\tag{2}
$$

where $\Gamma(.,.)$ is the incomplete gamma function and $\Gamma(.)$ is the gamma function.

We note that the non-cooperative false alarm probability depends on the position of SU. Thus, we can drop the index k in Eq. (2) and we get the missing probability perceived by the i-th SU, namely [4, 7]

$$
P_{mis,i} = 1 - P_{det,i}
\tag{3}
$$

It is obvious that the reduction of the missing probability will decrease the interference on the PU and increase the probability of its detection. Within a coalition the SUs minimize their missing probabilities. We assume that in each coalition C an SU is selected as coalition head. It collects the sensing bits from the coalition's members and acts as the head to form a coalition. Thus, assuming Rayleigh fading and BPSK modulation within each coalition, the probability of reporting error between an SU and the coalition head is given by [21]:

$$
P_{e,i,j} = \frac{1}{2} \left(1 - \sqrt{\frac{\overline{\gamma}_{i,j}}{2 + \overline{\gamma}_{i,j}}}\right)
\tag{4}
$$

where $\overline{\gamma}_{i,j} = \frac{P_i h_{i,j}}{\sigma^2}$ is the average SNR between SU i and the coalition head j inside coalition C with P_i the transmit power of SU i, σ^2 the Gaussian noise and $h_{i,j} = \frac{\kappa}{d_{i,j}^\mu}$ the path loss between SU i and coalition head j. We assume that any SU can be selected as a coalition head within a coalition. However, we suppose that within a coalition C, the SU having the lowest missing probability $P_{mis,j}$ is chosen as a coalition head. It means that the coalition head j should not risk sending his local sensing bit over the fading channel and thus it will serve as a fusion center for the other SUs in the coalition.

We assume that each SU can be chosen as the head of a coalition. As a criterion of the head selection, we suppose that with the coalition the SU with the lowest non-cooperative missing probability can be the head of the coalition. Then the coalition head $j \in C$ of a coalition C is given by Eq. (3). In a coalition formed of SUs by collaborative sensing, the missing and false alarm probabilities of a coalition C possessing a coalition head j are, respectively, given by [17]:

$$Q_{mis,C} = \prod_{i \in C} [P_{mis,i}(1 - P_{e,i,j}) + (1 - P_{mis,i})P_{e,i,j}] \tag{5}$$

$$Q_{fal,C} = 1 - \prod_{i \in C} [(1 - P_{fal})(1 - P_{e,i,j}) + P_{fal}P_{e,i,j}] \tag{6}$$

where $P_{fal}, P_{mis,i}$, and $P_{e,i,j}$ are given for an SU and coalition head $j, j \in C$ by Eqs. (2), (3), and (4), respectively.

We can see from both the above mentioned equations that as the number of SUs per coalition is increased, the missing probability will decrease while the probability of false alarm will increase. This means that in collaborative spectrum sensing it has a major impact on the collaboration strategies of each SU. It must be considered in a distributed algorithm planned for coalition formation.

3   Coalition Formation among Secondary Users of Incomplete Information

In this section, we model the problem of coalition formation among SUs of incomplete information.

For the purpose of deriving a distributed algorithm for forming the coalition of SUs, we use a cooperative game theory [13]. Our problem can be modeled as (Ω, u) coalitional game, where Ω is the set of players (the SUs) and u is the utility function or value of a coalition.

We assume that the value $u(C)$ of a coalition $C \subset \Omega$ must capture the trade off between the probability of false alarm. Then, $u(C)$ must be an increasing function of the detection probability $Q_{d,C} = 1 - Q_{mis,C}$ within coalition C. Also, $u(C)$ must be a decreasing function of the false alarm probability $Q_{fal,C}$. A utility function $u(C)$ is given by

$$u(C) = Q_{d,C} - Cost(Q_{fal,C}) = (1 - Q_{mis,C}) - Cost(Q_{fal,C}) \tag{7}$$

where $Cost(Q_{fal,C}$ is a cost function of the false alarm, $Q_{mis,C}$ is the missing probability of coalition C given by Eq. (6).

For mathematically modeling the cooperation problem, we refer to the coalitional game theory. Thus, we provide the following definition [6]:

The problem can be formulated with the help of using a cooperative game theory [13]. More formally, we have a (Ω, u) coalition game, where Ω is the set of players (the SBSs) and u is the utility function or the value of the coalition.

Following the coalition game of Harsanyi [8], a possible definition for a Bayesian game [6] is as follows.

Definition 1 (Bayesian game). *A Bayesian game G is a strategic-form game with incomplete information, which can be described as follows:*

$$\mathcal{G} = \langle \Omega, \{\mathcal{T}_n, \mathcal{A}_n, \rho_n, u_n\}_{n \in \mathcal{N}} \rangle \tag{8}$$

which consists of

- *a player set:* $\Omega = \{1, \ldots, N\}$
- *a type set:* $\mathcal{T}_n (\mathcal{T} = \mathcal{T}_1 \times \mathcal{T}_2 \times \cdots \times \mathcal{T}_N)$
- *an action set:* $\mathcal{A}_n (\mathcal{A} = \mathcal{A}_1 \times \mathcal{A}_2 \times \cdots \mathcal{A}_N)$
- *a probability function set:* $\rho_n : \mathcal{T}_n \to \mathcal{F}(\mathcal{T}_{-n})$
- *a payoff function set:* $u_n : \mathcal{A} \times \mathcal{T} \to \mathcal{R}$, *where* $u_n(a, \tau)$ *is the the payoff of player* n *when action profile is* $a \in \mathcal{A}$ *and type profile is* $\tau \in \mathcal{T}$.

The set of strategies depends on the type of the player. Additionally, we assume the type of the player is relevant to his decision. The decision is dependent on information which it possesses. A strategy for the player is a function mapping its type set into it action set. The probability function ρ_n represents the conditional probability $\rho_n(-\tau_n | \tau_n)$ that is assigned to the type of profile $\tau_n \in \mathcal{T}_{-n}$ by the given τ_n.

The payoff function of player n is a function of strategy profile $s(.) = \{s_1(.), \ldots, s_N(.)\}$ and the type profile $\tau = \{\tau_1, \ldots, \tau_N\}$ of all players in the game and is given by

$$u_k(s(\tau), \tau) = u_n(s_1(\tau_1), \ldots, s_N(\tau_N), \tau_1, \ldots, \tau_N) \tag{9}$$

We recall that in a trategic-form game with complete information, each player chooses one action. In a Bayesian game each player chooses a set or collection of actions (strategy $s_n(.)$).

A definition for a payoff of player in the Bayesian game as follows:

Definition 2. *The player's payoff in a Bayesian game is given by*

$$u_n(\tilde{s}_n(\tau_n), s_{-n}(\tau_{-n}), \tau) = u_n(s_1(\tau_1), \ldots, \tilde{s}_n(\tau_n),$$
$$s_{n+1}(\tau_{n+1}), \ldots, s_N(\tau_N), \tau) \tag{10}$$

where $\tilde{s}_n(.), s_{-n}(.)$ *denotes the strategy profile where all players play* $s(.)$ *except player* n.

Next, we define the Bayesian equilibrium (BE) as follows:

Definition 3 (Bayesian equilibrium). *The strategy profile* $s^*(.)$ *is a Bayesian equilibrium (BE), if for all* $n \in \mathcal{N}$, *and for all* $s_n(.) \in S_n$ *and* $s_{-n}(.) \in S_{-n}$

$$E_\tau \left[u_n(s_n^*(\tau_{-n}, \tau) \right] \geq E_\tau \left[u_n(s_n(\tau_n, s_{-n}^*(\tau_{-n}), \tau) \right] \tag{11}$$

where

$$E_\tau \left[u_n(x_n(\tau_n), x_{-n}(\tau_{-n}), \tau) \right] \triangleq \sum_{\tau_{-n} \in \mathcal{T}_{-n}} p_n(\tau_{-n} \mid \tau_n)$$

$$u_n(x_n(\tau_n), x_{-n}(\tau_{-n}), \tau) \tag{12}$$

is the expected payoff of player n, which is averaged over the joint distribution of all players' types.

Based on these concerns, it is important to say, that the utility of coalition C is equal to the utility of each SU in the coalition. Thus, the used (Ω, u) coalitional game model has a non-transferable utility. We assume that in the coalitional game the stability of the grand coalition of all the players is generally assumed and the grand coalition maximizes the utilities of the players. Then, player i may to choose the randomized strategy s which maximizes his expected utility. Informally, we could provide a Nash equilibrium here.

Assuming the perfect coalition of SBS C_{per}, the false alarm probability is given by

$$Q_{fal,C_{per}} = 1 - \prod_{i \in C_{per}} (1 - P_{fal}) = 1 - (1 - P_{fal})^{|C_{per}|} \tag{13}$$

4   Coalition Formation

In this section, we propose a distributed algorithm for coalition formation of SUs with incomplete information in the CRN.

As mentioned previously, the performance of a coalition formation algorithm among the SUs with incomplete information will depend on the false alarm constraint α and the false alarm probability P_{fal}. This can be seen, as the false alarm probability decreases,

> **procedure** *SUs coalition formation*;
> **begin**
> $i := 0;\ |\ C_{per}\ | := 1;$
>   *compute* $Q_{fal,C}$;
>   **while** $(|\ C_{per}\ | < M_{max}$ **and** $Q_{fal,C} > Q_{fal,C_{per}})$ **do**
>   **begin**
> $i := i + 1;$
>   **if** *BE exists for given* $|\ C_{per}\ |$ **then**
>   **begin**
> $|\ C_{per}\ | := i;$ *compute* $Q_{fal,C}$;
>   *find the coalition head*;
>   **end**;
>   **end**;
> **end**;

Fig. 2. Coalition formation algorithm

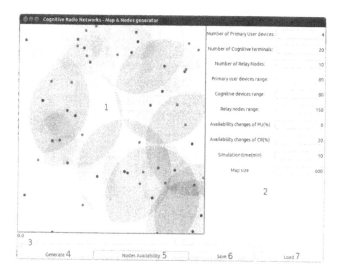

Fig. 3. A toolkit of our simulation program

the possibilities for collaboration is increased. A maximum number of SUs per coalition is given by [17]:

$$M_{max} = \frac{\log(1 - \alpha)}{\log(1 - P_{fal})} \tag{14}$$

And the false alarm probability in a perfect coalition of SUs $| C_{per} |$ is given by

$$Q_{fal, C_{per}} = 1 - \prod_{i \in C_{per}} (1 - P_{fal}) = 1 - (1 - P_{fal})^{|C_{per}|} \tag{15}$$

Fig. 2. shows the pseudo-code of the proposed algorithm. Initially, each SU creates a single-membered coalition. In each iteration, one coalition attempts to improve the utility function of its member by making a coalition. The coalition formation algorithm is terminated whenever a Bayesian equilibrium (BE) has been reached.

5   Simulation Results

A simulation was used to confirm the above given algorithm for the coalition formation among the SUs with incomplete information. In Fig. 3, we present a toolkit of our simulation study. The simulation of the CR network has a four square with the PU at the center. Each square is equal to 1 km × 1 km. In each square 4 SBSs deployed. On this square the SUs are randomly deployed around the PU. We set the time bandwidth product $m = 5$ [7], the PU transmit power $P_{PU} = 100$ mW, the SU transmit power $P_i = 10$ mW and the noise level $\sigma^2 = -90$ dBm. We set $\mu = 3$ and $\kappa = 1$ We assumed that the maximum constraint on the false alarm is given by $\alpha = 0.1$ as has been recommended by the IEEE 802.22 [10].

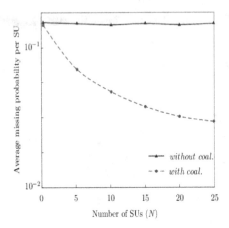

Fig. 4. Average missing probabilities versus number of SUs

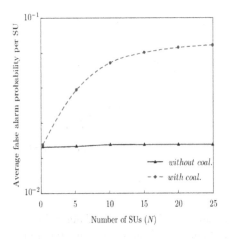

Fig. 5. Average false alarm probabilities versus number of SUs

Fig. 4 shows the average missing probabilities achieved per single SU for a different network size for the network with noncooperative game and with a formed coalition of the SUs. It can be seen in Fig. 4 that the proposed algorithm provides an improvement in the average missing probability of up to 60% reduction in comparison with the network without coalition formation. This is an advantage which is especially important in a network with a large number of SUs.

Fig. 5 presents the average false alarm probabilities per SU versus the number of SUs for both networks: without cooperation and with the coalition. These probabilities are averaged over random locations of the SUs as well as range detection thresholds, which does not violate the false alarm constraint. It can be seen that the obtained false alarm probabilities in the network with the coalition formation achieved by the proposed algorithm are worse than for the network without cooperative spectrum sensing.

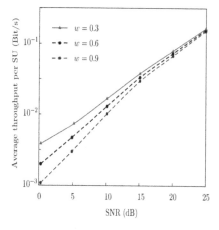

Fig. 6. Maximum achieved throughput for various value of the utility contribution weighting factor w

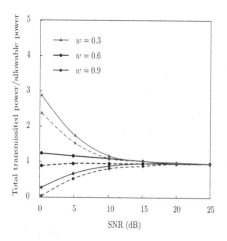

Fig. 7. Total transmitted power relative to the assumed power limit for various values of the utility contribution weighting factor. Dashed graphs are obtained for coalition of four SUs.

The result of this is that the network with centralized solution obtains a better missing probability. The formed coalition compensates this performance downfall by decreasing the corresponding cost of achieved average false alarm probability.

Fig. 6 presents the results of the average transmissions of SUs coalition in terms of achieved throughput per SU (the player's achieved throughput in the coalition divided by the total available bandwidth) for various values of the utility-contribution weighting factor w ($w \in [0.3, 0.6, 0.9]$), and for assumed SNR value. Here, it can be seen that the transmitted power exceeds the power limit for small SNR value and for $w = 0.3$. Thus, a higher throughput is achieved due to the lack of noise. If $w = 0.9$, the opposite results are obtained because the SU is forced to be more power efficient. By assuming $w = 0.6$ in the game, we obtain the optimal curve of the achieved throughput and the bandwidth.

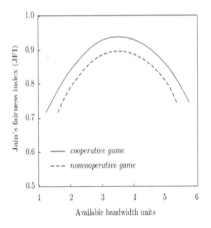

Fig. 8. Jain's fairness index (JFI) for available bandwidth units for cooperative and noncooperative games

Fig. 7. shows the total transmitted power relative to the assumed power limit for coalition of two and four SUs in dependence of various values of the utility contribution weighting factor w ($w \in [0.3, 0.6, 0.9]$) and for assumed SNR value. The curves in Fig. 7 lead us to conclusion that by choosing the appropriate value of the utility contribution weighting factor for given coalition, the SUs can obtain the ideal transmitted power.

Fig. 8 presents the Jain's fairness index (JFI) for available bandwidth units in coalition games. The results show that all cooperation games take the maximum values of the JFI index. We recall that Jain's fairness index is defined as [19]

$$JFI = \frac{(\sum_{i}^{N_b} x_i)^2}{N_b \sum_{i=1}^{N_b} x_i^2}$$

where x_i denotes a bandwidth unit and N_b is the number of all bandwidth units. The results show that all cooperation games take the maximum values of the JFI index.

6   Conclusions

In this paper, we proposed a new method for coalition formation among the SUs with incomplete information in a cognitive radio network. We introduced a novel distributed algorithm for coalition formation with incomplete information in these networks. The proposed coalition formation algorithm is based on simple rules that enable SUs in the CRN to cooperate in order to improve their effectiveness. We showed that a maximum number of SUs per coalition exists in order to maintain their stability. The provided simulation results showed that the proposed algorithm can adapt to changes in the network topology. Also, these results indicate that the average missing probability per SU is smaller by 60% than in the CRN without the proposed algorithm. Finally, the obtained results showed that through the proposed algorithm the SUs can self-organize and the CRN achieves better performance than the CRN without this algorithm.

References

[1] Akyildiz, I.F., Lee, W.Y., Vuran, M.C., Mohanty, S.: Next Generation Dynamic Spectrum Access/cognitive Radio Wireless Networks: A Survey. Computer Networks 50(13), 2127–2159 (2006)

[2] Akyildiz, I.F., Lo, B.F., Balakrishnan, R.: Cooperative Spectrum Sensing in Cognitive Radio Networks: A Survey. Physical Communication 4, 40–62 (2011)

[3] Altman, E., Debbah, M., Silva, A.: Game Theoretic Approach for Routing in Dense Ad Hoc Networks. In: Stochastic Network Workshop, Edinburg, UK (July 2007)

[4] Digham, F.F., Alouini, M.S., Simon, M.K.: On the Energy Detection of Unknown Signals over Fading Channels. In: Proc. Int. Conf. on Communications Alaska, USA, pp. 3575–3579 (2003)

[5] First Report and Order, Federal Communications Commission Std. FCC 02-48 (February 2002)

[6] Fudenberg, D., Tirole, J.: Game Theory. MIT Press, Cambridge (1991)

[7] Ghasemi, A., Sousa, E.S.: Collaborative Spectrum Sensing for Opportunistic Access in Fading Environments. In: IEEE Symp. New Frontiers in Dynamic Spectrum Access Networks, Baltimore, USA, pp. 131–136 (2005)

[8] Harsanyi, J.C., Selten, R.: A Generalized Nash Solution for Two-Person Bargaining Games with Incomplete Information. Management Science 18(5), 80–106 (1972)

[9] Haykin, S.: Cognitive Radio: Brain-empowered Wireless Communications. IEEE J. Select Areas Commun. 23(2), 201–220 (2005)

[10] IEEE 802.22, Cognitive Wireless Regional Network - Functional Requirements, 902.22-06/0089r3, Technical Report (June 2006)

[11] Lasaulce, S., Debbah, M., Altman, E.: Methodologies for Analyzing Equilibria in Wireless Games. IEEE Signal Processing Magazine 26(5), 41–52 (2009)

[12] Martyna, J.: The Use of Shapley Value to Power Allocation Games in Cognitive Radio Networks. In: Jiang, H., Ding, W., Ali, M., Wu, X. (eds.) IEA/AIE 2012. LNCS, vol. 7345, pp. 627–636. Springer, Heidelberg (2012)

[13] Myerson, R.B.: Game Theory, Analysis of Conflict. Harvard University Press, Cambridge (1991)

[14] Mathur, S., Sankaranarayanan, L.: Coalitional Games in Gaussian Interference Interference Channels. In: Proc. of IEEE ISIT, Seattle, WA, pp. 2210–2214 (2006)

[15] Mitola III, J., Maguire Jr., G.Q.: Cognitive Radio: Making Software Radios More Personal. IEEE Personal Communication Magazine 6(4), 13–18 (1999)

[16] Perlaza, S.M., Lasaulce, S., Debbah, M., Chaufray, J.M.: Game Theory for Dynamic Spectrum Sharing. In: Zhang, Y., et al. (eds.) Cognitive Radio Networks: Architecture, Protocols and Standards. Taylor and Francis Group, Auerbach Publications, Boca Raton (2010)

[17] Saad, W., Han, Z., Debbah, M., Hjorungnes, A.: Coalitional Games for Distributed Collaborative Spectrum Sensing in Cognitive Radio Networks. In: IEEE INFOCOM, pp. 2114–2122 (2009)

[18] Saad, W., Han, Z., Basar, T., Hjorungnes, A., Song, J.B.: Hedonic Coalition Formation Games for Secondary Base Station Cooperation in Cognitive Radio Networks. In: IEEE Wireless Communication and Networking Conference (WCNC), pp. 1–6 (2010)

[19] Vassaki, S., Panagopoulos, A., Constantinou, P.: Game-theoretic Approach of Fair Bandwidth Allocation in DVB-RCS Networks. In: Int. Workshop on Satellite and Space Communications, IWSSC 2009, Italy, pp. 321–325 (2009)

[20] Visotsky, E., Kuffner, S., Peterson, R.: On Collaborative Detection of TV Transmission in Support of Dynamic Spectrum Sensing. In: IEEE Symp. New Frontiers in Dynamic Spectrum Access Networks, Baltimore, USA, pp. 338–356 (2005)

[21] Zhang, W., Letaief, K.B.: Cooperative Spectrum Sensing with Transmit and Relay Diversity in Cognitive Networks. IEEE Trans. Wireless Commun. 7(12), 4761–4766 (2008)

MOVEDETECT – Secure Detection, Localization and Classification in Wireless Sensor Networks

Benjamin Langmann[1], Michael Niedermeier[2], Hermann de Meer[2],
Carsten Buschmann[3], Michael Koch[4], Dennis Pfisterer[5], Stefan Fischer[5],
and Klaus Hartmann[1]

[1] Center for Sensor Systems (ZESS), University of Siegen, Paul-Bonatz-Str. 9-11,
57068 Siegen, Germany
{langmann,hartmann}@zess.uni-siegen.de
[2] Department of Computer Networks and Computer Communcation,
University of Passau, 94032 Passau, Germany
{michael.niedermeier,hermann.demeer}@uni-passau.de
[3] Coalesenses GmbH, Maria-Goeppert-Str. 1, 23562 Lübeck, Germany
buschmann@coalesenses.com
[4] SINUS Messtechnik GmbH, Föpplstr. 13, 04347 Leipzig, Germany
michael.koch@sinusmess.de
[5] Institute of Telematics, University of Lübeck, 23562 Lübeck, Germany
{pfisterer,fischer}@itm.uni-luebeck.de

Abstract. In this paper a secure wireless sensor network (WSN) developed within the MOVEDETECT project is presented. The goal of the project was to design, implement and demonstrate a secure WSN for the protection of critical infrastructure. In order to provide a reliable service, the system must detect any kind of tampering with the sensor nodes, prevent eavesdropping and manipulation of the communication as well as detect, track and classify intruders in the protected region. Therefore based on previous experiences, a real-world WSN was developed, which addresses practical issues like water proofing, energy consumption, sensor deployment and visualization of the WSN state, but also provides a unique security concept, a interesting combination of sensors and sophisticated sensor data processing and analysis. The system was evaluated by examining firstly the sensors and the sensor processing algorithms and then conducting realistic field test.

Keywords: Wireless sensor network, Detection, Security, Functional safety, Networking.

1 Introduction

Wireless sensor networks (WSN) have attracted attention of many researches of different disciplines in the past driven by the increasing availability of microprocessors with wireless communication. In general, a WSN consists mainly of a

S. Balandin et al. (Eds.): NEW2AN/ruSMART 2013, LNCS 8121, pp. 284–297, 2013.
© Springer-Verlag Berlin Heidelberg 2013

number of sensors nodes. However, management and routing nodes are possible as well and hence different network structures have been proposed, e.g. mesh or hierarchical structures. Usually, each sensor node has only very limited resources (computational power, energy supply, memory, communication bandwidth) and its purpose is the decentralized and reliable acquisition and transmission of data. Higher reasoning is either performed by dedicated processing units or outside the WSN. The applications of WSN are located mostly in the safety and security area ranging from military and police tasks to structural and health monitoring.

Within the MOVEDETECT project a demonstrator for a real-world WSN was developed to be used for the protection of critical infrastructure or small key areas and its design, implementation and evaluation is described in this paper. In addition to the practical applicability which includes water proofing, a low energy consumption and methods for the sensor deployment, the system must provide secure and reliable communication and of course accurate detection, tracking as well as classification of intruders. Therefore, a combination of techniques to ensure the systems integrity and confidentiality were applied and a sophisticated distributed approach for sensor data processing and analysis was developed. In addition, reporting and logging mechanisms allow users to constantly monitor activities inside the WSN, both in real-time and time-delayed. Moreover, the usability of the system is not impaired by the inherent complexity of the WSN, as it can be configured and operated from a central command center. The combination of these features creates a uniquely secure, flexible and usable system.

The remainder of this paper is structured as follows: Section 2 introduces related work on WSNs to protect critical infrastructure. Section 3 details the requirements of the proposed system. Section 4 covers all aspects of the system development and Section 5 demonstrates the abilities of the deployed sensors. Section 6 discusses the results of two field tests and the paper closes with a conclusion and an outlook of possible future work in Section 7.

2   Related Work

The work related to the presented project is manifold, as MOVEDETECT combines multiple research fields, e.g. communication security, functional safety but also energy efficient communication and the fusion of WSN data. Therefore, this section covers firstly the state-of-the-art in conventional (non-networked) solutions for secure infrastructures. Secondly, related work on systems that rely on wireless communication and work in critical environments is described. Finally, recent projects that partly cover the goals of MOVEDETECT are introduced and compared.

Conventional solutions to secure critical infrastructures range from very simple setups involving mechanical barriers, like doors, fences and trenches or security personnel [1] to complex – however non-networked – equipment, including e.g. passive infrared (PIR), sound and seismic sensors or video surveillance. While these devices allow detecting certain influences, they lack the possibility to cooperatively analyze a situation [2], as no data transfer between them is possible.

However, over the last years, advances in the area of networked sensor technologies have opened up new application areas for these sensor devices [3]. With new technological paradigms, like the Internet of Things (IoT) or Automated Living, the use of WSNs is rapidly growing as networked sensors are vital in these scenarios. However, as future technologies more and more depend on sensor networks, the requirements for these – now critical infrastructures – rise in a similar way, especially in the fields of security and safety. An already realized example in this area is the SmartSantander project, where more than 10,000 IoT devices are deployed to build a so-called "Smart City". In such scenarios, strict security requirements apply, as personal and critical information is transferred. While there are research efforts in the areas of access protection, secure reprogramming over wireless connections and secure communication, a complete security solution for a wireless network does not yet exist. This example already characterizes several important aspects of WSN security: Due to the usage in unsupervised or even hostile environments while performing critical tasks, it is important to provide measures ensuring integrity and confidentiality of data (communication security) and protection of the devices against malfunctions that either occur randomly or due to tampering [4,5].

FleGSens [6] is a project for secure area monitoring using WSNs. The system is able to detect movement in critical areas and considers authenticity and data integrity of alarm signals. Other security issues, like jamming and functional safety, are not addressed. FleGSens uses a flat network hierarchy, which limits system scalability due to message collision. The used detection algorithm only considers the position of trespassers and – in contrast to MOVEDETECT – does not include a classification algorithm.

The POmSe ("Personen- und Objektdetektion mit mobilen Sensoren") project [7], evaluates the general applicability of WSNs for the protection of critical infrastructures and environments with a focus on the applicability for trespasser detection due to their inherent advantages compared to conventional security techniques (like e.g. security doors, fences, etc.). In contrast to MOVEDETECT, POmSe performs a general analysis of WSNs including the required security features in critical applications. However, the development, application and evaluation of a WSN is not included in the project, neither is a classification of trespassing objects performed.

A complementary project, named "Personen- und Objekterkennung basierend auf Trittschall (POT)" [8], evaluates the suitability of seismic sensors for the detection, classification and localization of humans and animals. It turns out that a reliable detection algorithm can be provided whereas the exact spatial localization of a person strongly depends on the environmental conditions as, e.g. the type and humidity of the surrounding soil.

In [9] a WSN for security in medical environments concentrating of high reliability and low delays is presented. A WSN to support safety in the transport domain is outlined in [10]. Here the cost is the most important design objective in order to increase the acceptance of the WSN. An approach focusing on the fusion of different sensors (underground, above ground and air) is introduced in [11] for border control. Moreover, in [11] an approach aiming at the detection of intrusions at borders is introduced. Lastly, in [12] methods to reduce the power consumption in WSNs including cameras and PIR sensors are discussed.

As shown here, the main difference between the presented related works is the implementation of a comprehensive security solution covering all necessary aspects, while still maintaining a high usability despite the high system complexity. In the following, first the requirements that originate from the usage scenario of the MOVEDETECT system are presented before the system development is described.

3 Requirements

The main functional goals of the system are defined by three tasks of the WSN: i) detection, ii) localization including tracking and iii) classification of objects in a predefined area. The system must be able to detect and locate persons or cars with certain accuracy. In our case, all values for deviations or timings are chosen in a way that the goal on the user's side – which is to be able to organize a well-directed response to a penetration of the surveillance area – is possible. For the given case, a deviation of ≤ 5 meters is therefore considered acceptable.

In addition, it is required to calculate the trajectory of the trespassing object to indicate its direction. In order to provide not only accurate but also timely data, the system is required to fulfill certain real-time criteria. The delay between object-entry and the signaling of the event to the user was fixed at a maximum delay of 3 seconds. Due to the special hardware used in WSNs, several additional system requirements have to be fulfilled. Among them is primarily a reliable, scalable and efficient communication architecture ensuring that all events occurring inside the monitored area are not only registered, but also that the messages of these events are transferred reliably to the base station [13].

The security of the system is of major importance, especially when the unattended operation of the system is taken into account. The system's overall security concept therefore comprises not only IT security measures but also functional safety aspects. It is required to protect the system from coincidental or targeted physical damage and hardware failures. Additionally, energy efficiency has to be carefully considered in both the hardware and software design, because the sensor nodes are powered with batteries and the intended operating time is two weeks.

4 WSN System Design

This section presents the general structure of the system to give an overview of how MOVEDETECT works. As depicted in Fig. 1a, the system is based on a hierarchical structure that consists of a single command center, several clusterheads on the intermediate level and for each clusterhead multiple sensor nodes on the bottom level. In the prototype system, which is described later on, a total number of 100 sensor nodes is used (10 clusterheads with 10 sensor nodes each). Due to the system's scalability, the system can however also be employed using a much larger number of nodes.

The sensor nodes are equipped with different sensor combinations and are described in the following section. An evaluation of the individual sensors can be found in Section 5.

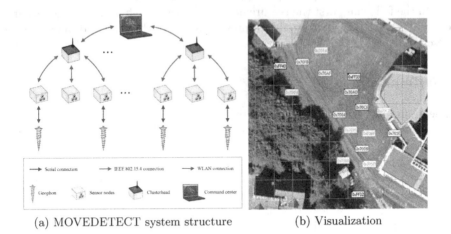

(a) MOVEDETECT system structure   (b) Visualization

Fig. 1. Structure of the MOVEDETECT WSN and the visualization software Spyglass

4.1   Hardware Design

In this section, the hardware concept is described in the order of the previously shown three levels: command center, clusterheads, and sensornodes.

The command center aggregates data from the clusterheads and displays the information to the user (cf. Fig. 1b). This information is presented using the detection algorithms described in Section 4.3 and delivered to a network visualization software called *Spyglass* [14]. The existing open-source Spyglass was extended to have a control window that allows to send commands to the network, e.g. restarting of nodes, sending messages to a number of nodes, or reprogramming nodes.

Each clusterhead consist of two components: an embedded-PC board (ARM-based CPU with 800 MHz, 512 MB RAM, 512 MB flash memory, power requirement of about 1.5 W) and an iSense sensor node. The embedded PC maintains the WiFi connection to the command center, pre-processes sensor data and performs network management tasks. In order to communicate with the sensor nodes, an iSense sensor network device was attached to the CPU board as depicted in Fig. 2a.

Several different configurations of sensor nodes have been designed, which differ in the type of attached sensors and housing (Fig. 2b) but all include several iSense modules. The Core Module includes a Jennic JN5148 32-bit RISC controller and an IEEE 802.15.4 compliant radio interface. The radio chip operates

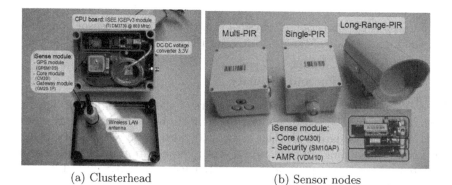

(a) Clusterhead   (b) Sensor nodes

Fig. 2. Hardware components of the WSN

at a frequency of 2.4 GHz, offers 16 different radio channels, provides a data transfer rate of 250 kB/s and includes a hardware AES engine as well as an ultra-stable real-time clock (RTC) (typical 6 ppm). The second type of housing contains a long-range Siemens IS392 PIR sensor and an accelerometer.

In the sensor network, the previously described four types of sensors are utilized for various purposes. The general detection of activity in the observed area is performed by PIR sensors, which can be differentiated into three types: Single-PIRs (Panasonic AMN34111) with a range up to 10 m, Multi-PIRs (AMN31111 + 2 × AMN33111) with a range of approximately 5 m and Long-Range-PIRs (Siemens IS392) with a range up to 50 m. With the Multi-PIRs, the directions of moving objects can be estimated and Long-Range-PIRs play a significant role in the detection of new objects entering the surveillance area. Fig. 3 shows detection results for these different PIR sensors.

The second type of sensors used for detection purposes are geophones. They are comprised of three seismic capsules (SM-24, Sensor Nederland) in an orthogonal arrangement, an analog signal conditioning part and a microcontroller-based (ARM-Cortex M3 MCU) signal processing unit. The complete geophone is housed in a protection enclosure and is connected to the sensor node via a communication cable that also provides the required power. For a sufficient coupling to the ground, a soil drilling tool in combination with a screw-shaped housing is used.

The iSense Vehicle Detection Modules are used for object classification, since only moving metallic objects influence their measurements. They are based on a 2-axis Phillips KMZ52 anisotropic magneto-resistive sensor bridge that is combined with two amplifier stages as well as circuitry for de-gaussing and earth-magnetic field compensation. It exploits the fact that large ferro-magnetic objects distort the earth-magnetic field to detect such objects by observing field changes. To achieve a detection range of more than 5 m, it amplifies the bridge output by a factor of approximately 40,000 and features a sensitivity of 786.2 mV/(kA/m) at a bandwidth of 1 kHz. Because the module typically has a current consumption of 20 − 25 mA during operation, it is activated after observing PIR sensor events within less than 180 ms.

The last sensor is an accelerometer, which is part of every sensor node. Its sole purpose is the detection of physical tampering with the sensor nodes, since even small movements of the device trigger the accelerometer.

4.2   Communication and Security

The communication and security architecture of the WSN is only outlined due to its complexity. The sensor nodes communicate based on the energy-efficient IEEE 802.15.4 standard with their associated clusterhead while applying adaptive frequency hopping. In order to prevent messages from being forged or manipulated, the message payload is encrypted using the AES-CBC-128 cipher, which is supported by a crypto co-processor available in all sensor nodes, which makes it very fast and energy-efficient. In addition, a timestamp, a CBC-MAC AES, and a sequence number is added to the payload. These are used to achieve resistance against replay attacks, delayed massage sending, or manipulation of the message order. The system also uses an advanced key management approach. The clusterheads pre-process and compact the data coming from the sensor nodes and communicate with the command center via TCP/IP-based connections over WiFi.

The integrity of the WSN is ensured on all levels. The sensor nodes and the clusterheads analyze the sensor data as well as test their CPU, RAM and firmware at start-up for manipulations, send heard beat messages regularly and observe their accelerometer.

4.3   Detection, Classification and Tracking of Objects

In order to achieve a robustness and reliability suitable for long term surveillance, the detection, tracking and classification methods are implemented in a simple yet effective way. After the WSN is started, the detection algorithm first remains in a waiting state, until a significant amount of sensor events are registered, which indicates an object entering the area. If the amount of events remains low, it is assumed that these are caused by random noise or other sources like wind. Additionally, it is required that there is a local accumulation of events. The amount of random noise depends on the sensor type as well as on other factors like the position of the sensor, the time of the day and the weather conditions. If the sensor data satisfies these conditions, it is inferred that an object is inside the surveillance area and an initial position is determined. Afterwards, only sensor events in a vicinity of the current position of the detected object are handled, until the object leaves the surveillance area. Each sensor event suggests possible locations of the object. For PIR sensors this is a cone of a specific length whereas for all other sensors, this is a radial symmetric area around the sensor node with a certain radius. The current position of the object is then determined by averaging the suggested positions of the sensor events while factoring in the previous position of the object.

The tracking simply consists of a concatenation of the determined object positions. The classification of the object (person, person carrying metal object, car or unknown) is performed by an analysis of the AMR and geophone events. Magnetic sensor events in the vicinity of the object suggest a car or a motorbike and geophone events indicate footsteps, but there need to be an accumulation of these events, which exceeds the noise level. Experiments show that the AMR sensors are surprisingly affected by wind (possibly due to movement of cables or the sensor node itself). An object in the surveillance area is considered to have left this area if the last known position is near the border of the area and further sensor events are below the detection level for some time.

5   Sensor Evaluation

The sensors employed by the sensor nodes need to be analyzed in order to achieve a desired quality of the surveillance, since datasheets often do not provide information in the way or detail required. Additionally, the preprocessing algorithms of the geophones and PIR sensors need to be evaluated and characterized. In the following an excerpt of the sensor related experiments is given.

For the PIR sensors, the region in which objects produce sensor events must be known in order to perform a localization of intruding objects. It is common to conduct simple walking and driving tests for this purpose, for which a path is defined and repeatedly driven or walked, respectively. In Fig. 3 the results for normal walking are given. Green marks the detection area specified in the datasheet, while red marks the region in which sensor events actually occur in practice. In Fig. 3e the joint detection region of a Multi-PIR sensor node is visualized. Here green marks the region where the two spot type PIR sensors (AMN33111) are sensitive and blue the region of the standard PIR type (AMN31111).

In Fig. 3d, the exemplary behavior of a Multi-PIR sensor node is shown for a walk-by test, in which a person crosses the detection area from right to left with normal walking speed and vice versa. It was found that the direction of the object can be determined reliably simply using the 3-PIR sensors included in a Multi-PIR sensor node. To do so, the time at which each PIR sensor activates is measured and compared. This is not possible using Single-PIR sensors, but promises more accurate localization and tracking of objects.

In order to be able to detect intruders and therefore to provide a basis for tracking, a robust algorithm is required, which filters the incoming signal at the geophone to isolate those representing human footsteps. To overcome the problem of varying environmental conditions (mostly soil quality and humidity as well as ground coverage with plants) and of disturbing seismic noise, a procedure with an adaptive threshold was developed and implemented. It is based on the so-called "sta/lta - picking", which is used in earthquake location scenarios [15]. The decision whether there is an event is made by comparing the ratio of a short and a long time average of the seismic signal with an empiric constant. By introducing an adaptive constant an accommodation to varying conditions is achieved. An example is shown in Fig. 4a.

The output event of a geophone reliably signals a person within the detection range. By calculating the geometric center of gravity of one or more geophones

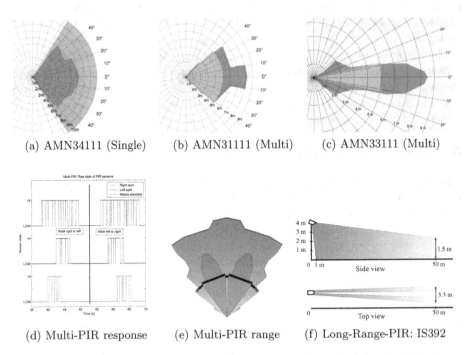

(a) AMN34111 (Single)   (b) AMN31111 (Multi)   (c) AMN33111 (Multi)

(d) Multi-PIR response   (e) Multi-PIR range   (f) Long-Range-PIR: IS392

Fig. 3. Experimental evaluation of the detection ranges of the different PIR sensor types. Green marks the detection area specified in the datasheet and red the actual detection area for walks with normal speed.

reporting events with similar timestamps, it is possible to locate (and track) an intruder within the sensor network.

For the vehicle detection based on the AMR sensors, the activation procedure, the sensor range and detection algorithms were tested prior to the implementation of the final system. The according test setup is shown in Fig. 4b. Two sensor nodes with a vehicle detection module are positioned at opposite sides of a road in a distance of 8 m from each other. A sensor node with a PIR sensor is placed next to the road 15 m before the AMR nodes. A vehicle then passes the installation at speeds of 50 km/h or 5 km/h, respectively. Once a sensor event of the PIR sensor is triggered, the sensor node sends a wireless message to the two AMR sensor nodes. This simulates a delay as it would occur in the final installation, where the communication would work via the clusterhead, instead of directly between the sensor nodes. Upon reception of that message, the AMR nodes activate their vehicle detection modules, de-gauss and calibrate the sensors to compensate for the static earth magnetic field, start sampling the two sensor module channels, and forward the live data to a forth sensor node that is connected to a PC to record timestamps and sensor data for all three nodes.

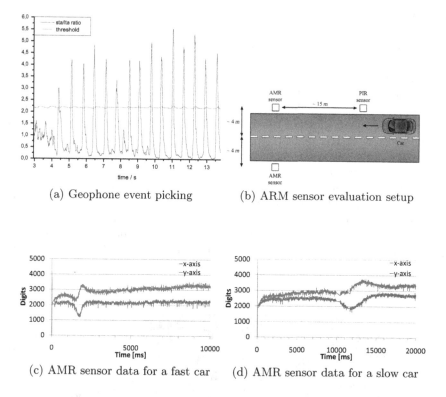

(a) Geophone event picking

(b) ARM sensor evaluation setup

(c) AMR sensor data for a fast car

(d) AMR sensor data for a slow car

Fig. 4. Results of the geophone event picking algorithm for footstep detection, test setup for the AMR sensor evaluation and AMR sensor data for fast and a slow car passing by

Fig. 4c shows the sensor readings of one of the AMR nodes for a vehicle speed of 50 km/h. The time axis of the diagram is set to start at 0 at the time when the PIR sensor event occurred. The y-axis shows the sensor signal as ADC digits, where 1 digit represents 0.59 mV or a field change of 786.2 mV/(kA/m). It is visible that the sensor signal starts around $t = 0.2$ s, as the calibration process commences before that.

The AMR sensor starts up and calibrates fast enough to be in normal operation by the time the vehicle passes by the sensor (characteristic signal shape between $t = 1$ s and $t = 2.3$ s). Consequently, the system would still work properly if the PIR sensor is placed at a distance of only 3 m to the AMR sensors.

Figure 4d shows the sensor readings of one of the AMR sensors for a vehicle speed of 5 km/h. As expected, the signal occurs much later, between $t = 9$ s and $t = 16$ s. The challenge is to develop a detection algorithm that detects vehicles from both the signal patterns, while still being robust against the signal drift that occurs when no vehicles passes.

6   Field Tests of the WSN

In the course of the project MOVEDETECT two main field tests of the whole WSN were conducted. Fig. 5a shows an illustration of the first testing terrain painted by the Spyglass application and it consisted of 100 sensor nodes and 10 clusterheads organized in 10 clusters, where one cluster was reserved for energy consumption monitoring. The surveillance area was a field with a path through and had a maximum width of 125 m and a length of 50 m at the largest extend. The nodes were distributed arbitrarily in distances of 5 m to 10 m from each other. After this initial setup, those 10 sensor nodes nearest to a clusterhead form one sensor cluster.

After the WSN was fully configured, several aspects of the whole system and their interactions had to be verified: First, the general system functions, like network connectivity, message sending, relaying and receiving as well as displaying of system events at the command center. Second, the detection, localization and classification had to be assessed and third a security evaluation to test the WSN's security and safety features was performed.

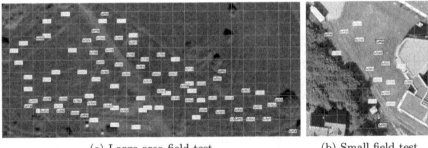

(a) Large area field test   (b) Small field test

(c) Example of a walking test

Fig. 5. Spyglass renderings of the maps for the field tests performed within the scope of the MOVEDETECT project (grid size is 5 m × 5 m)

Table 1. Event symbols and classification colors in the visualization using Spyglass

Symbol	Meaning	Symbol	Meaning
	Geophon event		AMR event
	Longrange-PIR event		Multi-PIR event
	Accelerometer event		Single-PIR event
	Node communication issue		Node failure
	Energy drain alert		

Color	Classification	Color	Classification
	Human		Vehicle
	Human with metal object		Unknown

In the second field test of the WSN a total number of 17 sensor nodes partitioned in 2 clusters were utilized. The surveillance area was a small field surrounded by bushes and buildings with a path going through it. Again, a set of walking and driving detection test were performed.

In order to verify the functions assessed in the field trials, multiple verification techniques were used. For the detection, localization and classification, a Mobotix IP-camera was used in combination with the Spyglass user interface. By doing so, it is possible to document if objects were on the one hand detected and localized and on the other hand if the classification works correctly and the real-time requirement of the system is fulfilled. As an example, Fig. 5c shows one of the walking tests done during the second field trial. As the figure depicts, the person walking on the grass was detected and classified correctly, which is indicated by the green field in the middle of the grid. Similar results were also achieved for other walking styles, like crawling or running, as well as tests with other trespassing objects, e.g. cars (the classification changed in those tests respectively). Furthermore, it was evaluated if a classification of a person carrying metal objects is possible, which would be especially useful to detect weapons like firearms. While the detection works with large ferromagnetic objects, like e.g. fire extinguishers or crowbars, as long as the distance between the object and the sensor node is ≤ 2 m, it is not possible for firearms. This is due to the low mass of ferromagnetic parts in modern firearms, resulting in a very low amplitude of the AMR sensor's readings, which does not trigger an event. All classification grid colors and symbols used in Spyglass are shown in Table 1.

Moreover, the varying weather conditions during the field trials, ranging from sunshine to thunderstorms, proved the system's adaptability and its resistance against hazardous environments.

7   Conclusion and Future Work

The MOVEDETECT system is a WSN solution capable of monitoring and securing critical infrastructures as was demonstrated in this paper. It provides security by using different sensors to detect, locate, track and classify various types of trespassers while operating in a secure and safe way. This is achieved by employing several functional safety and security features to guarantee high levels of confidentiality, integrity and availability. The system operates in real-time using efficient detection algorithms and a network hierarchy. MOVEDETECT enables its user to keep track of the events inside the sensor network by logging all events in a database for later analysis and, at the same time, reporting it in real-time to the command center. Moreover, the system uses a scalable, hierarchic network structure making it easily extendable for a wider area, if necessary. To adapt the system to changing usage conditions, it is possible to dynamically reconfigure the whole system in the field by using an over-the-air-programming (OTAP) function. This unique combination of features makes the system ideal for unattended operations in hazardous environments.

While the described features already enable a secure detection of trespassing objects, additional hardware, e.g. video cameras, could be included in the concept in order to document trespassers in future work. These devices could be integrated as a type of additional sensor into the existing sensor nodes and be activated if an object is detected by the other sensors. Of course, these new nodes would need to be carefully analyzed, especially their energy consumption and network traffic generation would need to be carefully balanced with the achievable visual quality. Another possible method is to treat the sensor nodes with visual capabilities separately and not to integrate them into the existing network hierarchy. The visual data would then be sent directly to the command center using WiFi.

Acknowledgments. The work was funded by and organized in cooperation with the German Federal Office for Information Security (BSI - "Bundesamt für Sicherheit in der Informationstechnik") within the project MOVEDETECT. We also thank the Federal Police of Germany, in particular Mr. Vehrkamp, for their support and numerous constructive discussions.

References

1. Critical Infrastructure Assurance Office: Practices for securing critical information assets (January 2000)
2. Intanagonwiwat, C., Govindan, R., Estrin, D.: Directed diffusion: a scalable and robust communication paradigm for sensor networks. In: Proceedings of the ACM Mobi-Com 2000, Boston, MA, pp. 56–67 (2000)
3. Akyildiz, I.F., Su, W., Sankarasubramaniam, Y., Cayirci, E.: Wireless sensor networks: a survey. Computer Networks 38, 393–422 (2002)

4. Roman, R., Zhou, J., López, J.: On the security of wireless sensor networks. In: Gervasi, O., Gavrilova, M.L., Kumar, V., Laganá, A., Lee, H.P., Mun, Y., Taniar, D., Tan, C.J.K. (eds.) ICCSA 2005. LNCS, vol. 3482, pp. 681–690. Springer, Heidelberg (2005)

5. Perrig, A., Stankovic, J., Wagner, D.: Security in wireless sensor networks. Communications of the ACM 47(6), 53–57 (2004)

6. Rothenpieler, P., Krüger, D., Pfisterer, D., Fischer, S., Dudek, D., Haas, C., Zitterbart, M.: Flegsens – secure area monitoring using wireless sensor networks. In: Proceedings of the 4th Safety and Security Systems in Europe (2009)

7. Bonitz, F., Ghobadi, S.E., Hartmann, K., Hauff, H., Herrmann, R., Kargel, C., Löpprich, O.E., Heckmann, D., Maisch, M.M., Seidl, A., de Meer, H., Ruser, H., Sachs, J., Wenzl, K.: Personen- und objektdetektion mit mobilen sensoren - abschlussbericht zum archbeitspacket 3 (teil b) - bericht zum meilenstein 5. End report (September 2010)

8. Koch, M., Hubert, C.: Personen- und objekterkennung basierend auf trittschall (pot). Technical report (2011)

9. Kaseva, V., Hämäläinen, T.D., Hännikäinen, M.: A wireless sensor network for hospital security: from user requirements to pilot deployment. EURASIP J. Wirel. Commun. Netw. 2011, 17:1–17:15 (2011)

10. Bohli, J.M., Hessler, A., Ugus, O., Westhoff, D.: A secure and resilient wsn roadside architecture for intelligent transport systems. In: Proceedings of the First ACM Conference on Wireless Network Security, WiSec 2008, pp. 161–171. ACM, New York (2008)

11. Sun, Z., Wang, P., Vuran, M.C., Al-Rodhaan, M.A., Al-Dhelaan, A.M., Akyildiz, I.F.: Bordersense: Border patrol through advanced wireless sensor networks. Ad Hoc Netw. 9(3), 468–477 (2011)

12. Magno, M., Marinkovic, S., Brunelli, D., Benini, L., Popovici, E.: Combined methods to extend the lifetime of power hungry wsn with multimodal sensors and nanopower wakeups. In: 2012 8th International on Wireless Communications and Mobile Computing Conference (IWCMC), pp. 112–117 (2012)

13. Estrin, D., Govindan, R., Heidemann, J., Kumar, S.: Next century challenges: Scalable coordination in sensor networks. In: Proceedings of International Conference on Mobile Computing and Networks (MobiCom 1999), Seattle, WA, USA (August 1999)

14. Institute of Telematics, University of Lübeck, G.: Spyglass, a modular and extensible visualization framework for wirelesssensor networks (2006), https://github.com/itm/spyglass

15. Havskov, J.: Instrumentation in Earthquake Seismology. Modern Approaches in Geophysics. Springer (2004)

Synchronization for Cooperative MIMO in Wireless Sensor Networks

Marco A.M. Marinho, Edison Pignaton de Freitas,
João Paulo Carvalho Lustosa da Costa, and Rafael Timóteo de Sousa Júnior

Universidade de Brasília,
Laboratory of Array Signal Processing,
P.O. Box 4386, Zip Code 70.919-970, Brasilia - DF
marco.marinho@ieee.org, {edisonpignaton,desousa}@unb.br,
joaopaulo.dacosta@ene.unb.br

Abstract. The application of Wireless Sensor Networks (WSNs) is hindered by the limited energy budget available for the member nodes. Energy aware solutions have been proposed for all tasks involved in WSNs, such as processing, routing, cluster formation and communication. With communication being responsible for a large part of the energetic demand of WSNs energy efficient communication is paramount. The application of MIMO (Multiple-Input Multiple-Output) techniques in WSNs emerges as a efficient alternative for long range communications, however, MIMO communication require precise synchronization in order to achieve good performance. In this paper the problem of transmission synchronization for WSNs employing Cooperative MIMO is studied, the main problems and limitations are highlighted and a synchronization method is proposed.

Keywords: multiple-in multiple-out (MIMO) systems, wireless sensor networks (WSNs), synchronization.

1 Introduction

Recently wireless sensor networks have emerged as the tool of choice for a large number of emerging applications. Their usage ranges from military applications, such as battlefield surveillance and targeting, to health care applications, such as automating drug applications in hospitals [1]. However, the large scale application of WSNs is still hindered by the limited energy budget available to the nodes that compose such network and due to the fact that, since WSNs are first choices for deployment in harsh and hard to reach environments, replacing individual nodes, or their batteries, may become unpractical. Extensive research has been conducted with the aim of maximizing the energy efficiency of WSNs [2].

Solutions aiming to minimize energy consumption in WSNs have been proposed trough different layers, with energy efficiency being analyzed for all tasks involved in WSNs. Energy efficient protocols for medium access control have

S. Balandin et al. (Eds.): NEW2AN/ruSMART 2013, LNCS 8121, pp. 298–311, 2013.

been proposed on [3, 4], while many proposals focus on enhancing energy efficiency in the network layer, by means of energy efficient routing protocols [5, 6]. Other proposals consider energy aware processing approaches for communications among sensor nodes in WSN, as presented in [7], as well as alternatives solutions on the physical layer [8, 9].

Since communication is responsible for a large part of the energetic demand in WSNs, special attention has been given to the study of energy efficient communication methods, with multi-hop communication being a widely used technique to obtain improved energy efficiency and maximize network life time by spreading energy consumption over different nodes [10]. Multi-hop takes advantage of the cooperative nature of WSNs in order to split the distance involved in communication by employing intermediary nodes to forward data packets. Since free space loss is not linear, splitting the distance results in reduced power demand, thus minimizing energy consumption. However care must be taken when applying multi-hop in order to avoid reduced energy efficiency, as presented in [10].

Also taking advantage of the cooperative nature of WSNs, the formation of Cooperative MIMO clusters have been proposed. In this context, the work presented in [11] proposes a cooperative MIMO system used in the communication amongst the sensor nodes, in [12] cooperative MIMO transmissions are studied, and (Single-Input Multiple-Output) SIMO and (Multiple-Input Single-Output) MISO cases are taken into account. In [13] a energy analysis considering single-hop, multi-hop and cooperative MIMO is presented. Results obtained in these works show that cooperative MIMO is only advantageous when long distances are involved. Nevertheless, the advantages of cooperative MIMO are not restricted to energy efficiency. The faster data rates achievable with MIMO system allow the interaction between fast moving mobile as well as traditional static nodes [14, 15]. Cooperative MIMO also allows the application of antenna array techniques such as beam forming or direction of arrival estimation can be employed.

The application of Cooperative MIMO results in reduced hardware complexity on single nodes, allowing the application of the technique with minimal to no modifications in the hardware of existing WSNs. This complexity is transfered to the software responsible for managing the communication involving a large number of nodes. One of the most critical aspects of successful MIMO communications is the proper synchronization between the nodes involved.

In this paper, we analyze the behavior of a simple synchronization mechanism for WSNs employing cooperative MIMO. The efficiency of the proposed methods is studied by means of simulations. The remainder of this paper is divided into six sections. In section 2 the principles of MIMO communication are presented. Traditional equalization methods are introduced and their performances are compared with standard SISO communications. In section 3 the cooperative MIMO transmission technique for WSNs is briefly presented. In section 4 two algorithms for synchronizing Cooperative MIMO clusters are presented. In section 5 simulation results are shown and discussed. Conclusions are drawn in section 6.

2   MIMO Communications

MIMO communications consist of the use of multiple antennas for data transmission and reception. The use of multiple antennas to achieve benefits in various aspects of communication such as a lower bit error ratio (BER) or increased throughput.

This work focuses on MIMO techniques to achieve spatial multiplexing. Spatial multiplexing is used to transmit parallel bit streams simultaneously over the same frequency. MIMO also results in the array gain phenomenon, which is the increase of effective received power, due to multiple copies of the signal being received on different antennas.

Consider a sequence of symbols

$$s = [s_1, s_2, ..., s_N], \tag{1}$$

that needs to be transmitted over a wireless channel. The channel is assumed to be flat fading, which means that channel impulse response is constant over the frequency domain, also equivalent to considering the transmitted signal to be narrow-band. The impulse response between antennas is assumed to be uncorrelated and constant over a transmission period.

A technique called V-BLAST [16] which is employed for MIMO communications in this work, is of particular interest. In a normal transmission at each time slot a single symbol would be transmitted over the channel while, in the case of V-BLAST transmission, the symbols are grouped into multiple parallel streams. In the case shown in Figure 1, groups of size Q , and transmitted over the same time slot.

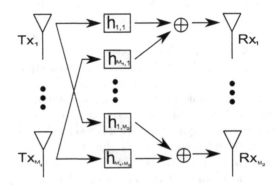

Fig. 1. Example of a Q by Q MIMO system

The received signal at a given receiving antenna x_i at a given time slot can be modeled as

$$x_i = \sum_{k=1}^{Q} h_{k,i} \cdot s_k + n_i, \tag{2}$$

where $h_{k,i}$ represents the complex impulse response of channel between transmit antenna k and receive antenna i and s_k is the symbol transmitted by the k-th antenna. n_i is the noise present at the i-th receiving antenna during sampling. Equation 2 can be rewritten in matrix form as

$$x_i = [h_{1,i}, h_{2,i}, ..., h_{Q,i}] \begin{bmatrix} s_1 \\ s_2 \\ \vdots \\ s_Q \end{bmatrix} + n_i. \tag{3}$$

Equivalently a matrix representation for the signals received at all receiving antennas can be written as

$$\begin{bmatrix} x_1 \\ x_2 \\ \vdots \\ x_Q \end{bmatrix} = \begin{bmatrix} h_{1,1} & h_{2,1} & \cdots & h_{Q,1} \\ h_{1,2} & h_{2,2} & \cdots & h_{Q,2} \\ \vdots & \vdots & \cdots & \vdots \\ h_{1,Q} & h_{2,Q} & \cdots & h_{Q,Q} \end{bmatrix} \cdot \begin{bmatrix} s_1 \\ s_2 \\ \vdots \\ s_Q \end{bmatrix} + \begin{bmatrix} n_1 \\ n_2 \\ \vdots \\ n_Q \end{bmatrix}, \tag{4}$$

$$\Updownarrow$$

$$x = Hs + n, \tag{5}$$

The first step necessary in order to estimate the transmitted symbols is to estimate the channel matrix H. An estimate $\hat{H}$ can be obtained by transmitting a set of pilot symbols vectors $P = [p_1, p_2, \ldots, p_U] \in \mathbb{C}^{Q \times U}$ where $p_i \in \mathbb{C}^{Q \times 1}$ and $U > Q$

$$\hat{H} = XP^\dagger, \tag{6}$$

here $P^\dagger = P^H(PP^H)^{-1}$ is known as the right pseudo inverse of matrix P and the operator ^{H} denotes the conjugate transposition. For a more detailed discussion on trade offs and optimal pilot symbol selection for MIMO channel estimation the reader may refer to [17, 18].

Once the channel matrix estimate $\hat{H}$ has been obtained the receiver needs to equalize the received symbols in order to obtain an estimate of the transmitted symbols, various methods exist for performing this equalization, here the Zero Forcing, Minimum Mean Square Error (MMSE) and Maximum Likelihood (ML) methods are analyzed.

The Zero Forcing method consists of finding a matrix W that satisfies $WH = I$, where I is an identity matrix. This matrix in given by

$$W = (\hat{H}^H \hat{H})^{-1} \hat{H}^H, \tag{7}$$

as Equation 7 shows, calculation W is equivalent to calculating the left pseudo inverse of $\hat{H}$. An estimate of the transmitted symbols is given by

$$\hat{S} = WHS + WN, \tag{8}$$

Equation 8 shows that depending on the structure of W the received noise might be amplified at equalization, thus degrading the estimate of the transmitted signals.

MMSE equalization tries to solve to problem the of noise amplification by taking into account the noise when calculating the equalizer. MMSE tries to find a matrix W that minimizes the criterion

$$E\left\{[WX - S][WX - S]^{\mathrm{H}}\right\},\tag{9}$$

where W is obtained by

$$W = (\hat{H}^{\mathrm{H}}\hat{H} + \mathrm{N}_0 I)^{-1}\hat{H}^{\mathrm{H}},\tag{10}$$

where N_0 is the power of the received noise. Notice that in the absence of noise Equation 10 reduces to 7.

Finally, ML equalization tries to find a matrix $\hat{S}$ that minimizes the criterion

$$Err = \left|X - \hat{H}\hat{S}\right|^2,\tag{11}$$

this is done numerically by testing all possible combinations of $\hat{S}$ and deciding on the one which leads to the minimum Err. Computationally efficient alternatives exist for the ML method such as spherical decoding.

Figure 2 shows a comparison between standard SISO systems and a 2×2 MIMO configuration using the equalization methods discussed previously. The ML equalization method is clearly the most efficient in terms of minimizing the bit error rate (BER) of the received bit stream, thus it is the method of choice for the remainder of this work.

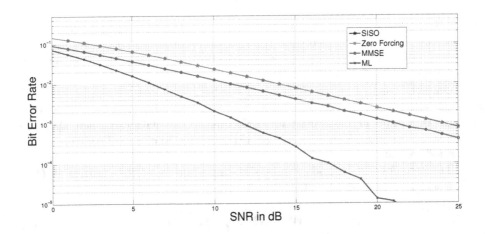

Fig. 2. Performance comparison between standard SISO systems and 2×2 MIMO systems using Zero Forcing, MMSE and ML equalization

3   Cooperative MIMO

Wireless sensor networks are cooperative by nature, taking advantage of this behavior a cooperative MIMO approach can be implemented in order to minimize
the energy spent with communication between nodes. As opposed to traditional
MIMO systems, where a set of antenna is present at the transmitter and at
the receiver, the cooperative MIMO utilizes a virtual MIMO approach, where
the multiple antennas involved are present at different systems (different nodes).
This avoids the increased hardware complexity involved, which is specially important in WSNs due to their limitations in term of size and hardware complexity. The additional complexity is transferred to the communication protocol.
Figure 3 presents the steps involved in a cooperative MIMO communication.

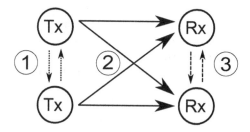

Fig. 3. Steps involved in a cooperative MIMO transmission

The first step represented by ① consists of synchronization and exchanging
data that needs to be transmitted, if both sensors need to transmit data this
exchange is not necessary, as each sensor can transmit its own data. Note that
since WSNs usually operate at low data rates the synchronization does not need
to be extremely precise as the symbol duration is usually long enough so that
small or even moderate offset in transmission instants does not result in errors.
The same can be said for the synchronization in the reception. Small offsets in
the sampling instant in the reception will not interfere with the overall system
performance. On ② both sensors transmit different symbols at the same time slot
according to the BLAST architecture discussed above. Space time block codes
(STBCs) such as [19][20] can be chosen according to the necessary or expected
behavior of the network. Finally on ③ the receiving sensors sample and quantify
the received symbols and exchange the quantified data so that the originally
transmitted symbols can be extracted. If the data is destined to only one sensor
of the receiving cluster this exchange becomes uni directional. Another option
is to exchange only a portion of the received information so that every sensor is
responsible for part of the decoding, alleviating the computational burden of a
single node.

4   Cooperative MIMO Synchronization

The problem of network wide synchronization in WSNs has been extensively studied [21–24]. Algorithms have been proposed in order to keep a common clock across the entire network. Most solutions suggest keeping a relationship between clocks across the network instead of trying to forcefully synchronizing clocks across all nodes. For networks relying on GPS synchronization it has been shown that very precise synchronization can be achieved, with variations being kept as small as 200 ns [25]. However, relying on GPS receivers results in increased energy demand and in a WSN that is no longer self contained. Broadcast synchronization schemes capable of achieving 1 μs of accuracy have been proposed [26].

For networks operating at a 256 kbps rate and using BPSK modulation the resulting symbol duration is approximately 4 μs, a synchronization error of 1 μs represents 25 % of the symbol duration. MIMO communications demand precise synchronization in order to achieve proper decoding, it is clear that another mean of synchronizing transmitting nodes and receiving nodes is necessary.

A simple method of synchronization is to over sample a received tonal wave and compare it with a reference wave kept internally. This can be done by using a sliding matched filter to digitally find the delay, in samples, that results in maximum correlation with the received wave. Since the networks are assumed to already be synchronized with a maximum error of 1 μs the range of comparison is reduced.

The first proposed method consists in scheduling a tonal transmission between a pair or tonal broadcast to a group of nodes, the sampling on the receiving nodes will start at the scheduled time, and the clock error can be compensated. To avoid problems created by sampling with a difference of more than a period of the tonal wave the time length of the tonal transmission needs to be known to the receiving nodes, this way, if a signal with less than the expected length is received the receiver can compensate by starting sampling earlier or later, and adjust its internal clock accordingly. Figure 4 presents the proposed mechanism. The sampling synchronization error d can be compensated prior to applying the sliding correlator achieve precise synchronization.

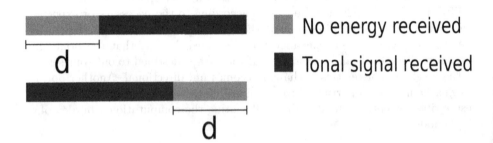

Fig. 4. Illustration of a sampling synchronization error

Once sampling is synchronized a finner synchronization can be done by applying a matched filter, the maximum theoretical error in the absence of noise is this case is equivalent to the sampling interval employed by the receiver. Figure 5 depicts how the sliding correlator is applied in order to achieve total synchronization.

Fig. 5. Illustration of sliding correlation synchronization

It is important to notice that this synchronization does not take into account the propagation time of the transmitted wave, this correction can be done by measuring the approximate distance between a pair of sensors. This can be done, for example, by means of the RSSI. Another alternative is to employ DOA techniques such as MUSIC [27] or ESPRIT [28] to map relative sensor positions in the network.

Finally Figure 6 presents the steps involved in achieving complete synchronization between a pair of nodes.

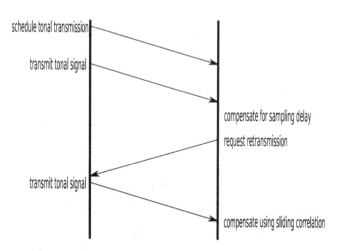

Fig. 6. Steps necessary for synchronization

A second and more robust method is to apply subspace methods used for parameters estimation in order to estimate the delay of the received wave with regard to the reference wave. This technique was first proposed in the context of achieving precise synchronization for GPS receivers [29] and its usage can be extended to the problem at hand. The time MUSIC consists of substituting the common antenna array configuration know in direction of arrival (DOA) estimation problems for a correlator bank configuration. The received signal is correlated to a set of forward and backward delayed replicas and a delay spectrum can be obtained similar to the spatial spectrum obtained in the original MUSIC.

The correlator bank transforms the input according to

$$x = Y, \tag{12}$$

where x is a vector containing the received signal and Y is a matrix with its rows containing the cross correlation between the received signal and its respective correlator. With Y an estimate of the covariance of the received signal trough the bank can be obtained by

$$R_{YY} = \mathrm{E}\{Y \times Y^{\mathrm{H}}\}, \tag{13}$$

where E is the expectation operator. The covariance matrix can be decomposed using the eigendecomposition, yielding

$$R_{YY} = \Sigma \Lambda \Sigma^{-1}. \tag{14}$$

By removing the eigenvector related to the strongest eigenvalue an estimate of the noise subspace Q_n can be obtained. Finally a delay spectrum can be obtained by

$$P(d) = \frac{a(d) \times a(d)^{\mathrm{H}}}{a(d)^{\mathrm{H}} \times Q_n \times Q_n^{\mathrm{H}} \times a(d)},$$

where $a(d)$ is the cross correlation between a signal that would be received with a given delay d and the bank of correlators.

5   Simulation Results and Discussion

In order to verify the precision of the synchronization possible with the proposed first method numerical simulations were performed. The first result that needs to be analyzed is the behavior of the proposed method in the presence of noise. Figure 7 depicts the synchronization precision achieved in seconds in relation to different levels of white Gaussian noise. The transmitted signal has a frequency of 2.4 GHz, 5000 samples are used for the sliding correlator and the tonal wave is sampled at twice the Nyquist rate.

It is possible to notice that even for low SNR scenarios the proposed method is capable of achieving synchronization in the order of 10^{-8} seconds, two order of magnitude superior to what is currently achievable in network synchronization methods [21–24].

Another important factor that needs to be analyzed is the performance of the proposed method for different numbers of samples. For this simulation the SNR is kept fixed at -15 dB and the sampling rate is twice the Nyquist rate. Sample size ranges from 100 to 10000.

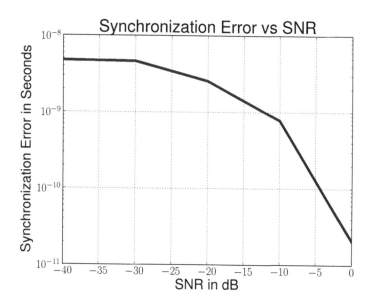

Fig. 7. Results for synchronization error in seconds versus SNR

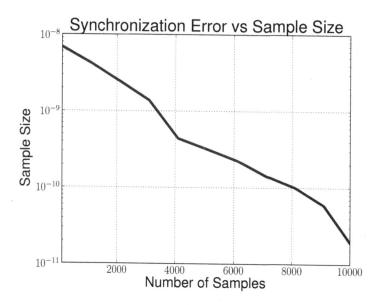

Fig. 8. Results for synchronization error in seconds versus sample size

Figure 8 presents the synchronization results for different sample sizes. It is possible to notice that, as expected, increased sample sizes result in increased accuracy. However, this comes at the cost of more time and energy being spent to perform synchronization and at the cost of increased computational complexity, since correlation complexity increases with the increased number of samples.

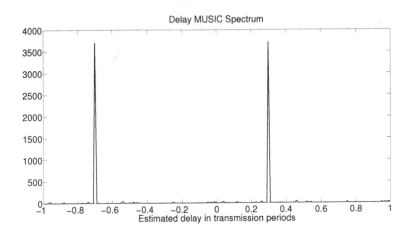

Fig. 9. Delay MUSIC spectrum

Figure 9 presents the spectrum obtained with the delay MUSIC technique. Notice that this technique is also unable to tell delays that are more than one transmission period apart, since their correlation is the same. Thus the proposed step of measuring the received energy is also necessary.

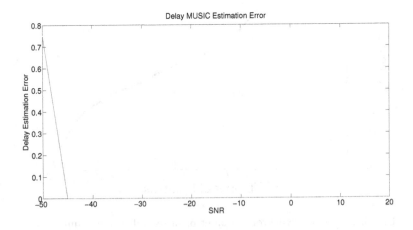

Fig. 10. Delay MUSIC mean error

Figure 10 presents the delay estimation error for the delay MUSIC technique, notice that this technique is extremely robust to noise in the transmission, allowing very precise estimation even at an SNR greatly below the noise floor. This however, comes at the cost of increased computational complexity, since it is necessary to calculate the covariance matrix of the signal after the correlator bank and its eigendecomposition. Also, a search mechanism needs to be employed to find the peaks over the obtained delay spectrum.

Note that transmission rates in WSNs are usually very low, with 2000 kb/s being considered a very high rate, and only achievable at small distances between nodes under low SNR. Rates ranging between 80 kb/s - 250 kb/s are typical data rates for WSNs in operation today [30, 31]. High data rate systems usually employ modulations with large constellations, resulting in increased symbol duration. This allows the proposed techniques to be efficient in allowing communications for even such networks.

6 Conclusion

This paper presents a initial approach for precise sensor synchronization in WSNs in order to employ Cooperative MIMO communications. Cooperative MIMO communications are capable of providing enhanced energy efficiency in WSNs by providing improved long range communications. However, since MIMO communications involve the decoding of multiple symbols transmitted at the same time slot, precise synchronization is required for proper decoding. Taking advantage of existing network synchronization protocols a method is proposed in order to achieve synchronization compatible with the transmission rate of WSNs in operation today. Simulation results corroborate that the proposed techniques are capable of such synchronization. Further study is planed in enhanced synchronization methods, employing reduced sample sizes or taking advantage of usual SISO traffic to avoid overhead specific to node synchronization.

References

1. Akyildiz, I., Su, W., Sankarasubramaniam, Y., Cayirci, E.: Wireless sensor networks: a survey. Computer Networks 38, 393–422 (2002)
2. Dietrich, I., Dressler, F.: On the Lifetime of Wireless Sensor Networks. ACM Transactions on Sensor Networks 5 (2009)
3. Cho, S., Chandrakasan, A.: Energy-efficient protocols for low duty cycle wireless microsensor. In: Proceedings of the 33rd Annual Hawaii International Conference on System Sciences (2000)
4. Lettieri, P., Srivastava, M.B.: Adaptive frame length control for improving wireless link throughput, range and energy efficiency. In: INFOCOM 1998, Proceedings of IEEESeventeenth Annual Joint Conference of the IEEE Computer and Communications Societies (1998)
5. Ganesan, D., Govindan, R., Shenker, S., Estrin, D.: Highly-resilient, energy-efficient multipath routing in wireless sensor networks. ACM SIGMOBILE Mobile Computing and Communications Review 5(4), 1125 (2001)

6. Heinzelman, W.R., Chandrakasan, A., Balakrishnan, H.: Energy-efficient communication protocol forwireless microsensor networks. In: Proceedings of the 33rd Hawaii International Conference on System Sciences (2000)
7. Weiser, M., Welch, B., Demers, A., Shenker, S.: Scheduling for reduced cpu energy. In: Proceedings of 1st USENIX Symposium on Operating System Design and Implementation, pp. 13–23 (November 1994)
8. Shih, E., Calhoun, B.H., Cho, S., Chandrakasan, A.P.: Energy-efficient link layer for wireless microsensor networks. In: Proceedings. IEEE Computer Society Workshop on VLSI (2001)
9. Shih, E., Cho, S., Ickes, N., Min, R., Sinha, A., Wang, A.C.A.: Physical layer driven protocol and algorithm design for energy-efficient wireless sensor networks. In: Proceedings of ACM MobiCom 2001 (2001)
10. Chen, C., Ma, J., Yu, K.: Designing energy-efficient wireless sensor networks with mobile sinks. In: Sensys 2006 (2006)
11. Bravos, G.N., Kanatas, A.G.: Combining MIMO and Multihop Based Transmissions on Energy Efficient Sensor Networks. In: Procedings of the Program for European Wireless (2007)
12. Cui, S., Goldsmith, A.J.: Energy-efficiency of mimo and cooperative mimo techniques in sensor networks. IEEE Journal on Selected Areas in Communications (2004)
13. de Freitas, E.P., da Costa, J.P.C.L., de Almeida, A.L.F., Marinho, M.: Applying MIMO techniques to minimize energy consumption for long distances communications in wireless sensor networks. In: Andreev, S., Balandin, S., Koucheryavy, Y. (eds.) NEW2AN/ruSMART 2012. LNCS, vol. 7469, pp. 379–390. Springer, Heidelberg (2012)
14. Marinho, M.A.M., de Freitas, E.P., da Costa, J.P.C.L., de Almeida, A.L.F., de Sousa Jr., R.T.: Using mimo techniques to enhance communication among static and mobile nodes in wireless sensor networks (2013)
15. Marinho, M.A.M., de Freitas, E.P., da Costa, J.P.C.L., de Almeida, A.L.F., de Sousa Jr., R.T.: Using cooperative mimo techniques and uav relay networks to support connectivity in sparse wireless sensor networks (2013)
16. Wolniansky, P.W., Foschini, G.J., Golden, G.D., Valenzuela, R.A.: V-BLAST: An architecture for realizing very high data rates over the rich-scattering wireless channel. In: Signals, Systems and Electronics (1998)
17. Biguesh, M., Gershman, A.B.: Training-based MIMO channel estimation: A study of estimator tradeoffs and optimal training signals. IEEE Transactions on Signal Processing 54, 884–893 (2006)
18. Trepkowski, R.: Channel estimation strategies for coded MIMO systems. Master's thesis, Virginia Polytechnic Institute and State University
19. Jafarkhani, H.: A Quasi-Orthogonal Space-Time Block Code. IEEE Transactions on Communications 49 (2001)
20. Alamouti, S.M.: A simple transmit diversity technique for wireless communications. IEEE Journal on Selected Areas in Communications 16 (1998)
21. Ganeriwal, S., Ganesan, D., Shim, H., Tsiatsis, V., Srivastava, B.: Estimating clock uncertainty for efficient duty-cycling in sensor networks. In: Proceedings of the Third International Conference on Embedded Networked Sensor Systems, Sensys (2005)
22. Lucarelli, D., Wang, I.-J.: Decentralized synchronization protocols with nearest neighbor communication. In: Proceedings of the Sensys (2004)

23. Wener-Allen, G., Tewari, G., Patel, A., Welsh, M., Nagpal, R.: Firefly inspired sensor network synchronicity with realistic radio effects. In: Proceedings of the Third International Conference on Embedded Networked Sensor Systems, Sensys (2005)
24. Ganeriwal, S., Kumar, R., Srivastava, M.: Timing-sync protocol for sensor networks. In: Proceedings of the Sensys (2003)
25. Mannermaa, J., Kalliomaki, K., Mansten, T., Turunen, S.: Timing performance of varios gps receivers. In: Proceedings of the 1999 Joint Meeting of the European Frequency and Time Forum and the IEEE International Frequency Control Symposium (1999)
26. Elson, J., Girod, L., Estrin, D.: Fine-grained network time synchronization using reference broadcasts. In: Proceedings of the Fifth Symposium on Operating Systems Design and Implementation (2002)
27. Schmidt, R.O.: Multiple emitter location and signal parameter estimation. IEEE Transactions on Antennas and Propagation 34, 276–280 (1986)
28. Roy, R., Kailath, T.: ESPRIT - estimation of signal parameters via rotation invariance techniques. IEEE Transactions on Acoustics Speech and Signal Processing 17 (1989)
29. Selva-Vera, J.: Subspace Methods to Multipath Mitigation in a Navigation Receiver. In: Proceedings of the IEEE Vehicular Technology Conference, VTC (1999)
30. Zhu, N., Du, W., Navarro, D., Mieyeville, F., Connor, I.O.: High data rate wireless sensor networks research. Technical Report
31. Lanzisera, S., Mehta, A.M., Pister, K.S.J.: Reducing average power in wireless sensor networks through data rate adaptation. In: Proc. IEEE International Conference on Communications, pp. 480–485 (2009)

The Mobile Sensor Network Life-Time under Different Spurious Flows Intrusion

Andrey Koucheryavy, I. Bogdanov, and Alexander Paramonov

State University of Telecommunication, St.Petersburg, pr.Bolshevikov 22, Russia
akouch@mail.ru

Abstract. The network-based security defined by International Telecommunication Union (ITU) has specific features for Ubiquitous Sensor Networks (USN). An enormous number of sensor nodes and hard energy constraints define specific network-based security features of USNs. The network-based security features of USNs are considered based on ITU-T Recommendation X.1311. A new type of attacks on the energy system of a USN is proposed, which is based on the spurious flows generation. The sensor nodes respond to the spurious flows and the sensor network life-time decreases. The mobile sensor network behavior in case of spurious flows intrusion is investigated in the paper. A homogenous mobile sensor network and the LEACH-M cluster head selection algorithm are used in this study. The effect of Poisson and deterministic spurious flows intrusion is investigated. The sensor node moving speed influence on the sensor network life-time is analyzed too.

Keywords: mobile sensor network, spurious flow, intrusion, life-time.

1 Introduction

The ITU-T Recommendation Y.2701 [1] defines network-based security as "security of end user communications across multiple-network administration domain". The user network and P2P application capabilities on customer equipment are not considered on Recommendation Y.2701. The network-based security requirements in Recommendation Y.2701 developed for NGN (Next Generation Networks) in accordance with the NGN functional architecture [2]. The network-based security requirements include the whole NGN and the NGN interfaces: UNI (User Network Interface), NNI (Network Network Interface), ANI (Application Network Interface).

The network-based security for sensor networks has specific features. These features are covered in ITU-T Recommendation X.1311 [3]. First of all, we should note that an enormous number of sensor nodes and hard energy constraints are the most important features of sensor networks. The more than 64000 nodes can be organized on a single sensor network in accordance with ZigBee protocol specifications [4]. The energy consumption is the most important challenge for sensor network life-time and for modern wireless networks [5, 6, 7]. The connectivity, mobility and coverage can also considered as important parameters [8].

An enormous number of sensor nodes and high requirements to energy system define specific features for sensor network-based security. The specific

S. Balandin et al. (Eds.): NEW2AN/ruSMART 2013, LNCS 8121, pp. 312–317, 2013.

network-based security support topics for the sensor networks include in accordance with Recommendation X.1311:

- difficulty of using public key cryptosystems,
- vulnerability of sensor nodes,
- difficulty in obtaining post deployment knowledge,
- limited memory size, transmission power, and transmission bandwidth,
- single point of failure of a base station.

The last item relates to a cluster head node too.

Sensor nodes are very simple devices. Cloning is the one more specific threat for sensor networks [9]. The sensor network energy system is very vulnerable. The energy system conditions are directly connected to the sensor network life-time. For example, attacks to the sensor network energy system can be organized as the sensor nodes sleep deprivation [10]. In this paper, we propose one more type of attacks to the sensor network energy system, which is based on the generation of spurious flows. The sensor nodes respond to the spurious flows and the sensor network life-time decreases. The mobile sensor network behavior in the case of the spurious flows intrusion is investigated in the paper.

2   The Sensor Network Structure and Algorithms

Today, sensor networks are commonly referred to as Ubiquitous Sensor Networks (USNs). It's understandable that the 7 trillion wireless telecommunication unites could be form the network at 2017-2020 years in according with forecast [11]. There are many USN applications – everywhere, anywhere, anytime. The most important USN application include building and industrial automation, logistics, transportation, body and intra-body sensors and RFIDs, military engineering, agriculture, environment data [12,13]. Further, it can be growth of trees, growth of animals and so on [14]. The library network will be one more USN application [15].

An enormous number of sensor nodes requires development of USN hierarchy structures. The cluster hierarchy is the most widely used structure for USN. In this case, the cluster head should be selected from sensor nodes. The cluster head selection algorithm is one of the most important research topics in the USN area. The energy consumption reduction and the sensor network life-time growth should be supported by cluster head selection algorithm. The LEACH (Low Energy Adaptive Cluster Hierarchy) algorithm [16] is the main algorithm for cluster head selection. The LEACH version for mobile sensor networks LEACH-Mobile, in short "LEACH-M" [17] is a variant of LEACH, which support node mobility. In LEACH-M mobile nodes declare the membership of a cluster as they move, and to confirm whether a mobile sensor node is able to communicate with a specific cluster head within a time slot allocated in TDMA schedule. This ensures dynamic joining and leaving of non-cluster mobile nodes in the steady state phase, unlike [16] where cluster membership is fixed after cluster formation. However, in LEACH-M, clusters are dynamically formed every time the sensor moves, giving rise to chance of overhead in the cluster maintenance.

We will use algorithm LEACH-M further and the homogenous sensor network [18] in the investigation model. All sensors in the homogenous sensor network have the same parameters: initial energy, radius, moving speed etc.

3   The Investigation Model and Modeling Results

The 100 sensor nodes are distributed on the plane 200x200m randomly. The sensor node activity radius is 25 m. The sensor nodes average moving speed varies from 1 m/s (slowly walking pedestrian) up to 8 m/s (average car speed in the big town). The initial energy is 2J for each sensor node, the energy consumption for receiving is 50 nJ/bit, the energy consumption for transmitting is 50 nJ/bit and 100 pJ per square meter additionally. The sink located in the plane center. The cluster head selection algorithm is LEACH.

The spurious flow two types are considered. One of them is the Poisson flow, another flow type is deterministic. The spurious flows intensity varies from 1 event/s up to 10 events/s. The C#.NET is used for modeling.

The life-time in rounds as a function of the spurious flow intensity for different moving speeds of sensor nodes in case of deterministic (2 m/s) and Poisson (2, 4, 6, and 8 m/s) flows is shown in Fig. 1. First of all we note that spurious flows intrusion reduce the sensor network life-time. Furthermore, the type of spurious flows also affects the sensor network life-time. This is especially noticeable in the range from 1 to 2 event/s. In this range, deterministic flows cause a more dramatic decrease in the sensor network life-time than Poisson ones. Thus, it is possible to adjust the spurious flow intensity and the type of flows to maximize the impact of the attack on the sensor network life-time.

The life-time in rounds as a function of the average moving speed of sensor nodes for different intensities (1, 2, 3, 4, and 10 event/s) of Poisson flows is shown in Fig. 2. The life-time depends medium from sensor nodes average moving speed than spurious flows intensity is 1 event/s and weakly in case 2 event/s.

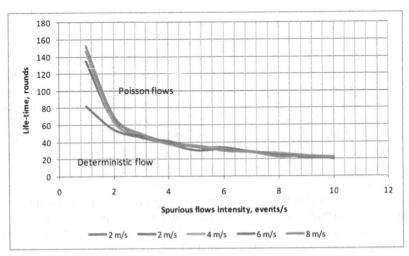

Fig. 1. The life-time in rounds as a function of the spurious flow intensity for different moving speeds of sensor nodes in case of deterministic (2 m/s) and Poisson (2, 4, 6, and 8 m/s) flows

The life-time is practically independent from sensor nodes average moving speed than spurious flows intensity is 3 events/s and more.

The life-time in rounds as a function of the average moving speed of sensor nodes for different intensities (1, 2, 3, 4, and 10 event/s) of deterministic flows is shown in Fig. 3. The life-time is practically independent from sensor nodes average moving speed in the all cases.

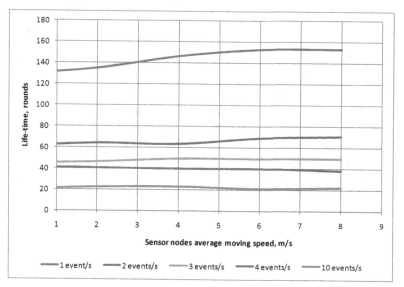

Fig. 2. The life-time in rounds as a function of the average moving speed of sensor nodes for different intensities of Poisson flows

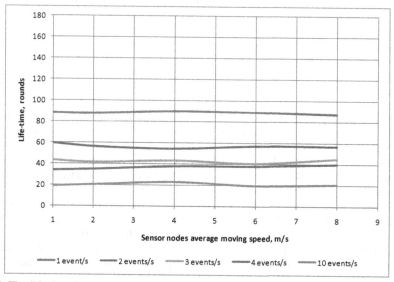

Fig. 3. The life-time in rounds as a function of the average moving speed of sensor nodes for different intensities of deterministic flows

4   Conclusions

A new attack type based on the different spurious flows generation is proposed. The spurious flows intrusion to the mobile ubiquitous sensor networks are investigated for Poisson and deterministic flows.

The spurious flow intrusion reduces the sensor network life-time. The flow type affects to the sensor network life-time. Deterministic flow cause a more noticeable decrease in the sensor network life-time than Poisson flow when the spurious flow intensity value interval is from 1 event/s up to 2 events/s. Hence, it is possible to adjust the spurious flow intensity and the type of flows to maximize the impact of the attack on the sensor network life-time.

The life-time depends medium from sensor nodes average moving speed than spurious flows intensity is 1 event/s. The life-time depends weakly from sensor nodes average moving speed than spurious flows intensity is 2 event/s. The life-time is practically independent from sensor nodes average moving speed than spurious flows intensity is 3 event/s and more. These results are valid in case of Poisson flow intrusion. The life-time is practically independent from sensor nodes average moving speed in the all investigated cases for deterministic flow intrusion.

References

1. Recommendation Y.2701. Security Requirements for NGN release 1. ITU-T, Geneva (April 2007)
2. Recommendation Y. 2012. Functional requirements and architecture of the NGN. ITU (2010)
3. Recommendation X.1311 Security Framework for Ubiquitous Sensor Networks. ITU-T, Geneva (February 2011)
4. http://www.zigbee.org
5. Younis, O., Fahmy, S.: Distributed clustering in ad-hoc sensor networks: A hybrid, energy-efficient approach. In: Proceedings, IEEE INFOCOM, Hong Kong, China (2004)
6. Andreev, S., Galinina, O., Koucheryavy, Y.: Energy-Efficient Client Relay Scheme for Machine-to-Machine Communication. In: IEEE Globecom 2011, Houston, TX, USA,
7. Aziz, A., Salim, A., Osamy, W.: Adaptive and Efficient Compressive Sensing based Technique for Routing in Wireless Sensor Networks. In: Proceedings, INTHITEN (IoT and its Enablers) Conference, June 3-4, St.Petersburg, State University of Telecommunication (2013)
8. Koucheryavy, A., Salim, A.: Prediction-based Clustering Algorithm for Mobile Wireless Sensor Networks. In: Proceedings, International Conference on Advanced Communication Technology, ICACT 2010, Phoenix Park, Korea (2010)
9. Bhattasali, T., Chaki, R.: A survey of recent intrusion detection systems for wireless sensor network. In: Wyld, D.C., Wozniak, M., Chaki, N., Meghanathan, N., Nagamalai, D. (eds.) CNSA 2011. CCIS, vol. 196, pp. 268–280. Springer, Heidelberg (2011)
10. Bhattassali, T., Chaki, R., Sanyal, S.: Sleep Deprivation Attack Detection in Wireless Sensor Networks. International Journal of Computer Applications 40(15) (February 2012)

11. Sorensen, L., Skouby, K.E.: Use scenarios 2020 – a worldwide wireless future. Visions and research directions for the Wireless World. Outlook. Wireless World Research Forum. 4 (July 2009)

12. Recommendation Y.2062. Framework of object-to-object communication using ubiquitous networking in NGN. ITU-T, Geneva (February 2012)

13. Kim, B.-T.: Broadband convergence Network (BcN) for Ubiquitous Korea Vision. In: Proceedings of the 7th International Conference on Advanced Communication Technology, ICACT 2005, Phoenix Park, Korea, February 21-23 (2005)

14. Marrocco, G.: Pervasive Electromagnetics: Sensing Paradigms by Passive RFID Technology. IEEE Wireless Communications 17(6) (December 2010)

15. Gorlatova, M., et al.: Energy Harvesting Active Networked Tags (EnHants) for Ubiquitous Object Networking. IEEE Wireless Communications 17(6) (December 2010)

16. Heinzelman, W., Chandrakasan, A., Balakrishnan, H.: Energy-efficient communication protocol for wireless microsensor networks. In: Proceedings 33rd Hawaii International Conference on System Sciences (HICSS), Wailea Maui, Hawaii, USA (January 2000)

17. Kim, D., Chung, Y.: Self-Organization Routing Protocol Supporting Mobile Nodes for Wireless Sensor Network. In: Proceedings of the First International Multi-Symposiums on Computer and Computational Sciences, vol. 2 (2006)

18. Koucheryavy, A., Salim, A.: Cluster head selection for homogeneous Wireless Sensor Networks. In: Proceedings of the International Conference on Advanced Communication Technology, ICACT 2009, Phoenix Park, Korea (2009)

19. Moltchanov, D., Koucheryavy, Y., Harju, J.: Performance response of wireless channels for quantitatively different loss and arrival statistics. Performance Evaluation 67(1), 1–27 (2010)

Internet Gateway Placement Optimization in Wireless Mesh Networks

Mojtaba Seyedzadegan, M. Othman, M.A. Borhanuddin, and S. Shamala

Universiti Putra Malaysia

Abstract. This paper elaborated on the importance of nodes degree and clustering for the efficient operation of Backbone Wireless Mesh Networks (BWMNs). A novel Zero-Degree (S) algorithm proposed for clustering the BWMN based on Wireless Mesh Routers (WMRs) degree/number of WMRs' connections, while ensuring Delay, Relay load and Cluster size constraints. It is shown that the performance of our algorithm outperforms the other alternatives. It places less Internet Gate-Ways (IGWs), and exhibits smooth and consistent performance when subject to various Quality of Service (QoS) constraints.

Keywords: WMN, Internet Gateway Placement, Optimization.

1 Introduction

Fast growing demand for high-speed connectivity from network customers due to rapid development of wireless technologies in the last few years, have encouraged the use of Wireless Mesh Networks (WMNs). Wireless mesh architecture offers the potential of providing a high-bandwidth network over a large coverage area at low costs.

One of the WMN promises is to provides a cost-effective alternative of high-speed Internet connectivity to wireless users by replacing conventional expensive cables. There are some typical services such as voice, messaging, e-mail, information services (e.g., news, stocks, weather, travel, sports, etc.), Internet fax, e-commerce, location-based services, health-care services, etc., that WMN could support. In addition, WMN could also support those applications that require high bandwidth communications such as online data, video broadcasting, video conference, and other multimedia services [1].

Performance of a wireless network evaluates by a key evaluation index which is "network capacity". It represents the long time achievable data transmission rate that can be supported by a network [2]. The capacity of wireless network depends on many aspects of the network: network architecture, routing strategy and radio interference model, communication paradigm, power and bandwidth constraints, etc. Among all of these factors, network architecture should be addressed carefully because it is especially critical. In order to maximize the network capacity and optimize the network performance, designing high bandwidth WMNs has been addressed in the past few years from different angles which

S. Balandin et al. (Eds.): NEW2AN/ruSMART 2013, LNCS 8121, pp. 318–331, 2013.

are location of the IGWs and the WMRs, the number of IGWs, the interface configuration and the channel assignment on each IGW and WMR, etc. Increasing demands also on the efficient resource initialization, fairness, and seamless handoff, can be seen to meet the Quality of Service (QoS) requirements of end users.

This paper proposes a novel algorithm, called "Zero-Degree (S)", for the IGW placement problem. Compared with existing algorithms, the new algorithm has the following advantages:

- guarantees to place a minimum number of IGWs to satisfy all the constraints;
- has competitive performance;
- has control of location of IGWs (i.e., aggregate or sparse) to distribute the IGWs in the locations that are closest to available wired network/Internet connection points;
- is easy to implement and use.

The rest of the paper is organized as follows: first, Overview of S-Model Architecture and S-Model Linear Program Formulation. This is followed by an overview of related works. Next, we discuss and detail the analysis of our Zero-Degree (S) algorithm. Performance evaluation and comparison to alternative approaches are performed in Section 6. Finally, we provide a description of the conclusion and future work.

2   S-Model Architecture and Linear Program Formulation

A S-Model architecture consisting of IGWs near each other and close to wired network/Internet connection points. Such a network architecture offers benefits like low wiring cost and efficient aggregation of equipments. More specifically, a S-Model architecture is suitable for the geographical restricted locations because: *(i)* it simplifies the wiring difficulties, and *(ii)* it can easily control, monitor, and maintain the most important equipments of a WMN like IGWs at same place. Therefore, in this paper, we focus on the S-Model architecture and formulate the IGW placement in such a way that all IGWs place in shortest distance from Internet connection points as well as each other. In other words, we can determine the IGWs positions for a given WMN to be organized into multiple clusters such that the head clusters or IGWs are aggregated and placed in shortest distance from Internet connection points.

Based on the specific S-Model architecture, and additional to the objectives discussed in Introduction Section, we further present the new objective in following way:

- **Minimizing the distance between IGWs:** The m number of IGWs in the network should be in shortest distance from each other. Wee need to determine the locations of IGW that results in minimum distance from each other to meet the constraint requirements.

We model the IGW placement problem as a Capacitated Facility Location Problem (CFLP)issue with multiple optimization objectives and additional constraints in the linear program. As discussed earlier, two optimization objectives, which are minimizing the number of IGWs and minimizing the number of IGW-WMR hops, must be achieved. Having these two goals in mind, we formulate the IGW placement problem as a multi-objective optimization linear program.

A binary variable I_i is used to indicate whether a WMR $(v_i \in V)$ is set up as an IGW or not.

$$I_i = \begin{cases} 1 \ if \ the \ WMR(v_i) is \ selected \ as \ an \ IGW \ v_i \in V \\ 0 \qquad\qquad\quad otherwise \end{cases} \tag{1}$$

To represent WMR assignment to IGWs, again another binary variable $\Gamma_{i,igw}$ is defined. It shows the relationship between WMRs (v_i) and IGWs (v_{igw}).

$$\Gamma_{i,igw} = \begin{cases} 1 \ if \ WMR(v_i) is \ assigned \ to \ the \ IGW \ (v_{igw}) \\ \quad v_i \in V, v_{igw} \in IGW, IGW \subset V \\ 0 \qquad\qquad\qquad otherwise \end{cases} \tag{2}$$

$h_{i,igw}$ represents the minimum number of hops between WMR $(v_i \in V)$ and IGW $(v_{igw} \in IGW)$. When a WMR $(v_i \in V)$ is assigned to an IGW $(v_{igw} \in IGW)$, the $h_{i,igw}$ acts as a IGW-WMR hop.

$h_{igw(i)}^{igw(j)}$ is used to indicate the hop distance between IGWs, where $igw(i) \in IGW$ and $igw(j) \in IGW$.

In addition, a binary variable $\lambda_{i,igw}^k$ is defined to identify whether the path from v_i to v_{igw} passes through node v_k.

The linear program searches the entire provided space to find the optimal results. Before start searching, it has to specify the upper bounds of the IGW-WMR hops, relay traffic, and total traffic, which are delay constraint D_{QoS} (i.e., $h_{i,igw}$), relay load constraint R_{QoS} (i.e., $W_{max}(v)$), and cluster size constraint C_{QoS} (i.e., $W_{max}(igw)$) respectively.

Our objectives function are formulated as follows:

$$min \sum_{i \in V} I_i \tag{3}$$

$$min \sum_{i \in V} \sum_{igw \in IGW} h_{i,igw} \cdot \Gamma_{i,igw} \tag{4}$$

subject to:

$$\forall v_i \in V : \sum_{v_{igw} \in IGW} \Gamma_{igw,i} = 1 \tag{5}$$

$$\forall v_i \in V, v_{igw} \in IGW : \Gamma_{i,j} \leq I_i \tag{6}$$

$$min \frac{1}{2} \sum_{i \in IGW} \sum_{j \in IGW} h_{igw(i)}^{igw(j)} \cdot I_i \Rightarrow WMR(v_i) = smallest \ \partial N(v_{igw}) \tag{7}$$

$$\forall v_i \in V : \sum_{v_{igw} \in IGW} \Gamma_{i,igw} \cdot h_{i,igw} \leq D_{QoS} \tag{8}$$

$$\forall v_k \in V, v_{igw} \in IGW : \sum_{v_i \in V} \lambda^k_{i,igw} \cdot T_t(v_k) \leq R_{QoS} \tag{9}$$

$$\forall v_{igw} \in IGW : \sum_{v_i \in V} \Gamma_{i,igw} \cdot T_l(v_i) \leq C_{QoS} \tag{10}$$

$$\forall v_{igw} \in IGW : I_i \in \{0,1\} \tag{11}$$

$$\forall v_i \in V, v_{igw} \in IGW : \Gamma_{i,igw} \in \{0,1\} \tag{12}$$

$$\forall v_k \in V, v_i \in V, v_{igw} \in IGW : \lambda^k_{i,igw} \in \{0,1\} \tag{13}$$

The first objective (3) means that a minimum number of nodes should be chosen as the IGW ($I_i = 1$) among $|V| = n$ nodes so that the objective of having the minimum number of IGWs is achieved. In the second objective (4), the total number of IGW-WMR hops is minimized while every WMR (v_i) is assigned to one of m IGWs (i.e., $\Gamma_{i,igw} = 1$).

The objectives are subject to the following conditions. Equation (5) ensures each WMR to be assigned to an IGW, which is the requirement of full coverage objective. Besides, it implies that each WMR is assigned to one IGW only. Inequality (6) denotes that an IGW has to be set up before WMRs are assigned. In other words, it guarantees that a WMR is only assigned to the node that has been selected as an IGW. In (7) the IGWs will be placed close to each other and to Internet/wired network connection points if the smallest degree of IGW's neighbour nodes are selected as the WMRs. "$\frac{1}{2}$" is used in (7) in order to remove the duplication of number of hops from igw_i to igw_j and igw_j to igw_i. Condition (8) means that the number of IGW-WMR hops is within the maximum hop allowed by QoS limitation (D_{QoS}). Inequality (9) ensures that all the traffic passing through WMR (v_k) is within its load capacity, which is R_{QoS}. Inequality (10) further ensures IGW traffic capacity (C_{QoS}) as the upper bound of the traffic in each cluster. The last three constraints define the binary values (0 or 1) for three variables I_i, $\Gamma_{i,igw}$, and $\lambda^k_{i,igw}$.

We formulated the IGW placement problem as a Multiple Objective Linear Program (MOLP). There are many methods for solving this type of problem in the literature. For instance, problems with multiple objectives can be reformulated as single-objective problems by either forming a weighted combination of the different objectives [3] or transferring some of the objectives to constraints. Therefore, after converting multiple objective into a single objective problem, some linear program solver such as Matlab or LP-solve [4] can be utilized to solve the problem.

3   Related Work

In the few past years, some researchers have begun to design BWMNs and studied the IGW placement problem. They formulated this problem using linear programming and solved it using different heuristic algorithms because of NP-hard specification of such a problem.

In [5], Chandra et al. formulated the Internet Transit Access Points (ITAPs) placement problem and developed algorithms to place ITAPs in the network. Their algorithm, called Greedy Placement, aimed to minimize the number of ITAPs while satisfying user's bandwidth variation demand and fault tolerance assurance. Greedy Placement is similar to BWMN IGW placement, but it considers only one constraint, which is users' bandwidth requirements.

Wong et al., [6] addressed the gateway placement problem. They considered both communication delay and energy consumption. For each problem, they proposed different heuristic approaches, but using the same strategy. Their algorithms can only be used for BWMNs that form a connected component. They also requires minimum of two hops for communication between at least one pair of nodes.

Clustering technique is widely used for facility location, location management, and routing in wireless networks. It can be applicable in IGW placement problem, too.

Bejerano [7] was the first researcher who started to develop the way of wireless mesh network design using clustering technique. He divided all the wireless network nodes into clusters and for each cluster, a node is selected as an access point. Bejerano breaks the problem into two sub-problems: (i) finding the minimal number of disjoint connected clusters that contain all the nodes, and satisfying the Delay constraint; (ii) dividing the clusters that violate the cluster size constraint. However, splitting a cluster without considering re-assigning those wireless mesh routers to existing clusters may create some unnecessary clusters and therefore increase significantly the number of clusters.

Bassam [8] tried to minimize the disadvantages of Bejerano's technique by combining the two sub-problems, where the spanning tree and cluster coverage evolve in parallel subject to satisfy the QoS constraints. This algorithm has two drawbacks: first, it can only be used for those BWMNs that form a connected component; second, it needs to set the initial radius size properly in order to create satisfactory results.

Another algorithm to be used for the BWMNs, that does not form a connected component, is presented by Maolin Tang [9]. The name of the algorithm is incremental clustering, which is an iterative algorithm. Unlike our proposed algorithm in this paper, in incremental clustering algorithm, R-step transitive closure should be built in every iteration and only the last step of transitive closure will be used for IGW selection. Using last step of transitive closure in order to select the IGWs may produce some nodes with zero connection due to unseen condition of network in middle steps of transitive closure. Therefore, those nodes with zero connection should be selected as IGW in the next iterations and consequently, number of IGWs will be increased.

In [10] two architectures are considered: tree-based and cluster-based. In tree-based, He proposed two algorithms based on Greedy Dominating Tree Set Partitioning (GDTSP), which is degree-based GDTSP and weight-based GDTSP. Both algorithms aim to minimize the number of IGWs as well as reduce IGW-WMR hops. Similar to [9], this approach uses the last step of R-step transitive closure and does not consider the IGWs location controlling.

Having reviewed the previous studies on IGW placement problem, we discovered a reason for the unnecessary placement of IGWs, that is, zero degree nodes or those WMRs with no connection to other nodes in the network. The proposed solution should prevent producing nodes with zero connections. The other problem realized is that no attention has been paid to controlling the IGWs location. In restricted geographical areas that IGWs should be placed in the locations that are closest to available wired network/Internet connection points, controlled distribution of IGWs is an important issue. Therefore, we introduce a new unique factor as an objective parameter called "IGW-IGW hop". This factor is for the purpose of comparing the location of each IGW to other IGWs. One of the main advantage of the proposed Zero-Degree (S) algorithm is that predicting the condition of next iterations and having the chance to prevent producing zero-degree nodes in order to form feasible clusters satisfying all QoS constraints. Distribute the IGWs in the locations that are closest to available wired network/Internet connection points is added advantage of Zero-Degree (S) algorithm.

4   Heuristic Zero-Degree (S) IGW Placement Approach

The IGW placement problem is NP-hard [10] and the search space of the linear program for finding optimum number of IGWs increases exponentially with increasing the number of nodes (i.e., $n = |V|$). Therefore, in order to obtain near-optimal solution, a heuristic algorithm should be developed for large scale networks. In this section we introduce an algorithm called *"Zero-Degree (S)"*. In this algorithm, the graph $G = (V, E)$ is divided into disjoint clusters and each cluster has a head-cluster which is IGW. The aim of this algorithm is (i) minimizing the number of clusters, and consequently, minimizing the number of IGWs, (ii) reducing the number of IGW-WMR hops, and accordingly, increasing the throughput performance of the network, and (iii) Aggregate distributing the IGWs in the locations that are closest to available Internet connection points while satisfying the constraints listed in section 2.

Figure 1 shows the steps of our *Zero-Degree (S)* algorithm, including network initializing and generating, S-Model IGW selection, and S-WMRs assigning and network updating.

4.1   System Initializing and Network Generating

In this step, network information such as number of nodes, connection range and minimum distance, location size, and QoS parameters are collected to generate the network graph $(G = (V, E))$. When all network information have been

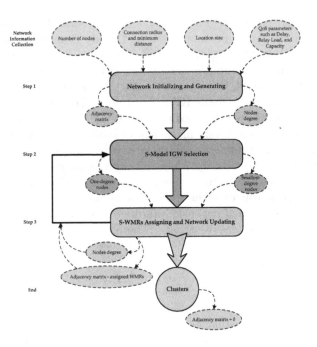

Fig. 1. Steps of Zero-Degree (S) Algorithm

collected, an adjacency matrix for all nodes in the graph ($G = (V, E)$) generates. The adjacency matrix consists of $(0, 1)$, 1 means there is a connection and
0 means there is no connection between two nodes. In the network graph, two
nodes are connected only when each node is located in the transmission range
of the other. Each node has a number of connections which is called "node degree". Nodes degree can be calculated using the adjacency matrix of the network
graph. The node degree and adjacency matrix is used to select the IGWs in the
following step.

4.2   S-Model IGW Selection

In this step, we assume that the largest degree node is the closest node to
available wired network/Internet connection points. Since largest degree node
will be selected as IGW, constraint 7, will be satisfied. An IGW is selected from
$G = (V, E)$ based on nodes degree. There is an algorithm to find an IGW in
S-Model architecture, which is shown in Algorithm 1. First, finding the largest-
degree node in adjacency matrix. Second, finding all other nodes with the same
degree as largest-degree node, if there is any. Otherwise, the only largest-degree
node will be the IGW. Bassam [8] and Maolin Tang [9] also select the largest-
degree nodes as IGW in their works. If there is more than one largest-degree
node, the algorithm will look for second hop nodes to find the nodes with only
one connection link in order to prevent producing zero degree/connection nodes

in the next iterations. In other words, it will look for nodes with degree equal to 1. Then, if there is the same situation for more than one node, it will look for third hop for the same reason as the second hop and so on (i.e., for forth, fifth, ... , and n^{th} hop. $n \leq D_{QoS}$). For the first attempt, a node with largest node-degree will be selected as an IGW, which is also connected to the largest number of single connection nodes, through single or multiple hop while satisfying all constraint such as QoS parameters. Thus, the selected IGW covers as much one degree nodes whom may convert to zero degree nodes in the next iterations as possible. If still there is some nodes with the same situation, IGW will be the node which has the smallest degree nodes among its neighbours. Node index will be used if two or more nodes have the same situation yet.

Algorithm 1. S-Model IGW Selection Algorithm

```
Input ← ⟨Adj_matrix, Delay, calculated_nodes_degree⟩
Output ← ⟨IGW⟩
if number of largest-degree nodes > 1 then
    for j = 1to number of largest-degree nodes do
        for k = 1to number of one-degree nodes do
            find number of connections from each largest-degree node to one-degree nodes via single or multiple hop,
            considering D_QoS
        end for
    end for
    if number of largest-degree nodes with highest number of connections to one-degree nodes > 1 then
        for i = 1to D_QoS do
            for j = 1to number of largest-degree nodes do
                find all the largest-degree nodes' neighbours degree
            end for
        end for
        if number of largest-degree nodes with smallest neighbours' nodes degree > 1 then
            IGW ← lowest index largest-degree node
        else
            IGW ← largest-degree node with smallest neighbours' nodes degree
        end if
    else
        IGW ← largest-degree node with highest number of connections to one-degree nodes
    end if
else
    IGW ← the only largest-degree node
end if
```

4.3 S-WMRs Assigning and Network Updating

When IGW is selected, a cluster is created by selecting a set of WMRs using S-WMRs assigning algorithm (Algorithm 2). Smallest degree model is used to select the nodes that have less connectivity links from one hop away nodes to D_{QoS} hop away nodes. This model, guarantees two objectives (i) the closer neighbouring WMRs to the IGW will be assigned before those which have far away. Thus, a minimum IGW-WMR hops objective will be met. (ii) the possibility of having the smallest degree nodes around assigned IGW will be reduced, and next IGW will be assigned close to selected IGW. Therefore, IGWs will be placed close to each other. Nodes with only one connectivity link in any hop, has the highest priority to be assigned because the possibility of having zero-degree nodes after deleting the assigned nodes from adjacency matrix and updating the nodes degree, will be decreased. After assigning WMRs to the IGW, assigned WMRs will be deleted from adjacency matrix. Finally, after deleting the assigned WMRs, nodes degree will be updated for next attempt.

At the same time, assigning procedure of any WMR to the IGW should satisfy the constraints imposed by cluster capacity and WMR traffic load. The traffic, local traffic (T_l) and relay traffic (T_r), that passes through a WMR must satisfy WMR traffic boundary. Furthermore, the entire traffic of a cluster should be less than maximal IGW traffic capacity.

Algorithm 2. S-WMRs Assigning Algorithm

```
Input ← ⟨Adj_matrix, Delay, Relay, Capacity, calculated_nodes_degree⟩
Output ← ⟨a formed_cluster⟩
true ← 1
while true do
    for j = 1to number of one-degree nodes do
        if the IGW has a connection with one-degree nodes in iᵗʰ hop via single or multiple hop then
            if cluster_size < C_QoS and relay_load < L_QoS then

                Cluster ← Cluster + iᵗʰ one-degree nodes
                Cluster ← Cluster + nodes in the path from the IGW to iᵗʰ one-degree nodes considering D_QoS
                Update cluster_size and relay_load
                true ← 1
            else
                break
            end if
        else
            true ← 0
        end if
    end for
    Adj_matrix ← Adj_matrix− assigned one-degree nodes
    update the nodes degree
    true ← 1
end while
if cluster_size < C_QoS and relay_load < L_QoS then
    assign a node with smallest node-degree from the first hop
    Adj_matrix ← Adj_matrix− assigned node
    update the nodes degree
    true ← 1
    go to beginning of the while loop
else
    break
end if
```

---

5   Zero-Degree (S) Algorithm

The Zero-Degree (S) algorithm (Algorithm 3) solves the IGW placement problem by iteratively identifying IGWs and assigning WMRs to identified IGWs considering node degrees. In each iteration, it identifies an IGW from the current unassigned WMRs set/Adj_matrix using $S - Model\_IGW\_Selection()$, which is provided in section (4.2), Algorithm 1, and then assigns WMRs to the identified IGW using $S - WMR\_Assigning()$, which is provided in section (4.3), Algorithm 2. Finally, it removes the assigned WMRs and IGW from Adj_matrix. The process is repeated until the Adj_matrix is empty. By the time Adj_matrix is empty, every WMR has been assigned to an IGW. Formed clusters will be specified in BWMN at the end of algorithm process. The Zero-Degree (S) IGW placement algorithm has a polynomial time complexity $O(n^2)$.

6   Performance Evaluation

In this section, a simulation-based analysis on our proposed heuristic IGW placement algorithm (i.e., Zero-Degree (S)) is performed. We evaluate the algorithm

Algorithm 3. Zero-Degree (S) Algorithm

```
Input ← ⟨Adj_matrix, Delay, Relay, Capacity⟩
Output ← ⟨formed_clusters⟩
Calculate the nodes degree
while Adj_matrix ≠ 0 do
    S − Model_IGW_Selection()  % Algorithm 1
    S − WMR_Assigning()  % Algorithm 2
    Delete IGW from Adj_matrix
end while
Plot the formed clusters
```

by comparing them with four top algorithms for the gateway placement problem. This four top algorithms are Incremental proposed by Maolin [9] similar to the one proposed by He in [10], Recursive algorithm proposed by Aoun et al. in [8], Iterative algorithm proposed by Bejerano in [7], and Augmenting algorithm similar to those proposed in [6] and [5].

MATLAB is used to randomly generate 30 topologies for 30 runs in any set ups in each evaluation. 200 WMRs are generated on a 10×10 plane. The connection radius is 1.0, and the minimum distance between any pair of WMRs is 0.5. The result is the average of 30 runs.

6.1 Effects of Hop-Based Delay

This section evaluates the effects of Delay constraint, which is hop-based, on the performance of five algorithms. In our set-up the relay load and IGW capacity constrains are relaxed. Delay constraint value vary from 1 to 10. As it is shown in Figure 2a, the performance of the Zero-Degree (S) is better than the other four algorithms. For $D = 4, 7, 9$, and 10, Iterative algorithm places the same number of IGWs required by Zero-Degree (S) while Incremental algorithm places the same number of IGWs required by Zero-Degree (S) when $D = 6$ and 7. Recursive algorithm has the worse performance compared to others.

IGW-WMR hop is a factor that only Recursive algorithm takes it into account among the top four algorithms in this paper. Therefore, Zero-Degree (S) will be evaluated against Recursive algorithm. In Figure 2b, both algorithms based on IGW-WMR hops are evaluated. Figure 2b shows that Zero-Degree (S) has better throughput performance results compared to Recursive algorithm for $D > 5$. Since, number of hops between IGW and its WMRs has direct effect on throughput performance, those algorithms with smaller number of IGW-WMR hops have better performance in terms of throughput in the network.

The IGW-IGW hop, which is introduced in this paper, is to provide a clear image of IGW location controlling technique. Figure 2c shows that Zero-Degree (S) can control the location of IGWs. Zero-Degree (S) located its IGWs in places close to available Internet points and more close to each other. The evaluation is based on largest number of hops between IGWs, using shortest path algorithm.

6.2 Effects of Relay Load

In this evaluation, the relay constraint varies from 1 to 15, IGW capacity is relaxed, and the delay constraint is fixed to 8. Figure 3a illustrates the results.

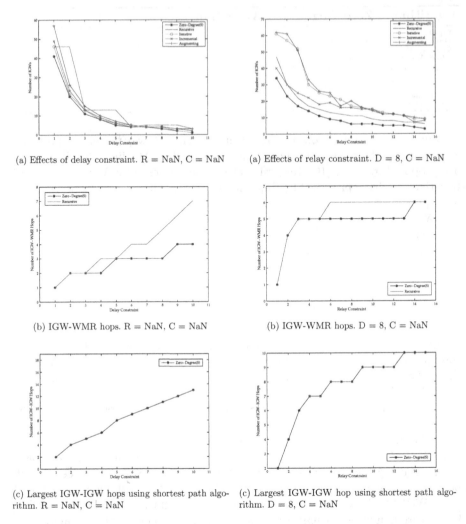

(a) Effects of delay constraint. R = NaN, C = NaN (a) Effects of relay constraint. D = 8, C = NaN

(b) IGW-WMR hops. R = NaN, C = NaN (b) IGW-WMR hops. D = 8, C = NaN

(c) Largest IGW-IGW hops using shortest path algo- (c) Largest IGW-IGW hop using shortest path algorithm. R = NaN, C = NaN rithm. D = 8, C = NaN

Fig. 2. Effects of Hop-Based Delay **Fig. 3.** Effects of Relay Load

The evaluation results show that Zero-Degree (S) performs best for all values of R. The effect of relay load constraint is mainly pronounced when it is very limited. For instance, the Iterative and Augmenting algorithms place twice the number of IGWs required by the Zero-Degree (S) algorithm when R = 1. In addition, when R exceeds 10, the number of required gateways by each algorithm remains constant.

For the IGW-WMR hop evaluation, Figure 3b shows that, Zero-Degree (S) has a better throughput performance for 5 < R < 14 and 6 < R < 14 compare to Recursive. WMRs in Zero-Degree (S) are closer to IGW compare to the other algorithm. Therefore, its throughput performance will be the best.

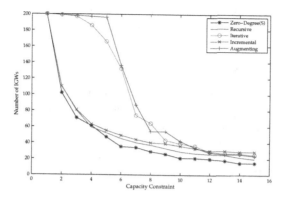

(a) Effects of capacity constraint. D = 8, R = NaN

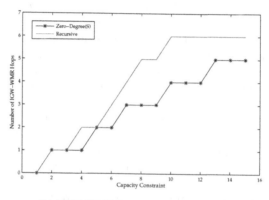

(b) IGW-WMR hops. D = 8, R = NaN

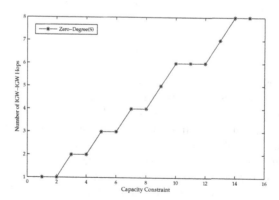

(c) Largest IGW-IGW hop using shortest path algorithm. D = 8, R = NaN

Fig. 4. Effects of Cluster Size

Figure 3c shows that Zero-Degree (S) algorithm is controlling the location of IGWs regardless of relay load effects. In Zero-Degree (S), IGWs are located in two hops away from each other to maximum 10 hops.

6.3 Effects of Cluster Size

This section evaluates the IGW capacity constraint. The IGW capacity varies from 1 to 15, delay constraint set to 8, and relay constraint is relaxed. Figure 4a shows that Zero-Degree (S) algorithm has better performance compare to other algorithms. It can be seen that the Iterative and Augmenting algorithms are heavily penalized when the cluster size constraint is strict. As C decreases, the number of required IGWs increases exponentially since each cluster is subdivided further as long as the cluster size constraint is violated [7].

On the other hand, the number of IGWs required by the Zero-Degree (S) algorithm increases almost linearly as the constraint on the cluster size becomes stricter. The reason is that Zero-Degree (S) algorithm has the chance to predict the next iterations conditions in order to prevent producing the zero-degree nodes and to form feasible clusters. As shown in Figure 4a, our Zero-Degree (S) algorithm produce much lower number of IGWs than the Augmenting and Iterative algorithms. For example, for $C = 6$, it require only 13% and even for $C = 7$, it require only 50% of the number of IGWs placed by the Iterative algorithm.

In Figure 4b, Zero-Degree (S) has the lowest number of IGW-WMR hops. It means that, WMRs in Zero-Degree (S) are closer to IGW in each cluster. Consequently, relayed traffics receive better services in terms of bandwidth and throughput. Thus, the overall throughput performance of the network in Zero-Degree (S) is the best compared to the other algorithm.

Figure 4c confirms the control advantage of Zero-Degree (S) algorithm on IGW locations, which is one of our objectives in this paper. Zero-Degree (S) aggregates the IGWs in a location close to available Internet points.

7 Conclusion and Future Work

IGWs are responsible for connecting the BWMNs to the Internet/wired backbone. Strategically placing the IGWs in a BWMN is critical to the WMN architecture. In order to solve the problem of IGWs placement in BWMNs, a novel algorithm is proposed in this paper. The new algorithm is involved in placing a minimum number of IGWs so that the QoS requirements are satisfied. Different from existing algorithms, this new algorithm incrementally identifies IGWs and prioritively assigns WMRs based on the computed degree of WMRs to identified IGWs. Performance evaluation results show that proposed algorithm outperforms other alternative algorithms by comparing the number of gateways placed in different scenarios. Furthermore, having control of the distribution of IGWs in order to locate them closest to available Internet/wired network connection points is an added advantage of this algorithm.

It is interesting to study the impact of topology changes and their effects on Zero-Degree (S) algorithm in future work. Thus, linear programming can be replaced by Particle Swarm Optimization (PSO). An application to use PSO is listed in [11].

Acknowledgments. This work was supported by Science Fund Grant, Ministry of Science, Technology and Innovation (MOSTI) Malaysia with consultancy from Telekom Malaysia (TM). Project no. 01-01-04-SF1462.

References

1. Nandiraju, N., Nandiraju, D., Santhanam, L., He, B., Wang, J., Agrawal, D.: Wireless Mesh Networks: Current Challenges and Future Directions of Web-In-The-Sky. IEEE Wireless Communications 14, 79 (2007), doi:10.1109/MWC.2007.4300987
2. Agrawal, D., Zeng, Q.: Introduction to Wireless and Mobile Systems. Thomson Engineering (2010)
3. Ralph, L.K., Raiffa, H.: Decisions with Multiple Objectives. Cambridge University Press (1993), doi:10.2277/0521438837
4. LP-Solve, http://lpsolve.sourceforge.net/ (access date: December 10, 2011)
5. Chandra, R., Qiu, L., Jain, K., Mahdian, M.: Optimizing the Placement of Internet TAPs in Wireless Neighborhood Networks. In: 12th IEEE International Conference on Network Protocols, ICNP 2004, pp. 271–282 (2004)
6. Wong, J.L., Jafari, R., Potkonjak, M.: Gateway Placement for Latency and Energy Efficient Data Aggregation. In: 29th IEEE International Annual on Local Computer Networks, pp. 490–497 (2004)
7. Bejerano, Y.: Efficient Integration of Multihop Wireless and Wired Networks with QoS Constraints. IEEE/ACM Transactions on Networking 12, 1064 (2004)
8. Aoun, B., Boutaba, R., Iraqi, Y., Gary, W.K.: Gateway Placement Optimization in Wireless Mesh Networks With QoS Constraints. IEEE Journal on Selected Areas in Communications 24, 2127 (2006)
9. Tang, M.: Gateways Placement in Backbone Wireless Mesh Networks. International Journal of Communications, Network and System Sciences 2, 44 (2009)
10. He, B.: Architecture Design and Performance Optimization of Wireless Mesh Networks. Ph.D. thesis, Computer Science and Engineering, College of Engineering, University of Cincinati, Ohio (2010)
11. Shi, Z., Beard, C.C., Mitchell, K.: Competition, Cooperation, and Optimization in Multi-Hop CSMA Networks with Correlated Traffic. International Journal of Next-Generation Computing 3 (2012)

Traffic Classification Approach Based on Support Vector Machine and Statistic Signature

Seonhwan Hwang[1], Keuchul Cho[1], Junhyung Kim[1], Youngmi Baek[1], Jeongbae Yun[1], and Kijun Han[2]

[1] The Graduate School of Electrical Engineering and Computer Science,
Kyungpook National University, 1370, Sankyuk-dong, Buk-gu, Daegu, 702-701, Korea
{shhwang,k5435n,jhkim,maya,jbyun}@netopia.knu.ac.kr
[2] The School of Computer Science and Engineering, Kyungpook National University, 1370, Sankyuk-dong, Buk-gu, Daegu, 702-701, Korea
kjhan@knu.ac.kr

Abstract. As network traffic is dramatically increasing, classification of application traffic becomes important for the effective use of network resources. Classification of network traffic using port-based or payload-based analysis is becoming increasingly difficult because of many peer-to-peer (P2P) applications using dynamic port numbers, masquerading techniques, and encryption. An alternative approach is to classify traffic by exploiting the distinctive characteristics of applications. In this paper, we propose a classification method of application traffic using statistic signatures based on SVM (Support Vector Machine). The statistic signatures, defined as a directional sequence of packet size in a flow, are collected for each application, and applications are classified by SVM mechanism.

Keywords: Traffic, Application, Classification, Statistic Signature, Support Vector Machine.

1   Introduction

Traffic classification plays an important role in common network management applications, such as intrusion detection and network monitoring. However, it is challenging to classify the applications associated with network connections according to their various characteristics and behaviors [1].

Former researches have proposed a number of methods to identify the application associated with a traffic flow. Port-based methods by examining TCP port numbers are simple because many well-known applications have specific port numbers assigned by IANA. However, the port-based classification is insufficient [2, 3, 4, 5], mainly because many applications use dynamic port-negotiation mechanisms to hide from firewalls and network security tools. Another approach is to inspect the payload of every packet. However, this approach has several problems. First, this method cannot be used if the payload is encrypted. Second, there are privacy concerns with examining user data. Third, there is a high storage and computational cost to study every packet that traverses a link.

S. Balandin et al. (Eds.): NEW2AN/ruSMART 2013, LNCS 8121, pp. 332–339, 2013.

To address these challenges, we propose a classification method of application traffic using SVM that only uses the size of the first few data packets of each connection.

The remainder of the paper is organized as follows. In Section 2, we describe related works. Section 3 presents our classification method that uses static signature based on SVM. Performance evaluation and discussion are given in Section 4. Section 5 proposes concludes the paper.

2   Related Works

Bernaille et al [1, 6] suggested traffic classification based on packet header trace technique, known as "on the fly" classification. They only utilized the first 5 significant packets (in both directions) of a TCP connection to classify, based on clustering. Usually, the packet header gives more information without inspecting payload. Their approach allows the unsupervised online classification of traffic, and therefore allows actions to be taken during the connections. They employed data clustering using only the sizes and directions of the first 5 packets. The direction is used as the signal (first and same-direction packets: positive; opposite direction packets: negative). By ignoring TCP handshake packets and ACKs with no payload, the clustering of only 5 packets allows the online classification of flows with a considerable hit ratio.

Support Vector Machine (SVM) developed by Vapnik in 1995 is a classification and regression prediction tool that uses machine learning theory to maximize predictive accuracy while automatically avoiding over-fit to the data. Fig. 1 shows an example of SVM using linear classifiers which could separate the data into two groups.

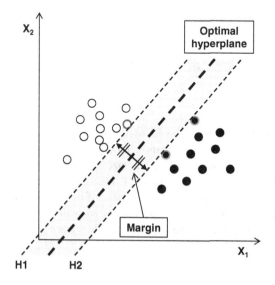

Fig. 1. An example of Support Vector Machine using linear classifier

3   Our Classification Method

In this section, we present our classification method that uses static signature based on SVM. Our classification mechanism works in two phases: SVM data set training phase and traffic classification phase as shown in Fig. 2. The training phase obtains models of application behaviors. The trace contains a representative sample of flow from all target applications. The classification phase associates a new flow with an application by using the group defined in the training phase.

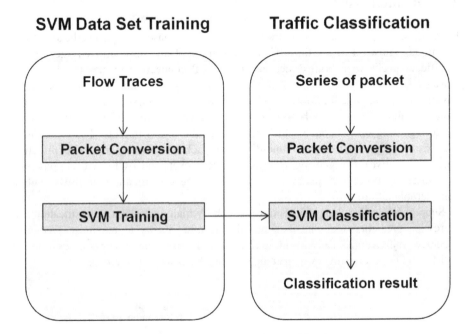

Fig. 2. Two Phases for Traffic Classification

3.1   Flow Trace

The flow traces are the input for the SVM. Flow is defined as a collection of both directional packets of 5-tuple information (source IP, destination IP, source port, destination port, protocol). We only use the size of the data packets, not using TCP control packets (SYN, Ack with no data, etc.). We analyze the trace and convert each flow into a spatial representation based on the size of its first N packets. Flow data set is generated by the size for the first N packets of each flow. Only one application is executed for collecting packet by the application. Flow trace server collects the first N packets of connection by the application.

3.2   Packet Conversion

The conversion module extracts the flow and the packet size. Statistic signature is made into vector which consists of payload size and direction of application packet. The analyzer filters out control traffic (the three packets of the TCP handshake) and stores the size of every packet in both directions of the connection. When it knows the size for the first N packets of the flow, it sends this information to the assignment module which associates the flow based on application descriptions.

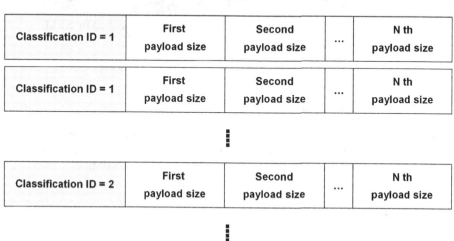

Fig. 3. SVM signature using the size of packet payload

Fig. 3 presents SVM database by the size of packet payload. Classification ID is whole number by application classification. Other fields composed the first N packet payload size by flow. And the database has enough statistics signature per application. Direction is represented by positive and negative, positive value represents sent packet from the client to the server, and negative from server to client in case of TCP. But, Because UDP classification is neither one thing nor the other, meaning of positive/negative is indicated to only the opposite direction. In UDP, the first packet is positive. And the second packet will be positive if it is the same direction. If not, it will be negative.

3.3   SVM Training

In our scheme, we use multi-class SVM for classifying many applications at the same time. In addition, we use Radial Basis Function (RBF) of non-linear way which is more accurate than liner classification. Fig. 4(a) presents an example of payload distribution by the first 2 packets of Nateon and uTorrent. After finishing the packet conversion phase, different application is divided to hyperplane at SVM training phase. Fig. 4(b) presents grouping result of two application distributions shown in Fig. 4(a) by RBF way.

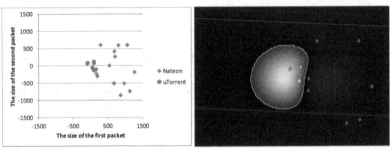

(a) Payload distribution of Nateon   (b) Two groups classified by SVM
and uTorrent.   RBF.

Fig. 4. An example of application classification by SVM

3.4   Traffic Classification

In traffic classification phase, we collect packet of various applications. The classifying module determines which application is most likely associated with the flow. Connection of each application is classified by multi-class SVM.

4   Simulations

We carried out simulations using statistic signature classification by SVM. All packets of simulation environment were generated by executing application in the test-bed. We used maximum 10-dimension vector for SVM training in simulation. In the simulation, we used 3 programs: Wireshark ver. 1.8.3 (Packet capture), C++ ver. 2008 (Signature conversion) and SVMmulticlass ver. 2.20 (Training & classification). Also, we used 3 types of kernel function in SVM: linear, radial basis function, sigmoid. Table 1 shows the number of applications and signatures, and Table 2 presents application list and type.

Table 1. Coverage of SVM classification

	Application	Training signature	Classification signature
Coverage	5	5,198	4,482

Table 2. Application list

Type	Application name
Messenger	Nateon
Webhard	Webhard
P2P	uTorrent
Game	League Of Legends(LOL)
Encrypted packet	SSH

Fig. 5, 6, 7, and 8 present the accuracy of our classifier using the first N packets of each flow by sigmoid, linear and RBF of multi-class SVM. Fig. 5 indicated that SVM sigmoid can only classify application packets of Nateon and LOL, but it is not appropriate to classify packets of Webhard, uTorrent and SSH, since the sigmoid classifier is not a good model to fit straight lines for traffic with similar packet size.

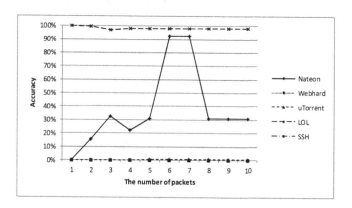

Fig. 5. Accuracy of SVM sigmoid classification

Fig. 6 indicated that when we use first 2 packets for signature capturing, the SVM linear can only classify application packets of Nateon and Webhard.   It is not appropriate to use the SVM linear classifier to separate packets of uTorrent, LOL and SSH. And when we use first 3~4 packets, the SVM linear classifier can roughly distinguish application packets of Nateon, uTorrent and SSH.   We can see that the linear classifier is not a good model to discriminate traffic with similar packet size.

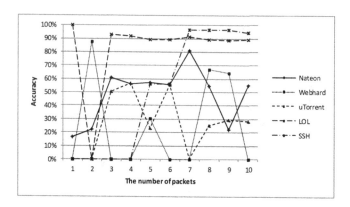

Fig. 6. Accuracy of SVM linear classification

Fig. 7 indicated that the SVM RBF classifier can successfully classify all 5 types of traffic. In special, it was shown that we could get very accurate results with the first 2~5 packets. Bu, the classification accuracy drops with more than 5 packets. This is because the packet size is different depending on the user application environments. It means when network applications communicate with each other, there is no significant difference in the packet size if less than 5 packets are used. After the fifth packet, the packet size becomes drastically different according to the operation of application.

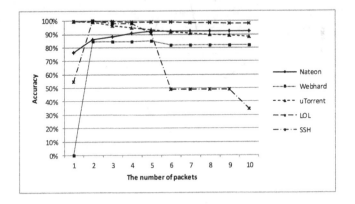

Fig. 7. Accuracy of SVM RBF classification

Fig. 8 shows the average accuracy of SVM RBF, linear, and sigmoid classifications. From this graph, we can see that the classification using RBF has the highest accuracy.

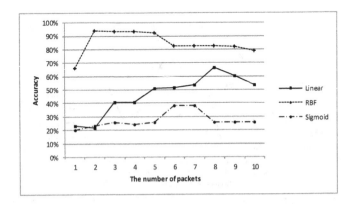

Fig. 8. Average accuracy of SVM RBF, linear, and sigmoid classifications

5   Conclusions

In this letter, we proposed a traffic classification scheme by using only payload size and direction of application packet without examining TCP port numbers and inspecting the payload of every packet. Our classification mechanism works in two phases: SVM data set training phase and traffic classification phase. The training phase obtains models of application behaviors. The trace contains a representative sample of flow from all target applications. We apply SVM techniques to a set of training data to group flows. The classification phase associates a new flow with an application by using the group defined in the training phase.

We evaluate three different classifier methods in SVM: sigmoid, linear and RBF. Simulation results showed that the RBF classifier can successfully classify 5 types of application traffic from Nateon and LOL, Webhard, uTorrent, and SSH, with the first 2~5 packets.

Acknowledgements. This work was supported by the IT R&D program of MOTIE/KEIT. [10041145, Self-Organized Software platform(SoSp) for Welfare Devices].

This research was supported by Basic Science Research Program through the National Research Foundation of Korea (NRF) funded by the Ministry of Education, Science and Technology (2011-0029034).

References

1. Bernaille, L., Teixeira, R., Akodkenou, I., Soule, A., Salamantian, K.: Traffic Classification On The Fly. ACM SIGCOMM Computer Communication Review 36(2), 23–26 (2006)
2. Karagiannis, T., Broido, A., Brownlee, N., Claffy, K., Faloutsos, M.: Is P2P dying or just hiding? In: IEEE Globecom (2004)
3. Roughan, M., Sen, S., Spatscheck, O., Duffield, N.: Class-of-service mapping for QoS: A statistical signature-based approach to IP traffic classification. In: Internet Measurement Conference (2004)
4. Moore, A., Zuev, D.: Internet traffic classification using bayesian analysis. In: ACM SIGMETRICS (2005)
5. Karagiannis, T., Papagiannaki, D., Faloutsos, M.: BLINC: Multilevel traffic classification in the dark. In: ACM SIGCOMM (2005)
6. Bernaille, L., Teixeira, R., Salamantian, K.: Early Application Identification. In: Second Conference on Future Networking Technologies (December 2006)
7. Callado, A., Kamienski, C., Szabó, G., Gerő, B.P., Kelner, J., Fernandes, S., Sadok, D.: A Survey on Internet Traffic Identification. IEEE Communications Surveys & Tutorials 11(3) (Third quarter 2009)
8. Vapnik, V.: The Nature of Statistical Learning Theory. Springer, NewYork (1995)

RPL Objective Function Impact on LLNs Topology and Performance

Agnieszka Brachman

Silesian University of Technology,
Institute of Informatics,
Akademicka 16, Gliwice, Poland
agnieszka.brachman@polsl.pl

Abstract. RPL is IPv6 Routing Protocol designed for Low Power and Lossy networks. RPL is a distance vector routing protocol that builds Directed Acyclic Graphs (DAGs), using implementation specific routing metrics and constraints. In the construction process of network topology, each node selects a set of potential parents towards the destination and associates itself to a preferred parent, basing on the outcome of Objective Function (OF). OF defines how one or more node and link metrics and constraints are used to compute the node rank.

The purpose of this paper is to analyse, how selection of Objective Function influences network topology. The absence of published papers concerning RPL Objective Function motivates author to focus on this topic. In addition, performance evaluation of network transmission when constructed using different OF is presented. Results where obtained using simulation environment prepared for this and further planned studies on RPL.

Keywords: RPL, objective function, routing, 6LoWPAN, LLN.

1 Introduction

Internet of Things (IoT) is considered the key component of the future Internet and our everyday life. Mobile communication giants expect that by 2020 billions of devices will be wirelessly connected [1]. Smart objects - computers equipped with a communication device, sensors or actuators, can be embedded in almost any object like meter, industry machinery, home equipment, light switches, engines and many others. It enables almost endless number of new applications in industrial, commercial and domestic areas such as smart cities, home and building automation, controlling and monitoring systems (lighting, ventilation, security systems, fire systems), security systems, habitat monitoring, environment monitoring, vehicular tracking, asset tracking systems, medical applications, Automated Meter Reading (AMR).

The prospective smart object networks consist of thousands of nodes, which are usually small and inexpensive devices. It implies many technical challenges in terms of assuring low power consumption and efficient algorithms, requiring

S. Balandin et al. (Eds.): NEW2AN/ruSMART 2013, LNCS 8121, pp. 340–351, 2013.

scarce amount of computing resources. Nodes may be designed to work for several years, usually in unattended manner, therefore should support some forms of auto-configuration and management. The wide range of applications and wireless communication medium cause that nodes often work in unfriendly and harsh environments. In order to reduce power consumption during data transmission, nodes operate on low power communication standards, these include wireless standards or Power Line Communication (PLC); such communication is unreliable and obstructed, therefore smart object networks are considered being lossy hence such networks are also referred to as Low power and Lossy Networks (LLNs).

Many already existing applications of such networks constitute non-IP-based architectures. The lack of an IP-based network architecture precludes from interoperating with the Internet. Since IP has proven itself highly scalable, stable and reliable technology, the protocols and solutions for introducing IP into LLNs, with regard to the aforementioned restrictions, are crucial.

The Internet Engineering Task Force (IETF) recognized this disconnect and chartered two working groups 6LoWPAN (IPv6 over Low power Wireless Personal Area Network) and RoLL (Routing over Low power and Lossy network) to "specify standards at various layers of the protocol stack with the goal of connecting Low-power and Lossy networks to the Internet" [2].

With RFC 4919 [3] Kushalnagar, et al. depicted the advantages of using IPv6 networking as well as problems and open issues associated with enabling IPv6 communication in a LoWPAN, compliant with LLN. Since the IEEE 802.15.4 standard [4] is commonly used for many smart object networks applications e.g. Wireless Sensor Networks (WSNs), detailed specification of the protocol for IPv6 transmission over IEEE 802.15.4 was soon provided in [5]. Furthermore problems regarding the fragmentation, header compression, IP Neighbour Discovery and Auto-Configuration were addressed in RFC 6282 [6] (updates RFC4944), RFC 4861 [7] (updated later by RFC 5942 [8]), RFC 6775 [9].

The IETF RoLL group evaluated the well-known routing protocols such as OSPF, AODV, OLSR however none of them was proved to provide the acceptable performance in the unique conditions that characterize LLNs. Several solutions have been proposed [10,11,12], however none of this protocols finished as a standard. In 2010 the first draft of Routing Protocol for Low Power and lossy networks referred to as RPL was announced. Two years later, the protocol specification has been published as RFC 6550 [13].

RPL is a distance vector routing protocol that builds Directed Acyclic Graphs (DAGs) using implementation specific routing metrics and constraints. The main features are:

- It supports different traffic patterns (multicast and unicast),
- It propagates the routing state, which causes quick reaction to routing changes,
- To achieve reliability, a node maintains a set of potential parents towards the destination instead of one,
- The protocol defines link's and node's metrics to compute a node rank and enable graph establishment in multiple applications.

In the construction process of network topology each node selects a set of potential parents towards the destination and associates itself to a preferred parent, basing on the outcome of Objective Function (OF). OF defines how one or more node's and link's metrics and constraints are used to compute node rank. The protocol defines rules how to use rank to select and optimize routes, with avoidance of creating loops. The RPL specification [13] defines a generic protocol but not OF itself. Adaptation of that single protocol, to the wide variety of network types is possible, because of the application of specific Objective Functions.

The purpose of this paper is to analyse, how selection of Objective Function influences the network topology. The absence of published papers concerning RPL Objective Function motivates author to focus on this topic. In addition performance evaluation of network transmission when established using different OF is presented. Results where obtained using a simulation environment prepared for this and further planned studies on the RPL protocol.

The remainder of this paper is organized as follows. Related work is referenced in section 2. In section 3 general requirements for RPL Objective Function are presented as well as known algorithms are described. Simple metrics provided by IETF RoLL working group are depicted and their usability discussed in section 4. In section 5 differences in the network topology are compared when using some basic objective functions, also description of simulation environment prepared for the purpose of further work is provided. Some basic performance evaluation of network implementing two different objective functions is presented. Finally, section 6 concludes the paper and discusses future work concerning more sophisticated OF evaluation in the presented simulation environment.

2 Related Work

Design of efficient Objective Function is still an open research issue. So far several drafts specifying OF have been proposed [14,15] along with the IETF RFC 6552 standard for Objective Function Zero (OF0) [16]. More details concerning aforementioned OFs are provided in section 3.

In [17] authors simulate and analyse the performance of the network formation process using ContikiRPL simulator. Among other parameters they verify how using two different Objective Functions influences average number of hops and average node energy. The observed differences are insignificant due to the choice of the OFs and their specific parametrization, which results in similar outcome when computing rank.

Some papers have been devoted to the comprehensive RPL studies and performance evaluation. In [18] authors provide detailed survey concerning RPL specification as well as the relevant research issues concerning RPL. Also main routing metrics are discussed along with the performance evaluation of RPL in different network scenarios is presented. Authors use Expected Number of Transmissions (ETX) OF (see section 3.1 described in [14], with improvement proposed in [15] and omit discussion concerning the OF details and influence on the network performance, hence emphasize its importance and lack of more complex proposals.

Among some other papers concerning RPL e.g. [19,20] authors rather focus themselves on the performance analysis of the RPL protocol. None of them discusses other than ETX Objective Function. In all presented papers authors use ContikiRPL implementation.

3 Objective Function Overview

RPL Objective Function determines, which parent will each node select, therefore it is directly responsible for the route establishment process and resultant topology. OF allows adapting RPL to a variety of network types. This section provides detailed description of known OFs.

3.1 EXT Objective Function

The ETX Objective Function (ETXOF) [14], is designed to select parent that provides delivery with the least number of transmissions. It does so by using the ETX metric defined in [21]. The ETX metric of a wireless link is the expected number of transmissions required to successfully transmit and acknowledge a packet on the link. It allows distinguishing less reliable paths that require larger number of packet transmissions from better ones. The RPL path evaluation using ETXOF results in selection of minimum-ETX paths, which are generally also the most energy-efficient.

Each node computes the ETX Path metric for a path to the root through each candidate neighbour n as $ETX(n) + MinPathETX(n)$, where $ETX(n)$ is ETX metric for the link to a neighbour n and $MinPathETX(n)$ is ETX metric advertised by that neighbour. ETX metric is defined as: $ETX = 1/(D_f * D_r)$, where D_f is the measured probability that a packet is received by the neighbour and D_r is the measured probability that the acknowledgement packet is successfully received.

Link performance affecting ETX metric can be highly variable. Such jitters, if reported and immediately used to change route, would cause network instability and routing oscillations. In [15] authors describe the Minimum Rank Objective Function with Hysteresis (MRHOF), an Objective Function that uses hysteresis while selecting the path with the smallest metric value. MRHOF can be used with any OF that uses metric that must be minimized. The use of MRHOF with the ETX metric allows RPL to find the stable minimum-ETX paths.

3.2 Objective Function Zero

RFC 6552 [16] specifies the default basic OF, namely Objective Function Zero (OF0), which operates on abstract information obtained using RPL DODAG Information Object (DIO) containers.

OF0 calculates a node rank by adding a strictly positive scalar *RankIncrease* to the rank of candidate neighbour. *RankIncrease* is based on a *StepOfRank* scalar that can vary from 1 (best case) to 9 (worst case) to represent the link properties. OF0 requires computing *StepOfRank*, that indicates the amount by which to increase the rank along a particular link. If a static metric is used, it results in rank being analogous to hop count, which favours paths with fewer but potentially longer hops, possibly of poorer connectivity [16]. Thus it is recommended to base the computation on dynamic link properties such as ETX.

OF0 definition allows stretching the *StepOfRank* through multiplying by a configurable factor *RankFactor*, however this may lead to network instability. Stretching the *StepOfRank* increases the network depth, however it may be necessary to distinguish links of different types.

OF0 is designed as default OF that will assure interoperation between different RPL implementations, therefore the standard refuses to specify how node and link metrics are transformed into the *RankIncrease* and other variables used.

4   Routing Metrics for Path Calculation

With the [21] IETF RoLL proposed some routing metrics that may be used for path calculation in LLNs. The RPL standard supports those metrics by serving DODAG Information Object (DIO) that advertises those data. The presented metrics may be used directly to construct Objective Function, however, as depicted in the following sections, using a single metric may lead to network imbalance and may highly reduce its performance and lifetime of some nodes often selected as parents, therefore usually combinations of those metrics should be used for the OF implementation.

4.1   Metric Categories

As stated in [21] routing metrics may be categorized according to the following characteristics:

- Link versus node metrics;
- Qualitative versus quantitative;
- Dynamic versus status.

Link metrics include throughput, delay, error bit rate and other channel characteristics. In the wireless environments, especially in LLNs those numbers may significantly differ among different network parts or at different daytime. Some nodes may experience higher bit error rate and/or throughput rate due to their position, lack of neighbours or weaker transmission unit.

Node metrics concern node itself and may refer to its energy capacity, expected lifetime, antenna sensitivity, ability to encode or encrypt transmitted data. In LLNs device diversity may by high in terms of resources, battery consumption, power supply and so on.

The qualitative metrics refer to the network characteristics e.g link reliability. The quantitative metrics include transmission channel parameters, which can be calculated or measured. The quantitative indicators are commonly believed to be objective, verifiable and process of their collection and assessment seems repeatable and strictly defined, in opposed to the qualitative metrics which may be subjective and more difficult to verify.

LLNs are rapidly changing environments in terms of transmission performance or energy level, therefore it seems quite obvious that to compensate this variability, related metrics should reflect those changes. However exchanging additional information concerning varying metrics, requires some additional computations, data transmission, which consumes bandwidth and energy. Therefore decision upon dynamic versus static as well as update frequency should be well considered. Metrics variation should be soothed to avoid routing oscillations.

4.2   Node and Link Metrics

A set of node's and link's routing metrics that can be exchanges between nodes and used as OF income, as defined in [21], is as follows:

- Node State and Attributes - this metric reports attributes as CPU overload, available memory, average workload.
- Node Energy - since energy is crucial resource in most LLNs, using information concerning the energy level may extend the network lifetime and improve the network stability. Also information regarding the device type , i.e. mains-powered, battery-powered, energy scavengers (solar panel, mechanical, etc.) may be exchanged.
- Node Hop Count - the metric simply reports the number of hops along the path from reporting node towards the root.
- Link Throughput - the metric reports recently estimated, actual throughput.
- Link Latency - this metric reports the path latency and may be used as an aggregated, additive metric.
- Link Reliability (LLQ/ETX) - this metric regards to the link reliability level and may be expressed in several ways such as packet error rate, bit error rate, mean time between failures. Standard specifies two link reliability metrics, namely the Link Quality Level (LLQ) and Expected Number of Transmissions (ETX). LQL is used to quantify the link reliability using a discrete value from 0 to 7. 0 indicates the unknown quality level, 1 refers to the highest value. ETX metric has already been explained in section 3.1. Although formula for calculating ETX is given in [21], the standard refuses to mandate the use of a specific formula.
- Link Color - this metric allows assigning 10-bit encoded color to links. Meaning of each color is implementation specific. It may be used for instance to indicate the links that support encryption or encoding.

5   Impact of the RPL Objective Function

To compare OF impact on the network topology and performance, two OFs have been selected that are based on the static and dynamic metrics. The first choice was OF0, which minimizes the number of hops. The second OF (LLQ OF) was based on the link reliability metric, assessed using nodes' distance. In wireless transmission, Received Signal Strength Indicator (RSSI) that can be used to estimate link quality, mostly depends on the distance between communicating nodes, therefore it was used as an indicator of link reliability in depicted simulation scenarios.

OF impact was evaluated using own simulation environment. Prepared testbed is a discrete-event simulator written in Python. For network graph generation networkx library was used. The purpose of this studies was to model network topology and behaviour using graphs to provide the wide range of possible analysis. Adapting the simulation environment to fully reflect RPL operation is work in progress.

5.1   Network Topology

Impact on the network topology was studied using different network sizes. Random geographic graph was generated. All nodes that stay within the selected range of each node, equal for all nodes, are its neighbours. Existing connectivity among nodes is presented on figures as graph edges. Fig. 1 presents the initial network state for 50 and 100 nodes. Networks consisting up to 1000 nodes are analysed. The node being in the middle is selected as a root.

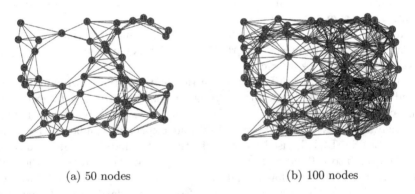

(a) 50 nodes   (b) 100 nodes

Fig. 1. All initial connections for random geographic graphs

When static Objective Function Zero, that aims minimize the number of hops, was used, the network topology is constant as long as all nodes are active. Nodes select its parents that provide shortest path towards the root in terms of hop counts. Fig. 2 presents network topology for 50 and 100 nodes when Objective Function Zero is used.

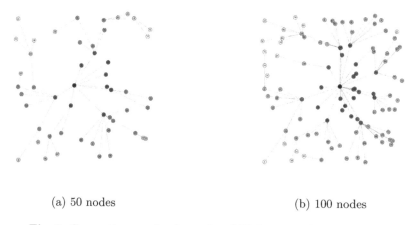

(a) 50 nodes   (b) 100 nodes

Fig. 2. Connection graph when using OF0 for network consisting of

When choosing OF0, average number of child devices is high for each node. There are two flaws of such configuration: if the transmission rate is high, efficient algorithms of traffic shaping and transmission scheduling are required. Distance between pairs of devices may be significant, which increases transmission failure rate. It is very efficient configuration when using multicast transmission from root device. OF0, for dense networks, results in very similar depth, regardless of node number. Fig. 3 presents the distribution of hop count for network consisting of 50, 100, 500 and 1000 nodes. Except for network with 50 nodes, none of the rest exceeds three hop counts.

When LLQ OF was used, routing process was captured in several phases. First step for network consisting of 50 nodes, is presented in fig. 4a. Nodes were associating with their parents, starting from nodes that were in the closest proximity to the root. The network configuration after second phase, when nodes can associate with closer parent, as long as the new parent has smaller or equal rank, is depicted in fig. 4b. The node cannot decrease its rank by more than one at the reconfiguration.

The collection of LLQ OF outcome is larger than in case of OF0, therefore the graph depth degree significantly increases. Such configuration requires much more transmissions when addressing single node however average "usage" of each node as a router is much lower. Selecting OF that tries to associate itself to the closest device results in the establishment of long communication chains. Fig. 5 presents hop count distribution for network configured using LLQ OF. If all nodes were battery-powered, such configuration would drain energy from closest-to-root nodes, that act as router for large part of the network, very fast. On the other hand small disctances result in lower transmission power and much lower rate of interferences, hence better transmission reliability.

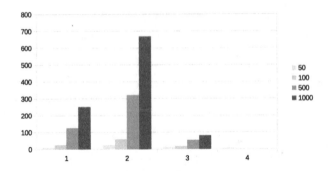

Fig. 3. Hop count histogram when using Objective Function Zero

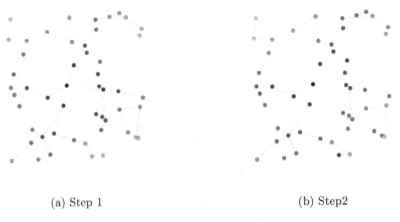

(a) Step 1   (b) Step2

Fig. 4. Connection graph when using LLQ OF

5.2   Performance Evaluation

Performance evaluation was tested for network consisting of 1000 nodes. The transmission speed in downlink direction is limited to 100 kbps, nodes can transmit uplink 5 kbps. Traffic shaping is performed on both ends. Root sends data with packet size randomly chosen between 1000 and 1280 bytes. Node responds with 1280 bytes packet. Only unicast transmission is used. Root sends 10, 100 or 1000 packets to randomly chosen nodes, only one session to each device can be established at the same time. Node response is randomly backoffed to simulate CSMA manner of accessing the medium. For each packet timeout how long to wait for response is calculated. Packet is retransmitted 5 times, if no answer is received. Packet error rate is set to 1%. At each simulation, the same topology is kept; only Objective Function and number of packets to send, are changed.

The presented results are preliminary. The implemented transmission module is very simple and assumes equal transmission parameters for each connection, which de facto would result in identical topology for both OFs, therefore more

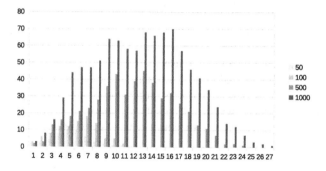

Fig. 5. Hop count histogram when using LLQ Objective Function

accurate results are desired, that require further development of simulation environment.

Table 1 compares time needed to complete task of sending required number of packets to randomly chosen recipients for OF0 and LLQ OF. For small number of packets, time required to accomplish the transaction is comparable for both topologies. When all nodes are addressed, transaction time for OF0 topology is almost three times faster than when using LLQ OF, mainly because of larger number of retransmissions performed in LLQ OF network, since probability of loosing the packet is the higher, the more hop counts.

Table 1. Time to complete number of tasks (Network: 1000 nodes, PER = 1%)

Packets:	10	100	1000
OF0 Time to complete:	0d 1:03:20	0d 4:33:20	0d 4:53:20
LLQ OF Time to complete:	0d 1:10:0	0d 4:50:0	0d 11:36:40

6 Conclusions

RPL is a very promising routing protocol for LLNs as it provides solution for many types of network applications. One of the most important issue regarding RPL is Objective Function, which highly influences routing behaviour and path selection process.

In this paper impact of choosing Objective Function on the network topology and transmission performance was analysed. Two OFs were evaluated: OF0 and LLQ OF, based on static and dynamic metric respectively.

The main conclusion is that OF determines average number of hop counts and child devices connected to each router as well as network stability. In battery-powered networks OF should not rely on static metric, and should not only minimize the hop counts, because such strategy leads to fast energy drain from nodes that are in the closest proximity to the network root. Combination or constrains concerning power metrics should be used.

On the other hand, when using wide range metric, to avoid increasing network depth and network instability, the realm of OF outcome should be limited and nodes should not be allowed to increase their rank with too large step. Communication in network with smaller hop count is faster, however average number of transmissions performed by each node is larger and transmission power is higher. It is impossible to choose which OF result in better outcome, the application of each OF is different, however to confirm this statement further studies are crucial.

Acknowledgements. This material is based upon work supported by the Polish National Science Centre under Grant No. N N516 479240.

References

1. Vasseur, J.P., Dunkels, A.: Interconnecting Smart Objects with IP: The Next Internet. Morgan Kaufmann Publishers Inc., San Francisco (2010)
2. Ko, J., Terzis, A., Dawson-Haggerty, S., Culler, D.E., Hui, J.W., Levis, P.: Connecting low-power and lossy networks to the internet. IEEE Communications Magazine 49(4), 96–101 (2011)
3. Kushalnagar, N., Montenegro, G., Schumacher, C.: IPv6 over Low-Power Wireless Personal Area Networks (6LoWPANs): Overview, Assumptions, Problem Statement, and Goals. RFC 4919 (Informational) (August 2007)
4. IEEE Std 802.15.4-2011: IEEE Standard for local and metropolitan area networks–part 15.4: Low-rate wireless personal area networks (LR-WPANs). IEEE Std 802.15.4-2011 (Revision of IEEE Std 802.15.4-2006) 1–314 (2011)
5. Montenegro, G., Kushalnagar, N., Hui, J., Culler, D.: Transmission of IPv6 Packets over IEEE 802.15.4 Networks. RFC 4944 (Proposed Standard), Updated by RFCs 6282, 6775 (September 2007)
6. Hui, J., Thubert, P.: Compression Format for IPv6 Datagrams over IEEE 802.15.4-Based Networks. RFC 6282 (Proposed Standard) (September 2011)
7. Narten, T., Nordmark, E., Simpson, W., Soliman, H.: Neighbor Discovery for IP version 6 (IPv6). RFC 4861 (Draft Standard), Updated by RFC 5942 (September 2007)
8. Singh, H., Beebee, W., Nordmark, E.: IPv6 Subnet Model: The Relationship between Links and Subnet Prefixes. RFC 5942 (Proposed Standard) (July 2010)
9. Shelby, Z., Chakrabarti, S., Nordmark, E., Bormann, C.: Neighbor Discovery Optimization for IPv6 over Low-Power Wireless Personal Area Networks (6LoWPANs). RFC 6775 (Proposed Standard) (November 2012)
10. Dawson-Haggerty, S., Tavakoli, A., Culler, D.: Hydro: A hybrid routing protocol for low-power and lossy networks. In: 2010 First IEEE International Conference on Smart Grid Communications (SmartGridComm), pp. 268–273 (2010)
11. Kim, K., Yoo, S., Park, J., Park, S., Lee, J.: Hierarchical routing over 6LoWPAN (HiLow). draft-deniel-6lowpan-hilow-hierarchical-routing-01.txt (2007)
12. Kim, K., Park, S., Chakeres, I.C.P.: Dynamic manet on-demand for 6LoWPAN (DYMO-low) Routing. draft-montenegro-6lowpan-dymo-low-routing-03.txt (2007)

13. Winter, T., Thubert, P., Brandt, A., Hui, J., Kelsey, R., Levis, P., Pister, K., Struik, R., Vasseur, J., Alexander, R.: RPL: IPv6 Routing Protocol for Low-Power and Lossy Networks. RFC 6550 (Proposed Standard) (March 2012)
14. Gnawali, O., Levis, P.: The ETX Objective Function for RPL. draft-gnawali-roll-etxof-01.txt (2010)
15. Gnawali, O., Levis, P.: The Minimum Rank Objective Function with Hysteresis. draft-gnawali-roll-minrank-hysteresis-of-02.txt (2011)
16. Thubert, P.: Objective Function Zero for the Routing Protocol for Low-Power and Lossy Networks (RPL). RFC 6552 (Proposed Standard) (March 2012)
17. Gaddour, O., Koubaa, A., Chaudhry, S., Tezeghdanti, M., Chaari, R., Abid, M.: Simulation and performance evaluation of DAG construction with RPL. In: 2012 Third International Conference on Communications and Networking (ComNet), pp. 1–8 (2012)
18. Gaddour, O., KoubíA, A.: Survey RPL in a nutshell: A survey. Comput. Netw. 56(14), 3163–3178 (2012)
19. Accettura, N., Grieco, L., Boggia, G., Camarda, P.: Performance analysis of the RPL routing protocol. In: 2011 IEEE International Conference on Mechatronics (ICM), pp. 767–772 (2011)
20. Tripathi, J., De Oliveira, J., Vasseur, J.P.: A performance evaluation study of RPL: Routing Protocol for Low power and Lossy Networks. In: 2010 44th Annual Conference on Information Sciences and Systems (CISS), pp. 1–6 (2010)
21. Vasseur, J., Kim, M., Pister, K., Dejean, N., Barthel, D.: Routing Metrics Used for Path Calculation in Low-Power and Lossy Networks. RFC 6551 (Proposed Standard) (March 2012)

Computing the Retransmission Timeout in CoAP

Ekaterina Balandina[1,2], Yevgeni Koucheryavy[2], and Andrei Gurtov[3]

[1] FRUCT Oy, Helsinki, Finland
Ekaterina.Balandina@fruct.org
[2] Tampere University of Technology, Tampere, Finland
yk@cs.tut.fi
[3] Department of Computer Science and Engineering, Aalto University, Helsinki, Finland
gurtov@hiit.fi

Abstract. The most prominent IT trend nowadays is connection of Wireless Sensor Networks (WSNs) with Internet service infrastructure. Interconnection of the millions of sensor and processing devices will create a tremendous traffic increase that can lead to congestion. In parallel to the development of new protocols for WSNs, e.g., Constrained Application Protocol (CoAP) there is plenty of research for new congestion control techniques (CC). This research shall carefully take into account all key restrictions of sensor networks, e.g., memory and power consumption, lousy paths and limited links throughput. This paper analyzes classical approach of definition of the retransmission timeout (RTO) estimate, proposed in RFC 6298, and compares it with the Eifel Retransmission Timer and the new ideas proposed in CoCoAP. Finally, we present our method for calculating RTO. Our approach could be seen as an extension of the classical TCP algorithm, where instead of constants that are used to take into account history of the current state we use a dynamically changing parameter. The value of this parameter is defined as a ratio between current sample of the round-trip time (RTT) and the RTO value.

Keywords: WSNs, Congestion Control, CoAP, RTO.

1 Introduction

Wireless sensor networks (WSN) are a fast growing technological domain and quite common phenomenon nowadays. Different types of WSNs are deployed for provision of a numerous services in various application areas, e.g., medicine, industrial, civil and army areas, etc [1]. The main challenge of WSN infrastructure is that most devices (sensors) in such networks have very limited memory, power and processing resources [2] and as a result require special resource-efficient software to be executed on top of them.

One of the key trends in today's ICT research is development of an efficient infrastructure for connecting WSNs with Internet services. It is important to mention here that most of classical Internet services are done without proper energy-awareness. At the same time sensor networks are characterized by short lifetime of sensors, inability to properly handle redundant traffic and network overloads. As a result, direct deployment

S. Balandin et al. (Eds.): NEW2AN/ruSMART 2013, LNCS 8121, pp. 352–362, 2013.
© Springer-Verlag Berlin Heidelberg 2013

of such services on sensor platforms will lead to unacceptable waste of energy, which cannot be delivered on the sensor side [3]. Moreover, complexity of Internet infrastructure and amount of network traffic will grow by orders of magnitude.

Constrained Application Protocol (CoAP) was proposed to address these demands [4]. It is light and efficient enough to work on constrained devices and provides good interface for the standard Internet services. CoAP protocol works on top of the unreliable UDP transport layer. A recent CoAP version has just a simple back-off mechanism that includes a timer and a retransmission counter. This is not sufficient for proper congestion management and there is a need in reliable congestion control mechanism that will address most of network overload scenarios.

In fact, high probability of congestion is in very core of most typical WSN use case scenarios. Service reliability in WSN is often based on sensing redundancy, when multiple sensors producing similar information in the same time and place. For example, in a forest fire and area lightening detection applications most likely a number of sensors will at the same time detect an event and start to signal about it. This may result in network overload for this network segment and even loss of important messages due to mutual blocking.

An important WSN design aspect is organization of sensor-scheduling activities that guarantee full area coverage while maximizing network lifetime [5]. Good overview of this problem with solution on how ensures full coverage of the monitored area by involving minimum number of sensors, which minimizes energy consumption and therefore extends network lifetime, was done by Lehsaini at el. [6]. Another key issue is efficient organization of data routing in WSNs [7].

However, proper control of network load cannot be provided without use of a congestion management mechanism [8]. Many approaches can be employed for this purpose [9], [10, [11]. In this study we decided to explore one of the most classical ideas, i.e., control congestion by choosing a policy for recalculating the retransmission timeout (RTO) values, which help to keep the transmissions on the required target level and also allows stopping transmission of flows when the network cannot deal with pushed amount of traffic. RTO is the time that elapses after a packet has been sent until the moment when sender will consider it to be lost and therefore a retransmission shall be initiated. The RTO could be seen as a prediction of the upper boundary of the round-trip time (RTT). Development of a method for accurate RTO prediction will provide the congestion control mechanism with a key tool to drive network via events of heavy load with minimal service degradation. It greatly influences reliable end-to-end performance. To evaluate importance of RTO it is enough to consider two opposite polices of recalculating RTO. A spurious timeout - too optimistic retransmission time can cause unnecessary traffic, which is reducing connection's effective throughput. A conservative retransmission timer causes long idle times before the lost packet is retransmitted during timeout period sensor is active and in vain spending energy.

The rest of the paper is structured as follows. Section 2 provides an overview of the Constrained Application Protocol. In Sections 3 and 4 we discuss the new method for RTO calculation. The developed analytic model of the new RTO calculation methods is presented in Section 5. Section 6 gives analysis and evaluation of RTO behavior in various scenarios. The paper is concludes by the list of main conclusions, acknowledgments and list of references.

2   Constrained Application Protocol (COAP)

Constrained Application Protocol was proposed by IETF group and in October 2012 its 12th draft was published. CoAP is a specific web protocol which is developed specifically for constrained nodes (sensors) and low-power networks that have high packet error rates and relatively small throughput (6lowPAN).

CoAP is designed specifically for machine-to-machine (M2M) applications that have embedded multicast support asynchronous message exchange and low overhead. Design of the interaction model of CoAP is very similar to the client/server model used in HTTP, but taking into account specifics of M2M applications, CoAP nodes can act either as clients or servers.

The send rate of CoAP protocol can be defined using a simplified version of TCP send rate formula, which has been derived by Padhye at el. in [12]. But for COAP case values of W_M and b are always equal to 1, so these parameters shall be excluded from calculation.

Unlike HTTP, the internal communication of CoAP is based on asynchronous datagram-oriented UDP transport layer. But it is important to remember that UDP does not provide internal mechanisms for congestion management and this is why one of the key requirements to ensure stable work of CoAP is to have own Congestion Control mechanism (CC). In this paper we propose an enhancement of the classical CC method, which is specifically adopted for CoAP. Current draft of CoAP specification states that CoAP has mechanisms for slowing down network overload, correct order of packets and check duplicates by using the exponential back-off mechanism and simple stop-and-wait mechanism [4]. This is achieved by strictly limiting number of simultaneous outstanding interactions to one, where the outstanding interaction is a confirmable message (CON) for which the acknowledgement (ACK) has not yet been received but is still expected. The second allowed case is when there was a request, for which neither a response nor ACK has yet been received, but is still expected. In fact both cases could occur at the same time, which is counted as one outstanding interaction. Message duplication detection is implemented for both confirmable and non-confirmable messages based on a simple idea of including message identification field to the message header and definition of the recipient endpoint.

Two parameters can be used for controlling the back-off mechanism - retransmission timeout (RTO) and retransmission counter. The initial value of retransmission timeout in CoAP is set to a random number within the interval of [ACK_TIMEOUT to ACK_TIMEOUT*ACK_RANDOM_FACTOR], where ACK_TIMEOUT and ACK_RANDOM_FACTOR are the transmission parameters, which default values are 2 sec. and 1.5 respectively. The initial value of the retransmission counter is always 0. The maximum number of retransmissions is defined by parameter MAX_RETRANSMIT and its default value is 4. The retransmission timer switch-on when CON has been sent, and the timer value doubles each time, when the timer expires and no ACK for the CON had been received. After four not successful attempts of retransmission, the sender shall close the session.

The following parameters are used for controlling the retransmission time in the current CoAP draft [4]:

- MAX_TRANSMIT_SPAN - is the maximum time from the first transmission of a confirmable message to its last retransmission.

$$MAX\_TRANSMIT\_SPAN = ACK\_TIMEOUT *$$
$$(2^{MAX\_RETRANSMIT} - 1) * ACK\_RANDOM\_FACTOR \qquad (1)$$

Default value for MAX_TRANSMIT_SPAN is 45 seconds.

- MAX_TRANSMIT_WAIT - is the maximum length of period of time that sender is capable to wait for ACK on the already sent confirmable message.

$$MAX\_TRANSMIT\_WAIT = ACK\_TIMEOUT *$$
$$(2^{MAX\_RETRANSMIT + 1} - 1) * ACK\_RANDOM\_FACTOR \qquad (2)$$

Default value for MAX_TRANSMIT_WAIT is 93 seconds.

- MAX_LATENCY - is the maximum time that will be needed for a datagram to be fully delivered to the destination. By default it is set to be 100 seconds.

- PROCESSING_DELAY - is the time required for a node to process a confirmable message and issue an acknowledgement. By default it is equal to ACK_TIMEOUT.

- MAX_RTT - is the maximum round-trip time calculated as follows:

$$MAX\_RTT = 2 * MAX\_LATENCY +$$
$$PROCESSING\_DELAY \qquad (3)$$

- EXCHANGE_LIFETIME - is the time from the moment when transmission of the confirmable message has started until the moment when an acknowledgement is no longer expected. As a result at the message layer information about this message exchange can be purged.

$$EXCHANGE\_LIFETIME = ACK\_TIMEOUT * (2^{MAX\_RETRANSMIT} - 1) *$$
$$ACK\_RANDOM\_FACTOR + 2 *$$
$$MAX\_LATENCY + PROCESSING\_DELAY \qquad (4)$$

For the default transmission parameters its value is 247 seconds.

Responsibility for congestion prediction, detection and control in CoAP networks is fully on clients' side. But potentially it is possible that the client will be hacked or broken, which will lead to abnormal behavior. To prevent any damage to the network in such cases, the server should have some mechanism that will limit the traffic data rate of nodes with abnormal behavior and this solution is discussed in the next section.

3   Enhancement of the Classical TCP Algorithm

The most classical definition of the algorithm for calculating value of the retransmission timer is proposed by RFC 6298 [13]. According to that definition the Transmission Control Protocol (TCP) uses an RTO to control reliability of data exchange. Calculation of new RTO value is done by an algorithm that is based on two variables: smoothed RTT (SRTT) and RTT variation (RTTVAR). SRTT can be seen as a mean to preserve history of RTT, its impact factor is constant and equals to 7/8. RTTVAR keeps the history of RTT variation, it is also constant and its impact factor is 3/4.

Before the first measurement of RTT is received RTO should be set to 1 second. After first RTT sample is received the following formulas are used for RTO calculation:

$$SRTT \leftarrow RTT$$
$$RTTVAR \leftarrow RTT/2$$
$$RTO \leftarrow SRTT + \max (G, K*RTTVAR) \tag{5}$$

After subsequent measurement is received the following formulas are applied:

$$RTTVAR \leftarrow (1 - \beta) * RTTVAR + \beta * |SRTT - RTT|$$
$$SRTT \leftarrow (1 - \alpha) * SRTT + \alpha * RTT$$
$$RTO \leftarrow SRTT + \max (G, K*RTTVAR) \tag{6}$$

In these equations α, β and K are constants and their values respectively are 1/4, 1/8 and 4. Value G defines the clock granularity and higher it is, more conservative is result RTO value. It is recommended to choose G value not greater than 100 ms [13]. At the same time G should be at least one order of magnitude smaller than the RTT [14].

In the later study we will primarily address the Machine-to-Machine (M2M) use case scenario. Shafiq at el. [15] illustrate the cumulative distribution functions (CDFs) of the median RTTs and packet loss ratio experienced by each device for smartphones and all M2M device categories. Based on this study we can say that if RTT varies between 500 ms and 2 seconds, G should be at least 50 ms and its maximum value in this case is 100 ms, to fulfill recommendation of RFC6298 and [15].

In RFC6298 it is underlined that SRTT and RTTVAR can be cleared if retransmission timeout expired several times and values of these variables became bogus.

4   CoCoAP Modifications

In addition to activities targeted in improvement of CoAP that resulted in releasing specification draft version 12, the same team proposed to take RTO estimation calculation algorithm [14] as a basis for an enhanced protocol solution on top of CoAP. The new protocol is named CoAP Simple Congestion Control/Advanced (CoCoAP).

The initial RTO estimate in this protocol is set to 2 seconds. Modification of RTO value is done by use of two mechanisms named "strong" and "weak" estimators. Both mechanisms implement the same algorithm, but have different sets of state variables. If the packet is received based on the initial transmission, i.e., without any retransmissions, then the "strong" estimator RTO calculation branch is in use. If the last packet was received as retransmissions were done then the "weak" estimator branch is used. It means that if we don't know for sure whether ACK is an acknowledgment of the initial message that was just delayed or it is already an acknowledgement of the retransmitted packet then the "weak" estimator is in use. If ACK came before retransmission timer expires, it means that real RTT sample was calculated and in this case "strong" estimator branch of RTO calculation algorithm shall be used. The last step is overall RTO estimate calculation that is an average of the currently calculated "weak" or "strong" value and the RTO overall value obtained on the previous step.

$$RTO\_overall = 0.5*RTO\_recent + 0.5*RTO\_overall \qquad (7)$$

CoCoAP support service provision for packets that don't need confirmation of delivery. Handling of such packets is provided by advanced part of the algorithm that defines a set of additional rules. For example, the date rate for sending non-confirmable messages must not exceed 1 Byte/s and at least 2 out of 16 consecutive messages sent to one endpoint must be confirmable. The full set of additional rules can be found in CoCoAP draft specification [14].

5   Analytic Model of Send Rate

The send rate is a key characteristic of network quality and it strongly dependents on frequency of congestion events as well as on speed of transmission recovery after such events. In this section we present derivation of an analytical characterization of CoAP send rate as a function of loss rate and RTT. This part of the study was partly inspired by work of Padhye at el. [12].

CoAP protocol detects packet loss when the retransmission timeout is expired, which in case of proper configuration of RTO happens if a packet or the corresponding ACK is lost. For simplicity reasons and to make it easy to read and understand the first steps of derivation process we adopting terminology and notations proposed in [12].

We would like to start from a general formula of send rate B_t. For any given time $t >$ 0 and N_t - number of packets transmitted during interval t, we can define the send rate as $B_t = N_t/t$ (the number of packets sent per unit of time).

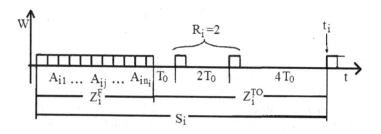

Fig. 1. CoAP performing scheme with timeout loss indications

As it was underlined before in Section 2, CoAP sends one packet and waits for the ACK. As a consequence it is enough for CoAP to support only one type of loss indication, which is based on occurrence of timeout event TO. Z^F_i is the time interval of normal transmission without packet loss. In practice it consists of n_i RTT intervals, as it is illustrated in Figure 1 by A intervals. This continues until the first timeout event (TO) occur. Then the protocol starts retransmission process, which on the sender side is associated to the sequence of timeouts Z^{TO}_i. Based on these notions the first full cycle of normal transmission and recovery for an error is defined as S_i:

$$S_i = Z^{TO}_i + Z^F_i \tag{8}$$

The number of packets sent during Z^{TO}_i is denoted as R_i. It counts the total number of packet transmissions in Z^{TO}_i. Parameter n_i is a number of packets sent during Z^F_i. Then we can define the total number of packets sent during S_i period (M_i) as follows:

$$M_i = n_i + R_i Z^F_i \tag{9}$$

So the most general definition of send rate B could be defined as a relation function of the total number of packets send (M) to the transmission periods (S).

$$B = \frac{E[M]}{E[S]} = \frac{E\left[\sum_{j=1}^{n_i} A_{ij}\right] + E[R]}{E\left[\sum_{j=1}^{n_i} A_{ij}\right] + E[Z^{TO}]} \tag{10}$$

Let's denote the average value of RTT as $E[r]$. Obviously when only one packet is allowed to be transmitted, the average value of A intervals denoted as $E[A]$ is:

$$E[A] = E[r] \tag{11}$$

Based on formulas 10 and 11 we can derive the the result formula for the send rate:

$$B = \frac{RTT + E[R]}{RTT + E[Z^{TO}]} \tag{12}$$

Timeout occurs k times for $k - 1$ consecutive losses, where the number k of timeout occurrences has a geometric distribution

$$P[R=k] = p^{k-1}(p-1) \tag{13}$$

Then the mean of R could be defined as:

$$E[R] = \sum_{k=1}^{n} P[R = k] = 1/(1-p) \tag{14}$$

The maximum number of retransmissions in CoAP is four. The duration of the sequence of four timeouts is denoted as L, where T_0 is the initial value of retransmission timer.

$$L = (2^4 - 1)T_0 \tag{15}$$

Now we can define the average time interval of timeout transmission after packet loss was detected: $E[Z^{TO}]$

$$E[Z^{TO}] = \sum_{k=1}^{4} L_k P[R = k] = T_0(1+2p+4p^2+8p^3-15p^4) \tag{16}$$

So finally we can calculate the send rate of CoAP protocol as:

$$B(p) = \frac{RTT+1/(1-p)}{RTT+T_0 f(p)} \tag{17}$$

where $f(p) = T_0(1+2p+4p^2+8p^3-15p^4)$

6   Evaluation

The proposed enhancement of CoCoAP protocol is based on combination of RTO estimation algorithm [13] and a set of the algorithm enhancements proposed in [14]. Also here we have used main findings of a survey of the current works involved in traffic analysis and modeling, network optimization and network anomaly detection for WSNs [16]. For our studies we also used the results by Ponmagal and Ramachandran [17] on wireless rate-control technique, whose link characteristics are identified by a variable link rate and burst transmission error. In addition the proposed CoCoAP enhancement utilizes ideas from the Eifel retransmission timer [18], [19].

The study of Eifel retransmission timer made an important conclusion on a role of "magic numbers" (α, β and K) for the performance of the algorithm. Estimator gains (values of α and β) are too high and the variation weight (K) is too low for the situations of the large senders load. They cause SRTT and RTTVAR to decay too quickly and RTO becomes too aggressive [14]. Based on these conclusions we propose to replace constant coefficients α and β to a coefficient γ that is defined as follows:

$$\gamma = RTT/RTO \tag{18}$$

As it was shown in the previous section, in current CoCoAP [13] RTO_overall is defined as an average of RTO recent (that is based on currently calculated "weak" or "strong" value) and RTO overall obtained on the previous step (see formula 7). In the proposed enhancement this rules works only if RTO recent was calculated based on "weak" estimate. In cases when RTO recent was calculated based on "strong" estimate then we assign RTO_overall = RTO_recent.

Figure 2 and Figure 3 compare the performance of the classical RTO estimation algorithm (blue dotted line), modified RTO estimation algorithm (red line) and the stair-step RTT function or saw-like RTT function respectively (green line).

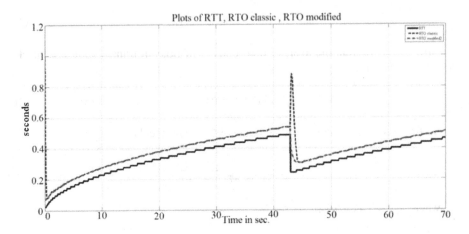

Fig. 2. Plots of the classical RTO estimation algorithm (blue), a modified RTO estimation algorithm (red) and with stair-step RTT function (black)

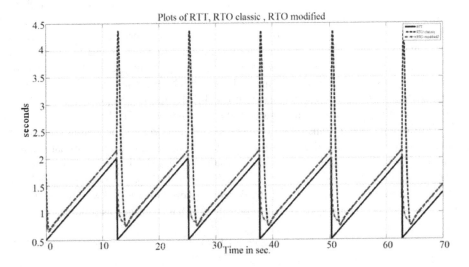

Fig. 3. Plots of the classical RTO estimation algorithm (blue), modified RTO estimation algorithm (red) and with a saw-like RTT function (black)

This result is based on intensive simulation and illustrates the same level of performance. In Figure 2, we can see a peak value for the classic RTO algorithm after approximately 42 seconds of simulation time. This peak is related to the decrease of RTT value, which first leads to a peak increase of RTO value as we can see. This happens as a difference of RTT and SRTT becomes negative, but formula (6) is calculating RTTVAR value using modulus that results in such an artificial increase of the RTO value. The modified algorithm adapts to this situation and reacts respectively by decreasing RTO prediction value [18].

Based on an intensive simulation studies, the following recommendations were generated. The first recommendation is that if RTT – SRTT < 0 then it is beneficial to skip the rule of changing RTTVAR, as it will allow faster adaptation to the current RTT values, while if RTT – SRTT > 0 we use the standard formula for changing RTTVAR as defined in [14]. The second recommendation is that when γ > 0.5 then it is better to change its value to 1-γ, as it means RTT is greater than ½ RTO, which could be only a consequence of the previous packet lost and it is better to slowdown adoption mechanism to avoid fast return to the too high level of data transmission.

7   Conclusion

The paper gives an overview of the existing congestion control solutions for wireless sensor networks, the CoAP protocol, and proposes the enhancement of CoCoAP protocol that combines best features of the prior art solutions and provides a set of new rules. The set of new rules is generated based on the results of intensive simulation tests that allowed us to create a new RTO estimation algorithm that works significantly better than the classical one. Our algorithm is slightly more aggressive, but at the same time more efficient in conditions of limited traffic fluctuations, as well as when some unpredictable events occur.

The performed analysis allowed us to derive an analytical model for CoAP congestion control behavior by defining the sending rate as a function of loss and RTT. In this model we use the COAP timeout mechanism as a tool of detecting packet losses.

Currently we are working on development of a more sophisticated model of the proposed algorithm. The implementation is targeted for Cooja simulator [20] that provides a good platform for this kind of study. We also plan to evaluate the new RTO mechanism using trace-driven simulations on real network data.

Acknowledgements. Authors are grateful for TiViT IoT SHOK program and GETA scholarship program of Tampere University of Technology that provided required support of this research. Also we want to acknowledge support MAMMOTH and SEMOHealth projects.

References

1. Akyildiz, I.F., Su, W., Sankarasubramaniam, Y., Cayirci, E.: Wireless sensor networks: a survey. Computer Networks 38(4), 393–422 (2002)
2. Cardei, M., Thai, M.T., Li, Y., Wu, W.: Energy-efficient target coverage in wireless sensor networks. In: IEEE INFOCOM 2005, pp. 1976–1984 (2005)
3. Cai, Y., Li, M., Shu, W., Wu, M.Y.: Acos: A precise energy-aware coverage control protocol for wireless sensor networks. International Journal Ad Hoc Sensor Wireless Networks 3(1), 77–98 (2007)
4. Shelby, Z., Hartke, K., Bormann, C., Frank, B.: Constrained Application Protocol (CoAP), draft-ietf-core-coap-12 (October 2012)
5. Mulligan, R., Ammari, H.M.: Coverage in Wireless Sensor Networks: A Survey. Network Protocols and Algorithms, Macrothing Institute 2(2), 27–53 (2010) ISSN 1943-3581

6. Lehsaini, M., Guyennet, H., Feham, M.: Cluster-based Energy-efficient k-Coverage for Wireless Sensor Networks. Network Protocols and Algorithms, Macrothing Institute 2(2), 89–106 (2010) ISSN 1943-3581

7. Ganesh, S., Amutha, R.: Efficient and Secure Routing Protocol for Wireless Sensor Networks through Optimal Power Control and Optimal Handoff-Based Recovery Mechanism. Journal of Computer Networks and Communications 2012 (2012), http://www.hindawi.com/journals/jcnc/2012/971685/ (retrived on May 08, 2013)

8. Gurtov, A.: TCP Performance in the Presence of Congestion and Corruption Losses. Master's Thesis, University of Helsinki, Department of Computer Science (December 2000)

9. Chen, L., Szymanski, B.K., Branch, J.W.: Auction-Based Congestion Management for Target Tracking in Wireless Sensor Networks. In: Proc. of the 7th IEEE PERCOM Conference, pp. 194–203 (March 2009)

10. Kumara, B., Naik, M.M.: Architecture for Node-level Congestion in WSN using Rate Optimization. IOSR Journal of Engineering 2(6), 30–34 (2012) ISSN 2250-3021, http://www.iosrjen.org/Papers/vol2_issue6%20%28part-3%29/E0263034.pdf (retrived on May 08, 2013)

11. Shaikh, F.K., Khelil, A., Ali, A., Suri, N.: Reliable congestion-aware information transport in wireless sensor networks. International Journal of Communication Networks and Distributed Systems 7(1/2), 135–152 (2011)

12. Padhye, J., Firoiu, V., Towsley, D.F., Kurose, J.F.: Modeling TCP Reno Performance: A Simple Model and Its Empirical Validation. IEEE/ACM Transactions on Networking 8(2), 133–145 (2000)

13. Paxson, V., Allman, M., Chu, J., Sargent, M.: Computing TCP's Retransmission Timer., RFC 6298 (June 2011)

14. Bormann, C.: CoAP Simple Congestion Control/Advanced., draft-bormann-core-cocoa-00 (August 2012)

15. Shafiq, M.Z., Ji, L., Liu, A.X., Pang, J., Wang, J.: A First Look at Cellular Machine-to-Machine Traffic - Large Scale Measurement and Characterization. In: SIGMETRICS 2012, London, UK (June 2012)

16. Wang, Q.: Traffic Analysis & Modeling in Wireless Sensor Networks and Their Applications on Network Optimization and Anomaly Detection. Network Protocols and Algorithms, Macrothing Institute 2(1), 74–92 (2010) ISSN 1943-3581

17. Ponmagal, R.S., Ramachandran, V.: Link Quality Estimated TCP for Wireless Sensor Networks. International Journal of Recent Trends in Engineering 1(1), 495–497 (2009)

18. Ludwig, R., Sklower, K.: The Eifel Retransmission Timer. ACM SIGCOMM Computer Communication Review 30(3), 17–27 (2000)

19. Ludwig, R., Gurtov, A.: The Eifel Response Algorithm for TCP, RFC 4015 (February 2005)

20. Official web page of Cooja simulator in Contiki OS, http://www.contiki-os.org/ (retrived on May 08, 2013)

Analytical Modeling of Playback Continuity in P2P Streaming Network with Latest First Download Strategy*

Yuliya Gaidamaka and Andrey Samuylov

Peoples' Friendship University of Russia, Telecommunication Systems Department
Ordzhonikidze str. 3
115419 Moscow, Russia
ygaidamaka@mail.ru, asam1988@gmail.com

Abstract. Nowadays peer-to-peer (P2P) overlay networks are widespread due to their numerous advantages. In P2P streaming networks the main measures to estimate the video playback quality are playback continuity and startup latency. Most analytical models suggest approximate formulas for playback quality measures or algorithms for different chunk download strategies. In this paper, we develop the discrete Markov chain model of the data exchange process between users in P2P streaming network with buffering mechanism. Unlike most of the existing models, we get the exact formulas to calculate the transition probability matrix for the discrete Markov chain. That let us develop a numerical method for the evaluation of playback continuity in P2P streaming network.

Keywords: P2P network, live streaming, buffer occupancy, playback continuity, Markov chain model.

1   Introduction

Mesh-based streaming P2P-networks became popular because of its scalability robustness, reliability and low cost of implementation. CoolStreaming, PPlive, PPstream, Gridmedia are examples for successful commercial P2P-streaming systems nowadays. The main measures to estimate the video playback quality are playback continuity or fluency (probability of continuous playback) and startup latency (expected time to start playback). These measures significantly depend on the data-driven download strategy used by a peer to select one or more chunks to download. Some models for P2P streaming system measures analyses were proposed in [1-8]. In [1-3] some overall conclusions about universal probability for all peers in the network

---

* This work was supported in part by the Russian Foundation for Basic Research (grant 13-07-00953) and the Ministry of education and science of Russia (projects 8.7962.2013, 14.U02.21.1874).

S. Balandin et al. (Eds.): NEW2AN/ruSMART 2013, LNCS 8121, pp. 363–370, 2013.

were obtained, in [4-7] the detailed distribution of what each peer has in its buffer was investigated for different download strategies, namely Rarest First, Greedy [4,5], and Latest Useful Chunk First strategy (LF) [6,7]. A similar comparison of different download strategies in VoD P2P networks and its impact on main performance measures was shown in [8], where they studied Rarest Random, Naive Sequential, Cascading, and Hybrid download strategies. In [4,6,8] a comparison was carried out using simple approximate formulas, but in [5,7] the significant error of such calculations was shown. Simulation based analysis of playback continuity for different download strategies, including mixed strategies, was presented in [3, 5], where a high computational complexity of the task was pointed out. So there are no analytical models available to compare the performance measures of streaming P2P-networks. Therefore in this paper we propose the method for finding playback continuity in terms of the Latest First strategy, which is nearly optimal in terms of peer upload capacity utilization and system throughput [6]. The advantage of LF strategy is its simplicity – a peer doesn't need to know the buffer maps of other peers, it selects a random peer in the network and downloads the chunk closest to the current server playback time. The contribution of this paper is developing of the discrete Markov chain model for changing buffer states proposed in [5] and developing exact formulas for calculating transition probability matrices in case of the LF download strategy.

The paper is organized as follows. The basic mathematical model is in Section 2. Section 3 goes into the details of how to calculate the transition probability matrices. Section 4 provides a numerical example to illustrate the formulas suggested in section 3. The conclusion of the paper is in Section 5.

2   Markov Chain Model

For a given network with N users and a single server, vector $\mathbf{x}(n)$ defines the state of each user (n-user), where $\mathbf{x}(n) = (x_0(n), x_1(n), \cdots, x_M(n))$ is the state of n-user's buffer. Here $x_m(n)$ is the state of n-user's buffer m-position: $x_m(n) = 1$, if n-user's buffer m-position is occupied with a chunk, otherwise $x_m(n) = 0$, $m = 0, \cdots, M$. Each user in the network uses buffer positions $m = 1, \cdots, M$ to store the downloaded chunks from the network, and uses 0-position only to download a chunk from the server. Thus, the oldest chunk in the buffer, which will be sent to the video player for playback during the next time slot, is located in M-position during the current time slot, and chunk in m-position will be sent to the player for playback in M-m time slots. Note that if during any time slot M-position is occupied, then n-user watches the video stream without any pause.

The state of the system is defined by $\mathbf{X} = (\mathbf{x}(n))_{n=1,\dots,N}$, where the n-th row of the matrix $\mathbf{X}$ corresponds to the buffer state of n-user, and dim $\mathbf{X} = N \cdot (M + 1)$. Therefore, the state space of the system is given by $X = \{\mathbf{X_0}\} \cup \{\mathbf{X} \in \{0,1\}^{N(M+1)} : \sum_{n=1}^{N} \mathbf{X}(n, 0) = 1\}$, where $\mathbf{X_0} \equiv \mathbf{0}$ and $|X| = N \cdot 2^{N \cdot M} + 1$.

Denote by $M^0(\mathbf{x}(n))$ and $M^1(\mathbf{x}(n))$ the set of all empty and occupied positions in n-user's buffer respectively, i.e. $M^0(\mathbf{x}(n)) = \{m: x_m(n) = 0, \ m = 1, \cdots, M\}$, $M^1(\mathbf{x}(n)) = \{m: x_m(n) = 1, \ m = 1, \cdots, M\}$, where $M^0(\mathbf{x}(n)) \cup M^1(\mathbf{x}(n)) = \{1, 2, \cdots, M\}$. Then $M^0(\mathbf{x}(n)) \cap M^1(\mathbf{x}(h))$ will be the set of all positions in buffer to which n-user can download a chunk from a target h-user, $n \neq h$. If $M^0(\mathbf{x}(n)) \cap M^1(\mathbf{x}(h)) \neq \emptyset$, then the index $m_\delta(\mathbf{x}(n), \mathbf{x}(h))$ of the position to which n-user can download a chunk from h-user is determined by the download strategy δ in use, i.e. $m_{LF}(\mathbf{x}(n), \mathbf{x}(h)) = \min\{m: m \in M^0(\mathbf{x}(n)) \cap M^1(\mathbf{x}(h))\}$, with $\delta = LF$ strategy.

Denote by $S\mathbf{x}(n)$ the shifting operator of vector $\mathbf{x}(n)$, meaning if $\mathbf{x}(n) = (x_0(n), x_1(n), \cdots, x_{M-1}(n), x_M(n))$, then $S\mathbf{x}(n) = (0, x_0(n), \cdots, x_{M-1}(n))$.

Let t_l be the shifting moment of buffer content. When constructing the model in discrete time it is assumed that if at the moment $t_l - 0$ a buffer is in the state $\mathbf{x}(n)$, then at the moment $t_l + 0$ it will be in the state $S\mathbf{x}(n)$.

According to the data distribution protocol in P2P live streaming networks with buffering mechanism, during the interval $[t_l, t_{l+1})$, which corresponds to the l-th time slot, the server and users interact in the following way.

First, every user shifts the contents of his buffer (i.e. chunk in buffer M-position is sent for playback and all the rest chunks are shifted towards the end of the buffer). Then the server chooses one user randomly and uploads a chunk for the current time slot to his buffer 0-position. After this, each user checks if he has any empty positions in his buffer. If there are empty positions, he chooses another user in attempt to download the missing chunks. Applying the LF strategy he will select the chunk with the minimum index among the available chunks to download. Otherwise, he will not perform any actions during this time slot.

Denote by $\mathbf{X}^l$ the network state at the moment $t_l - 0$. Note that $\{\mathbf{X}^l, \ l \geq 0\}$ forms a Markov chain over state space X. Generally speaking the Markov chain is decomposable, with one class of essential states.

Let $\pi^l(\mathbf{X})$ be the probability that Markov chain $\{\mathbf{X}^l, \ l \geq 0\}$ during l-th time slot is in state $\mathbf{X}$, i.e. $\pi^l(\mathbf{X}) = P\{\mathbf{X}^l = \mathbf{X}\}$ and $\Pi^{l,l+1}(\mathbf{X}, \mathbf{Y}) = P\{\mathbf{X}^{l+1} = \mathbf{Y} | \mathbf{X}^l = \mathbf{X}\}$ be the corresponding transition probability, $\mathbf{X}, \mathbf{Y} \in X, l \geq 0$.

The probability distribution $\pi^l(\mathbf{X})$ satisfies the Kolmogorov-Chapman equations:

$$\pi^{l+1}(\mathbf{Y}) = \sum_{\mathbf{X} \in X} \pi^l(\mathbf{X})\Pi^{l,l+1}(\mathbf{X}, \mathbf{Y}), \ \mathbf{Y} \in X, \ l \geq 0. \tag{1}$$

Denote by $p_1^l(n, m)$ the probability that m-position of n-buffer is occupied during l-th time slot:

$$P\{x_m^l(n) = 1\} = \sum_{\mathbf{X} \in X: x_m(n) = 1} \pi^l(\mathbf{X}) = p_1^l(n, m),$$

where $x_m^l(n)$ is an element of matrix $\mathbf{X}^l$.

Assume that the equilibrium distribution of the Markov chain $\{\mathbf{X}^l, \ l \geq 0\}$ exists. Denote by $p_1(n, m) = \lim_{l \to \infty} p_1^l(n, m)$ the probability that m-position of n-buffer is occupied. Then we have the following formula for the probability $V(n)$ that n-user is watching video without pauses during playback, i.e. probability of playback continuity,

$$V(n) = p_1 (n, M).$$ (2)

Thus, the probability of playback continuity is acquired by solving equation (1). In order to solve the equation we have to calculate the transition probability matrix $\Pi^{l,l+1}$. In the next section we propose the analytical formula for calculating the transitional matrix.

3   Transition Probability Matrix

In order to calculate the transition probability matrix $\Pi^{l,l+1}$ we introduce an auxiliary matrix $\mathbf{A} = \mathbf{X}^{l+1} - S\mathbf{X}^l$. Each element in matrix $\mathbf{A}$ shows the changes occurred to n-user's buffer m-position during l-th time slot, where $\mathbf{A}(n, m) = 0$ if nothing happened to n-user's buffer m-position, $\mathbf{A}(n, m) = 1$ if n-user successfully downloaded a chunk to his buffer m-position, and $\mathbf{A}(n, m) = -1$ if the corresponding chunk vanished from n-user's buffer.

Note that $\mathbf{A}(n, 0) = 1$ if n-user received a chunk directly from the server during l-th time slot, therefore $\sum_{n=1}^{N} \mathbf{A}(n, 0) \leq 1$.

If $\sum_{m=0}^{M} \mathbf{A}(n, m) = 0$, then during l-th time slot n-user did not download a single chunk. This might take place when there are no empty positions in n-user's buffer or n-user selected a user, who has no available chunks to download.

If $\sum_{m=1}^{M} \mathbf{A}(n, m) = 1$, then during l-th time slot n-user has successfully downloaded a chunk.

We now define the following sets of pairs $(\mathbf{X}, \mathbf{Y}) \subset X \times X$. The set $X_{-1} = \{(\mathbf{X}, \mathbf{Y}): \exists n \in \mathbb{N}, \exists m \in M\ Y(n, m) - SX(n, m) = -1\}$, which describes transitions from state $\mathbf{X}$ to state $\mathbf{Y}$, when at least one chunk has vanished from any buffer, and the set $X_{>1} = \{(\mathbf{X}, \mathbf{Y}): \exists n \in \mathbb{N}, \sum_{m \in M}(\mathbf{Y}(n, m) - SX(n, m)) \geq 2\}$ which describes transitions from state $\mathbf{X}$ to state $\mathbf{Y}$, when at least one of the users has downloaded more than one chunk. According to the protocol for the data distribution in P2P live streaming networks with buffering mechanism during the l-th time slot a user might not download any useful chunks or will download only one chunk – either from the server or from any other user. Hence, the sets X_{-1} and $X_{>1}$ contain the pairs of states, the transitions between which are not possible.

Next we define the probabilities of the following events, that occur when the system transits from state $\mathbf{X}$ to state $\mathbf{Y}$:

$P^s_{\mathbf{XY}}(n)$ - n-user received a chunk from the server;

$P^-_{\mathbf{XY}}(n)$ - n-user did not receive a single chunk – neither from the server, nor from any other user;

$P^+_{\mathbf{XY}}(n)$ - n-user received a chunk according to the LF strategy from another user.

One can prove, that these probabilities can be calculated using the formulas below.

$$P^s_{\mathbf{XY}}(n) = \frac{1}{N}$$ (3)

$$P_{XY}^-(n) = \frac{1}{N-1} \cdot \sum_{h \in N \setminus \{n\}} I\left(\sum_{m \in M} (1 - SX(n,m)) \cdot SX(h,m) = 0\right) \qquad (4)$$

$$P_{XY}^+(n) = \frac{1}{N-1} \times \sum_{m \in M \setminus \{0\}} [(Y(n,m) - SX(n,m)) \cdot$$

$$\cdot \sum_{h \in N \setminus \{n\}} \left(SX(h,m) \cdot I\left(\sum_{i=1}^{m-1} (1 - SX(n,i)) \cdot SX(h,i) = 0\right)\right)] \qquad (5)$$

The following proposition contains formulas for calculating the transition probability matrices $\Pi^{l,l+1}$.

Proposition. Let $\mathbf{X}^l = \mathbf{X}$ and $\mathbf{X}^{l+1} = \mathbf{Y}$. Then
$\Pi^{l,l+1}(\mathbf{X}, \mathbf{Y}) = 0$, if $\mathbf{X} = \mathbf{Y} = \mathbf{X}_0$;
$\Pi^{l,l+1}(\mathbf{X}, \mathbf{Y}) = 0$, if $(\mathbf{X}, \mathbf{Y}) \in X_{-1} \cup X_{>1}$;
$\Pi^{l,l+1}(\mathbf{X}, \mathbf{Y}) = \prod_{n \in N}(I(A(n,0) = 1) \cdot P_{XY}^s(n) +$
$+I(\sum_{m \in M} A(n,m) = 0) \cdot P_{XY}^-(n) +$

$$+I(\sum_{m \in M \setminus \{0\}} A(n,m) = 1) \cdot P_{XY}^+(n)), \text{ if } (\mathbf{X}, \mathbf{Y}) \in X \setminus \{X_{-1} \cup X_{>1}\}. \qquad (6)$$

According to formula (6), the transition probability matrix Π is independent of l, i.e. $\Pi^{l,l+1}(\mathbf{X}, \mathbf{Y}) = \Pi(\mathbf{X}, \mathbf{Y}) \; \forall l \geq 0, (\mathbf{X}, \mathbf{Y}) \in X \times X$. Thus, formula (6) allows to calculate matrix Π and equilibrium probabilities $\pi(\mathbf{X}), \mathbf{X} \in X$ of Markov chain $\{\mathbf{X}^l, l \geq 0\}$. After that by assuming independent and homogeneous peers in a symmetric network setting we can use formula (2) for evaluation of playback continuity in the P2P streaming network with high dimension state space.

4   Numerical Example

To illustrate the formulas derived in the previous section let us consider an example with N=4 peers in the network using the same size playback buffer of total length M+1=4 and chunk download strategy LF.

First, consider the case $(\mathbf{X}, \mathbf{Y}) \in X_{-1}$ depicted in Fig. 1, where matrix $\mathbf{A} = \mathbf{Y} -$

$$SX = \begin{pmatrix} 0 & 0 & -1 & 1 \\ 1 & 0 & 0 & 0 \\ 0 & 0 & 1 & 0 \\ 0 & 0 & 0 & 0 \end{pmatrix}.$$ The element $\mathbf{A}(1,3) = -1$, which means that the first user

(n=1) has lost a chunk from his second buffer position (m=2). This event is impossible, as $(\mathbf{X}, \mathbf{Y}) \in X_{-1}$, and therefore $\Pi(\mathbf{X}, \mathbf{Y}) = 0$.

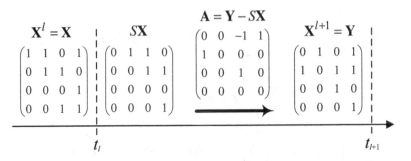

Fig. 1. Case $(\mathbf{X}, \mathbf{Y}) \in X_{-1}$

Consider the case $(\mathbf{X}, \mathbf{Y}) \in X_{>1}$ depicted in Fig. 2.

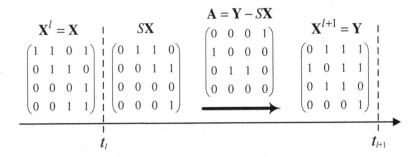

Fig. 2. Case $(\mathbf{X}, \mathbf{Y}) \in X_{>1}$

The third row of matrix **A** shows that the third user (n=3) has downloaded two chunks into his first and second buffer positions (m=1, 2). Here $\sum_{m \in M} \mathbf{A}(3, m) = 2 > 1$. This event is impossible as well, as $(\mathbf{X}, \mathbf{Y}) \in X_{>1}$ and therefore $\Pi(\mathbf{X}, \mathbf{Y}) = 0$.

Now we consider the case shown in Fig. 3, where $(\mathbf{X}, \mathbf{Y}) \notin X_{-1} \cup X_{>1}$.

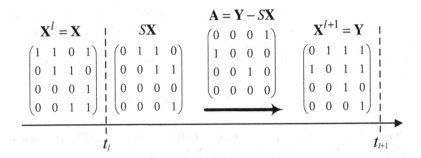

Fig. 3. Case $(\mathbf{X}, \mathbf{Y}) \in X \backslash \{X_{-1} \cup X_{>1}\}$

In this case the first user ($n=1$) has successfully downloaded a chunk into his third buffer position ($m=3$), the second user ($n=2$) received a chunk from the server, the third user ($n=3$) downloaded a chunk into his second buffer position ($m=2$), and the forth user ($n=4$) did not download anything. Next we have to calculate the transition probability $\Pi(\mathbf{X}, \mathbf{Y})$ for the states $\mathbf{X}$ and $\mathbf{Y}$ shown in Fig. 3. According to formula (6) the corresponding indicator function results are shown in Table 1.

As the first user ($n=1$) has downloaded a chunk into his third buffer position ($m=3$), then according to formula (5) $P_{XY}^+(1) = \frac{2}{3}$. According to formula (3) $P_{XY}^s(2) = \frac{1}{4}$, according to formula (5) $P_{XY}^+(3) = \frac{1}{3}$, and finally according to formula (4) $P_{XY}^-(4) = \frac{1}{3}$.

Table 1. The indicator functions values

	$n=1$	$n=2$	$n=3$	$n=4$
$I(A(n,0) = 1)$	0	1	0	0
$I\left(\sum_{m \in M} A(n,m) = 0\right)$	0	0	0	1
$I\left(\sum_{m \in M \setminus \{0\}} A(n,m) = 1\right)$	1	0	1	0

Hence in this example the transition probability is calculated as follows:
$$\Pi(\mathbf{X}, \mathbf{Y}) = P_{XY}^+(1) \cdot P_{XY}^s(2) \cdot P_{XY}^+(3) \cdot P_{XY}^-(4) = \frac{2}{3} \cdot \frac{1}{4} \cdot \frac{1}{3} \cdot \frac{1}{3} = \frac{1}{18}.$$

5 Conclusion

In this paper, we presented the exact formulas for calculating the transition probability matrix for the discrete Markov chain model of data exchange process between users of P2P streaming network. We assume that the derived exact formulas, unlike the approximate formulas in [4, 6], can be applied to compute the playback continuity for large networks. Simulating a small scale network [5, 7] shows that the results converge to the ones computed via the exact formulas in $2 \cdot 10^6$ time slots. The formulas were derived for LF strategy, corresponding formulas for Rarest First and Greedy strategies can be derived in the same manner.

Our further research will be devoted to the application of the exact formulas for cases of large scale networks, as well as the extension of the model taking into account playback lags, i.e. the time difference between the two peers playing the stream, that occur because of the transfer delays from server. The subject of further research is the comparison of Rarest First, LF, Greedy, and some mixed or hybrid strategies for streaming networks using the developed in this paper analytical tools.

We thank Prof. Konstantin Samouylov for comments that greatly improved the manuscript.

References

1. Kumar, R., Liu, Y., Ross, K.W.: Stochastic fluid theory for P2P streaming systems. In: Proc. of the IEEE INFOCOM, pp. 919–927 (2007)
2. Wu, D., Liu, Y., Ross, K.W.: Queuing Network Models for Multi-Channel Live Streaming Systems. In: Proc. of the 28th Conference on Computer Communications (IEEE Infocom 2009), Rio de Janeiro, Brazil, April 19-25, pp. 73–81 (2009)
3. Adamu, A., Gaidamaka, Y., Samuylov, A.: Analytical Modeling of P2PTV Network. In: Proc. of the 2d International Congress on Ultra Modern Telecommunications and Control Systems (IEEE ICUMT 2010), Moscow, Russia, October 18-20, pp. 1115–1120 (2010)
4. Zhou, Y., Chiu, D.M., Lui, J.C.S.: A Simple Model for Analyzing P2P Streaming Protocols. In: Proc. of the 15th IEEE International Conference on Network Protocols (ICNP 2007), Beijing, China, October 16-19, pp. 226–235 (2007)
5. Adamu, A., Gaidamaka, Y., Samuylov, A.: Discrete Markov Chain Model for Analyzing Probability Measures of P2P Streaming Network. In: Balandin, S., Koucheryavy, Y., Hu, H. (eds.) NEW2AN 2011 and ruSMART 2011. LNCS, vol. 6869, pp. 428–439. Springer, Heidelberg (2011)
6. Zhao, Y., Shen, H.: A simple analysis on P2P streaming with peer playback lags. In: Proc. of the 3rd International Conference on Communication Software and Networks (IEEE ICCSN 2011), Xi'an, China, May 27-29, pp. 396–400 (2011)
7. Gaidamaka, Y., Samuylov, A., Samouylov, K.: Mathematical Modeling and Performance Analysis of P2P Streaming Networks. In: Int. Conf. INTHITEN (INternet of THings and ITs ENablers), June 3-4, pp. 69–81. St Petersburg, Russia (2013)
8. Fan, B., Andersen, D., Kaminsky, M., Papagiannaki, K.: Balancing Throughput, Robustness, and In-Order Delivery in P2P VoD. In: Proc. ACM CoNEXT (December 2010)

Detection of Anomalous HTTP Requests Based on Advanced N-gram Model and Clustering Techniques

Mikhail Zolotukhin and Timo Hämäläinen

Department of Mathematical Information Technology,
University of Jyväskylä, Jyväskylä, FI-40014, Finland
{mikhail.m.zolotukhin,timo.t.hamalainen}@jyu.fi
https://www.jyu.fi/it/laitokset/mit/en/

Abstract. Nowadays HTTP servers and applications are some of the most popular targets for network attacks. In this research, we consider an algorithm for HTTP intrusions detection based on simple clustering algorithms and advanced processing of HTTP requests which allows the analysis of all queries at once and does not separate them by resource. The method proposed allows detection of HTTP intrusions in case of continuously updated web-applications and does not require a set of HTTP requests free of attacks to build the normal user behaviour model. The algorithm is tested using logs acquired from a large real-life web service and, as a result, all attacks from these logs are detected, while the number of false alarms remains zero.

Keywords: Intrusion detection, anomaly detection, n-gram, clustering.

1 Introduction

Modern networks are vulnerable to different types of intrusions which can exploit systems legitimate features as well as take advantage of their misconfigurations, programming mistakes or buffer overflows. Some of the most popular targets of such attacks are web servers and web based applications which use the HTTP protocol. In HTTP traffic, clients send and request information, using request messages. As a rule, the main part of each such request message is a string, which consists of the path to a web resource located on the server and of parameters which are processed by the corresponding web application. Attackers are able to manipulate such queries to inject malicious code into the query parameters to create requests that corrupt the server or collect confidential information from the server databases. In addition, a HTTP request message contains header fields, which define the operating parameters of an HTTP transaction. Such fields usually contain information about user agent, preferred response languages, connection type, referer, etc. The attacker can inject malicious code to these fields to construct various kinds of attacks based on HTTP response splitting or malicious redirecting [1]. Thus, the detection of intrusive HTTP requests is

S. Balandin et al. (Eds.): NEW2AN/ruSMART 2013, LNCS 8121, pp. 371–382, 2013.

one of the most important tasks of network security when deploying a HTTP service.

Intrusion Detection Systems (IDSs) are often used to ensure the security of web servers and web based applications. IDS analyzes HTTP requests, detects suspicious activities and determines suitable responses to these activities [2]. IDSs can differ in audit source location, detection method, behavior on detection, usage frequency, etc [3,4]. There are two best-known approaches for detecting attacks on web servers: misuse detection and anomaly detection [5,6].

The misuse detection approach is based on searching predefined attack signatures. It allows one to detect attacks very accurately but, on the other hand, makes this approach unsuitable for the detection of zero-day attacks. Such attacks are based on zero-day vulnerabilities, which are defects found in software. Exploits of these defects often appear even before any fix for those defects can be made available. Anomaly detection based IDSs analyze data and create normal user behavior patterns during training phase. Then they monitor the incoming data and compare it with the patterns of normal behavior. Any deviation from these models is considered as an attack, and therefore these systems can detect any kind of abnormal data. Thus, IDSs based on the anomaly detection approach are capable of detecting even previously unknown zero-day attacks [7]. However, due to the fact that not all anomalous data are malicious these IDSs have higher number of false alarms than misuse-based systems.

While there are many misuse based IDSs for HTTP traffic which are widely used in commercial intrusion detection in HTTP traffic, there is more scope for innovations for anomaly based IDSs. The development of anomaly based IDSs which detects all intrusive HTTP requests while the number of false alarms remains low is a topic of great interest nowadays. Different algorithms can be applied to find anomalous HTTP queries. For example, [8] considers a method of detecting web attacks which is based on dimensionality reduction by applying diffusion maps and on subsequent application of spectral clustering. Application of self-organizing map (SOM) and its growing hierarchical version (GHSOM) for finding intrusive HTTP queries are investigated in [9]. After extracting features from HTTP requests, the dimensionality of feature matrices obtained is reduced using GHSOM and anomalies are detected with clustering methods based on distance relationships of the map nodes and on the density structure of the map. In [10], a technique for reducing the number of GHSOMs is proposed. A framework for detecting anomalies in HTTP traffic using instance-based learning and one-class k-nearest neighbor classification is proposed in [11], where HTTP request messages are compared to their most similar instances learned in the training phase. If a new message deviates from the learned ones to a considerable degree it is considered to be anomalous.

One of the most important stages in algorithms for intrusive HTTP queries detection is transformation of queries to numeric feature vectors. A correct method of transformation can drastically simplify the process of finding anomalous requests. Some simple transformation methods are presented in [12], [11] and [13].

Length of requests and attributes, character distribution and data type are features which are used to extract information from HTTP queries in those papers. Preprocessing of HTTP requests in studies [8] and [9] is performed by applying an n-gram model. An n-gram is a sub-sequence of n overlapping items (characters, letters, words, etc) from a given sequence. N-gram models are more accurate than simple character distribution but require huge amounts of computing and memory resources even for small values of n. Some studies ([14],[16]) compare the usage of n-gram with finite automaton for detecting anomalies. According to the results presented in [14], the 4-gram model has the best performance among all tested n-grams with lengths from three to seven, and has accuracy similar to a deterministic finite automaton.

Despite the fact that there are plenty of different techniques of preprocessing HTTP requests they have several weaknesses. Some techniques produce high dimensional feature vectors which subsequently require huge amounts of time and computing resources to analyze them, e.g. n-gram [8],[9]. Methods which build grammar models ([14],[15],[16]) usually can be applied only if there is a set of HTTP requests free of attacks. However, in the real world, it can be difficult to obtain such a set for a particular web server. Another problem is, since different web resources most likely use different attributes, requests to some web application can be considered by the IDS as anomalies with respect to the rest of web resources. It is reasonable to separate requests to different web resources during the preprocessing stage and therefore obtain several feature matrices, one for each unique web resource ([12],[11],[13]). Thus, for huge HTTP services several thousands of matrices can probably be constructed, which is not efficient from the computing resources point-of-view. In addition, it is difficult to define normal user behavior for resources that have been requested few times only.

In this research, we consider an algorithm of processing HTTP queries which allows the analysis of all HTTP request messages at once and does not separate them by resource. Once requests have been processed, standard clustering techniques are employed to find anomalous entries in the feature matrix obtained. In addition, the algorithm proposed can be used to build the model of the normal behavior fast even in the case the set of HTTP requests free of attacks can not be extracted. The method proposed also allows the detection of HTTP intrusions in continuously updated web-applications. The algorithm is tested using logs acquired from a large real-life web service and, as a result, all attacks from these logs are detected, while the number of false alarms remains zero.

This paper is organized as follows. Section 2 introduces a method of preprocessing HTTP request messages which allows the extraction of the most important features from HTTP queries and header fields. Section 3 presents an algorithm which is based on simple clustering methods and can be used for detecting intrusive queries. The test settings and discussion of the experimental results are described in Section 4. Finally, Section 5 draws conclusions and outlines future work.

2   Features Extraction

In this study, a scheme which detects HTTP intrusions is based on several classifiers. In that scheme, the main classifier is responsible for detecting intrusions among HTTP queries sent to different web resources. In addition, several classifiers are employed for detecting code injections in HTTP header fields: each header type has its own classifiers. Different classifiers use different features extracted from queries and header fields. In this section, request messages are processed to extract the most valuable features.

2.1   Processing HTTP Queries

Let us consider the network activity logs of a web-service of some HTTP server. A request message contains a query which is a string consisting of a web resource URI and some attributes and different headers. Here is an example from an Apache server log file of a HTTP request stored in a combined log format [25]:

```
127.0.0.1 - frank [10/Oct/2000:13:55:36 -0700]
"GET /resource?parameter1=value1&parameter2=value2 HTTP/1.0" 200 2326
"http://www.example.com/start.html" "Mozilla/4.08 [en] (Win98; I ;Nav)"
```

In order to analyze such requests some features have to be extracted.

Features extraction from HTTP queries is based on an n-gram model. N-gram models are widely used in statistical natural language processing [17] and speech recognition [18]. An n-gram is a sub-sequence of n overlapping items (characters, letters, words, etc) from a given sequence. In order to process HTTP queries an n-gram character model is applied. Since, when making code injections, attackers use specific combinations of non-alphanumeric symbols, the usage of those symbols is of the most interest. In order to concentrate on them, all numbers and Latin letters are considered as the same character. Taking this fact into account, we divide the HTTP query into overlapping character subsequences, each of which contains n symbols. For the i-th HTTP query considered, we store all unique overlapping character subsequences obtained S_i^q as well as the vector which contains the frequencies of those sequences appearance f_i^q. In addition, we store the length of this query l_i^q. Thus, each HTTP query q_i is transformed to the following triplet:

$$q_i \rightarrow [S_i^q, f_i^q, l_i^q]. \tag{1}$$

2.2   Processing HTTP Header Fields

As a rule, HTTP header fields have a finite set of possible values, and therefore extracting simple features is enough to solve the problem of finding code injections in headers. In this research, we employ the idea, that attacks injected into header fields stand out by abnormal lengths and the distribution of non-alphanumeric symbols used in them as proposed in [19]. When analyzing a header field each alphanumeric symbol used in it is substituted by zero, whereas each non-alphanumeric symbol is replaced by the one. After that, a n-gram character model is applied to the bit sequence obtained. Similar to the processing of HTTP queries, for each header field h_{ij}, we store all unique overlapping character subsequences S_{ij}^h, the frequencies of those sequences appearance f_{ij}^h and length of the bit sequence l_{ij}^h. All different header types are supposed to be analyzed separately. Thus, the value h_{ij} of the j-th header in the i-th HTTP request message is transformed to the following triplet:

$$h_{ij} \rightarrow [S_{ij}^h, f_{ij}^h, l_{ij}^h]. \tag{2}$$

Despite the fact that the extraction of features described above is enough to find code injections in almost all header fields, there is an optional header field which allows the client to specify, for the server's benefit, the address of the document (or element within the document) from which the URI in the request was obtained. This header is called the referer and in this study we process values of this header type the same way as in HTTP queries, i.e. before applying n-gram character model all numbers and Latin letters are considered as the same character, whereas different non-alphanumeric symbols are considered as different.

3   Training the System and Detecting Intrusions

When applying the IDS anomaly detection approach to detect HTTP intrusions, we should first learn the normal user behavior patterns. The most intuitive way to obtain these patterns is to teach the system, using a training set which contains only legitimate HTTP requests. However, when deploying IDS for a real web service, it is difficult to find a set that is guaranteed to consist of only normal requests - there is always a chance that it contains several intrusive requests. In this research, it is assumed that most request messages sent to the web server in question are normal: i.e. they use legitimate features of the service, while the percentage of intrusive requests does not exceed ten percent.

Let us consider some network activity logs which contain request messages sent to the web server considered. After extracting features from those requests, as described in the previous section, the system is supposed to learn the normal user behavior to be able to detect attacks. As mentioned in the previous section, the classifiers we use for HTTP queries and the headers are different. For training those classifiers, corresponding features are used.

3.1   HTTP Queries Classifier

After obtaining triplets for each HTTP query in the training set, a list of all unique n-grams is built. For each n-gram from this list, we calculate the frequency of n-gram usage in the set, as follows:

$$F_k^q = \frac{N_k^q}{N^q}, \tag{3}$$

where F_k^q is the frequency of usage of the k-th n-gram, N_k^q is the number of queries where k-th n-gram is used and N^q is the total number of unique HTTP queries in the training set.

As a rule, there are several popular n-grams which are used for constructing almost every HTTP query. For example, if we use a 2-gram model, the pairs "alphanumeric symbol" and "/", "?" and "alphanumeric symbol" or "alphanumeric symbol" and "alphanumeric symbol" can be found almost in each query. Since the usage frequency of those popular n-grams is about one, they can be easily classified using simple clustering techniques: k-means [21] or single-linkage clustering [20] with a predefined number of clusters equal to two. The cluster with the biggest centroid value is considered as the one containing the most popular n-grams.

Let us consider the set of the most popular n-grams found as the centroid of normal user behavior. We denote this set as Ω. This means that if an HTTP query contains only n-grams from that set, this query will be classified as normal. However, if there are other n-grams in the query considered, it can be classified either as normal or intrusive depending on how far the set of n-grams contained in the query is from the normal set Ω. The distance from the normal set to the i-th query set S_i^q is supposed to be based on the ratios of frequency of appearance of abnormal n-grams in that query f_{ik}^q and frequency of usage of those n-grams in the training set F_k^q: $\sum_{k \notin \Omega} \frac{f_{ik}^q}{F_k^q}$. Despite the fact that such distance metric allows one to find the majority of abnormal HTTP queries, very long queries containing many abnormal n-grams are still close to the centroid of normal behavior Ω, because frequencies of appearance of those n-grams remain low. Thus, several dangerous attacks based on buffer overflows can be disguised and classified as normal. For this reason, the impact of the number of abnormal n-grams is supposed to be taken into account. We propose the following formula for measuring each abnormal n-gram impact $\alpha_k f_{ik}/F_k$, where α_{ik}^q is the number of appearance of the k-th abnormal request in the i-th query. Since $\alpha_{ik}^q = f_{ik}^q(l_i - n + 1)$, the distance between the i-th query n-grams set and the normal set Ω is calculated as follows:

$$D_i^q = \sum_{k \notin \Omega} \log \left(1 + (l_i^q - n + 1)\frac{(f_{ik}^q)^2}{F_k^q} \right), \tag{4}$$

where the logarithmic function is used to exclude huge outliers. For those HTTP queries which contain few abnormal n-grams, the distance is small while queries containing lots of such n-grams are distant from the normal behavior centroid Ω.

Thus, the HTTP query processing technique proposed concentrates only on combinations of non-alphanumeric symbols which are usually employed by attackers to build intrusive requests. In addition, the technique takes into consideration the frequencies of appearance of n-grams in each query separately and in the whole training set. Moreover, the distance metric takes into account the query lengths and the number of abnormal n-grams used in them.

Once the distances for all HTTP queries in the training set have been calculated they can be classified easily with the help of a clustering technique. In this research, we use DBSCAN which is one of the powerful density-based clustering algorithms for detecting outliers. It finds the number of clusters starting from the estimated density distribution of the corresponding nodes [22]. DBSCAN requires two parameters: the size of neighborhood ϵ and the minimum number of points required to form a cluster N_{min}. It starts with an arbitrary starting point that has not been visited. This points ϵ-neighborhood is found, and if it contains sufficiently many points (more than N_{min}), a cluster is started. Otherwise, the point is labeled as noise, although this point might later be discovered as a part of another point ϵ-environment and hence be made a part of a cluster. If a point is found to be a dense part of a cluster, its ϵ-neighborhood is also part of that cluster. The process continues until the density-connected cluster is completely found. Then, a new unvisited point is processed, leading to a discovery of a further cluster or noise. All cluster-less points are classified as anomalies.

Despite the fact that DBSCAN is widely used to detect outliers, there are some difficulties in finding its parameters (ϵ and N_{min}). There are several studies devoted to this problem [23], [24]. In this research, we choose ϵ equal to absolute deviation of distances values D^q:

$$\epsilon = \sum_i^N \left| D_i^q - \frac{1}{N} \sum_i^N D_i^q \right|. \tag{5}$$

The value of ϵ calculated in this way takes into account the noises and the core points. If the distance corresponds to a normal HTTP query it most likely has a lot of normal queries in such ϵ-neighborhood. The minimum number of points required to form a cluster N_{min} is calculated as follows:

$$N_{min} = max_{k \notin \Omega} F_k^q N^q, \tag{6}$$

where $max_{k \notin \Omega} F_k^q$ is the maximal frequency value from the cluster of abnormal n-grams. Thus, only D^q values corresponding to normal queries can form the cluster while all abnormal queries are classified as noise. Once the system has been trained, the following variables are stored: a set of normal n-grams Ω, the vector of frequencies of all unique n-grams usage in the training set F^q, $max_i D_i^q$ among the values classified as normal and DBSCAN parameter ϵ.

When a new HTTP query is sent to the server, it is first divided into n-grams. Then, the distance D^{nq} between the set of unique n-grams of this query and n-grams normal set Ω is calculated (4). If this distance is greater than the sum of $max_i D_i^q$ and ϵ, the query is classified as intrusive, otherwise it is considered as legitimate:

$$\text{HTTP query is} \begin{cases} \text{normal, if } D^{nq} \leq max_i D_i^q + \epsilon, \\ \text{intrusive, if } D^{nq} > max_i D_i^q + \epsilon. \end{cases} \tag{7}$$

3.2 HTTP Header Classifier

Injections in HTTP header fields are detected by using the simplified version of the technique we applied for classifying HTTP queries. In the training stage, we obtain the set of the most popular n-grams extracted for each header type separately. When a new HTTP request is sent to the server, its header values are divided into n-grams. Then, for each header type j the distance D_j^{nh} between the set of unique n-grams of this type and the corresponding set of popular n-grams is calculated (4). If at least for one header type the distance is greater than zero, the query is classified as intrusive, otherwise it is considered as legitimate:

$$\text{HTTP headers are} \quad \begin{cases} \text{normal, if } D_j^{nh} = 0, \quad \forall j \\ \text{intrusive, if } \exists j^* : D_{j^*}^{nh} > 0. \end{cases} \tag{8}$$

For the referer, we use the same method which is applied for queries.

3.3 Updating the System

Web-applications deployed on the server can change on a regular basis, which can cause noticeable changes in the HTTP requests which are sent to the web-server. This can lead to a situation where all new allowable requests will be classified as intrusions. For this reason, to be capable of classifying new requests the system should be retrained after a certain period of time or after processing a certain number of requests. When retraining the system, several requests from the training set are replaced with requests received during the detection stage using the-first-in-first-out strategy. After that all variables calculated during the training (F^q, Ω^q, ϵ, etc) must be updated. In addition, countermeasures are necessary against attackers who try to affect the training set by flooding the web-server with a large number of intrusions, for example by allowing a client to replace a configurable number of HTTP requests in the training set per time slot.

4 Simulation Results

The method proposed is tested using HTTP request logs acquired from a large real-life web service. The log files are acquired from several Apache servers and stored in a combined log format [25]. The logs contain requests from multiple web-resources.

In the first simulation, we test the algorithm described in the paper by applying n-gram models with different values of n. The training set contains 5000 request messages sent to different resources of the web server. Most of the requests in the set are normal, but some are HTTP attacks such as SQL injections, buffer overflows, directory traversal attacks, cross-site scripting, etc. After training the system, new requests are chosen from log files and classified one by one with the technique proposed in this study. The number of requests in the testing set is equal to 25000. The testing set contains 1240 intrusive HTTP queries and 498 header code injections. During the testing stage, the system is updated after each processing of 5000 requests.

To evaluate the performance of the proposed technique, the following characteristics are calculated in our test:

- True positive rate: the ratio of the number of correctly detected intrusions to the total number of intrusions in the testing set;
- False positive rate: the ratio of the number of normal requests classified as intrusions to the total number of normal requests in the testing set;
- Accuracy: the ratio of the total number of correctly detected requests to the total number of requests in the testing set;
- Precision: the ratio of the number of correctly detected intrusions to the number of requests classified as intrusions.

These performance metrics calculated after classifying all requests during testing are shown in Tables 1 and 2. In Table 1 the results of the application of different n-gram models to detect intrusive HTTP queries are presented, and Table 2 shows the performance of the scheme in detecting header field attacks. As one can see even for $n = 1$ the accuracy of the method is about 100% while the number of false alarms is very low. When $n = 2$ and when $n = 3$, both normal and intrusive queries are classified correctly. Table 2 presents the performance of the header injections detection for different n-gram models. Similar to classifying HTTP queries, the statistics obtained by applying 1-gram model can not properly describe header field values, especially for header fields in which different non-alphanumeric symbols can be used. However, with increasing n in n-gram model the accuracy of the method increases to 100% and all header field injections are detected.

Table 1. Performance of the method proposed for detecting intrusive HTTP queries using different n-gram models

N-gram model	True positive rate	False positive rate	Accuracy	Precision
1-gram	100.00 %	0.538 %	99.49 %	91.068 %
2-gram	100.00 %	0 %	100 %	100 %
3-gram	100.00 %	0 %	100 %	100 %

Table 2. Performance of the method proposed for detecting HTTP header fields injections

N-gram model	True positive rate	False positive rate	Accuracy	Precision
1-gram	66.87 %	0 %	99.46 %	100 %
2-gram	100.00 %	0 %	100 %	100 %
3-gram	100.00 %	0 %	100 %	100 %

The algorithm proposed in this research is compared with several web attack detection mechanisms. The part concerning the detection of intrusive HTTP queries is compared with k-nearest neighbor classification of queries character distribution [11], an n-gram model and GHSOMs [9], an n-gram model and diffusion maps [8]. The technique applied for the detection of injections in HTTP header fields is collated with k-nearest neighbor classification of headers character distributions [11] and statistical models based on the lengths of header fields and alphanumeric and non-alphanumeric symbols used in them [19]. We did not take into account techniques based on grammar models such as deterministic finite automaton and the nondeterministic finite automaton since they require a set of HTTP requests free of attacks during the training stage. For each method of the intrusive queries detection, two tests are carried out. In the first test, the same training and testing sets as described above are used. In the second test, we choose from log files one web resource to which the highest number of requests have been sent and use only corresponding queries in the training and testing sets to evaluate algorithms. The performance results are presented in Tables 3 and 4. As one can see, all algorithms show good results when only one web resource is explored (the accuracy is about 100 %). However, when the number of resources grows, the accuracy of other methods decreases while the performance of the technique proposed in this study remains the same. The comparison of algorithms for the header injections search is presented in Table 5. As one can notice, the algorithm described in this study outperforms analogues in terms of accuracy.

We implemented algorithms mentioned above in Matlab to compare time required to train the system and detect attacks. Training the system based on the method proposed is much faster compared to techniques based on n-gram and dimensionality reduction. Besides that, the rate of classifying new requests is few milliseconds which is enough for online intrusion detection.

Table 3. Performance of the method proposed for detecting intrusive HTTP queries compared with other web attacks techniques (one resource)

Algorithm	True positive rate	False positive rate	Accuracy	Precision
K-nearest neighbor	59.31 %	0.06 %	97.93 %	97.99 %
N-gram + GHSOM	92.51 %	0.19 %	99.45 %	96.21 %
N-gram + Diffusion Maps	98.72 %	0 %	99.94 %	100 %
Algorithm proposed	100 %	0 %	100 %	100 %

Table 4. Performance of the method proposed for detecting intrusive HTTP queries compared with other web attacks techniques (several resources)

Algorithm	True positive rate	False positive rate	Accuracy	Precision
K-nearest neighbor	55.51 %	2.05 %	97.79 %	59.12 %
N-gram + GHSOM	58.22 %	0.86 %	97.05 %	78.56 %
N-gram + Diffusion Maps	97.37 %	23.15 %	77.94 %	19.11 %
Algorithm proposed	100 %	0 %	100 %	100 %

Table 5. Performance of the method proposed for detecting HTTP headers injections compared with other web attacks techniques

Algorithm	True positive rate	False positive rate	Accuracy	Precision
K-nearest neighbor	70.39 %	0.35 %	99.05 %	81.01 %
Statistical models	94.62 %	0 %	99.90 %	100 %
Algorithm proposed	100 %	0 %	100 %	100 %

5 Conclusion

In this research, an algorithm for processing HTTP queries based on n-gram is proposed. The method allows at once allows the analysis of all HTTP request messages without separating them by resource. Simple clustering techniques are then employed to find anomalous entries in the feature matrix obtained. Code injections in HTTP header fields are detected by applying similar technique with some simplifications. The method proposed allows the detection of HTTP intrusions in case of continuously updated web-applications. The algorithm is tested using logs acquired from a large real-life web service. All attacks from these logs are detected, while the number of false alarms remains zero.

In the future, we are planning to continue studying techniques of extracting features from HTTP requests and of detecting intrusions based on the anomaly detection approach. Time reduction for training the system and classification of new requests are also of great interest. In addition, we are going to design a system that is capable of detecting complex intrusions, which can consist of several HTTP requests, e.g. requests aiming to find the most vulnerable features of a web application deployed on a server.

References

1. Klein, A.: Detecting and Preventing HTTP Response Splitting and HTTP Request Smuggling Attacks at the TCP Level. Tech. Note (August 2005), http://www.securityfocus.com/archive/1/408135
2. Axelsson, S.: Research in intrusion-detection systems: a survey. Department of Computer Engineering, Chalmers University of Technology, Goteborg, Sweden. Technical Report. pp. 98–117 (December 1998)
3. Patcha, A., Park, J.M.: An overview of anomaly detection techniques: Existing solutions and latest technological trends. Computer Networks: The International Journal of Computer and Telecommunications Networking 51(12) (August 2007)
4. Verwoerd, T., Hunt, R.: Intrusion detection techniques and approaches. Computer Communications - COMCOM 25(15), 1356–1365 (2002)
5. Kemmerer, R.A., Vigna, G.: Intrusion Detection: A Brief History and Overview. Computer 35, 27–30 (2002)
6. Gollmann, D.: Computer Security, 2nd edn. Wiley (2006)
7. Sriraghavan, R.G.: Data processing and anomaly detection in web-based applications. In: IEEE Workshop on Machine Learning for Signal Processing, MLSP 2008, pp. 187–192 (October 2008)

8. Sipola, T., Juvonen, A., Lehtonen, J.: Anomaly detection from network logs using diffusion maps. In: Iliadis, L., Jayne, C. (eds.) EANN/AIAI 2011, Part I. IFIP AICT, vol. 363, pp. 172–181. Springer, Heidelberg (2011)

9. Zolotukhin, M., Hämäläinen, T., Juvonen, A.: Growing Hierarchical Self-organizing Maps and Statistical Distribution Models for Online Detection of Web Attacks. In: Cordeiro, J., Krempels, K.-H. (eds.) WEBIST 2012. Lecture Notes in Business Information Processing, vol. 140, pp. 281–295. Springer, Heidelberg (2013)

10. Zolotukhin, M., Hämäläinen, T., Juvonen, A.: Online Anomaly Detection by Using N-gram Model and Growing Hierarchical Self-Organizing Maps. In: Proc. of the IWCMC (2012)

11. Kirchner, M.: A framework for detecting anomalies in HTTP traffic using instance-based learning and k-nearest neighbor classification. In: 2nd International Workshop on Security and Communication Networks (IWSCN), pp. 1–8 (May 2010)

12. Kruegel, C., Vigna, G.: Anomaly detection of web-based attacks. In: Proc. of the 10th ACM Conference on Computer and Communications Security, pp. 251–261 (2003)

13. Lucchese, L.: Data processing and anomaly detection in web-based applications. In: IEEE Workshop on Machine Learning for Signal Processing, MLSP 2008, pp. 187–192 (October 2008)

14. Lin, L., Leckie, C., Chenfeng, Z.: Comparative Analysis of HTTP Anomaly Detection Algorithms: DFA vs N-Grams. In: 4th International Conference on Network and System Security (NSS), pp. 113–119 (September 2010)

15. Ingham, K., Somayaji, A., Burge, J., Forrest, S.: Learning DFA representations of HTTP for protecting web applications. Computer Networks 51, 1239–1255 (2007)

16. Sun, M., Xuelei, H., Yang, J.: Grammar-Based Anomaly Methods for HTTP Attacks. In: Chinese Conference on Pattern Recognition, CCPR 2009, vol. 1-5 (November 2009)

17. Suen, C.Y.: N-Gram Statistics for Natural Language Understanding and Text Processing. IEEE Transactions on Pattern Analysis and Machine Intelligence PAMI-1(2), 164–172 (1979)

18. Hirsimaki, T., Pylkkonen, J., Kurimo, M.: Importance of High-Order N-Gram Models in Morph-Based Speech Recognition. IEEE Transactions on Audio, Speech, and Language Processing 17(4), 724–732 (2009)

19. Corona, I., Giacinto, G.: Detection of Server-side Web Attacks. In: Proc. of JMLR: Workshop on Applications of Pattern Analysis, pp. 160–166 (2010)

20. Jain, A., Murty, M., Flynn, P.: Data clustering: a review. ACM Computing Surveys 31(3), 264–323 (1999) ISSN 0360-0300

21. Xie, J.: A Simple and Fast Algorithm for Global K-means Clustering. In: Proc. of 2nd International Workshop Education Technology and Computer Science (ETCS), vol. 2, pp. 36–40 (March 2010)

22. Ester, M., Kriegel, H.-P., Sander, J., Xu, X.: A density-based algorithm for discovering clusters in large spatial databases with noise. In: Proc. of 2nd International Conference on Knowledge Discovery and Data Mining, pp. 226–231 (1996)

23. Kim, J.: The Anomaly Detection by Using DBSCAN Clustering with Multiple Parameters. In: Proc. of ICISA, pp. 1–5 (April 2011)

24. Smiti, A.: DBSCAN-GM: An improved clustering method based on Gaussian Means and DBSCAN techniques. In: Proc. of International Conference on Intelligent Engineering Systems (INES), pp. 573–578 (June 2012)

25. Apache 2.0 Documentation (2011), http://www.apache.org/

Queuing Model for SIP Server Hysteretic Overload Control with Bursty Traffic *

Pavel Abaev[1] and Rostislav Valerievich Razumchik[2]

[1] Telecommunication Systems Department,
Peoples' Friendship University of Russia,
Ordzhonikidze str. 3, 115419 Moscow, Russia
`pabaev@sci.pfu.edu.ru`
[2] Institute of Informatics Problems of
Russian Academy of Sciences,
Vavilova str., 44-2,
119333 Moscow, Russia
`rrazumchik@gmail.com`

Abstract. In this paper, we develop a mathematical model of a load control mechanism for SIP server signaling networks based on a hysteretic technique. We investigate loss-based overload control, as proposed in recent IETF documents. The queuing model takes into account three types of system state – normal load, overload, and discard. The hysteretic control is made possible by introducing two thresholds, L and H, in the buffer of total size R. We denote the mathematical model using the modified Kendall notation as an $MMPP|M|1|\langle L, H \rangle|R$ queue with hysteretic load control and bursty input flow. Algorithms for computation the key performance parameters of the system were obtained. A numerical example illustrating the control mechanism that minimizes the return time from overloading states satisfying the throttling and mean control cycle time constraints is also presented.

Keywords: SIP server, hop-by-hop overload control, loss-based overload control, hysteretic control, return time, queuing model, MMPP flow.

1 Introduction

In modern telecommunications networks quality of service depends on the timely and reliable delivery of signaling messages between network nodes. SIP is considered the primary signaling protocol for modern telecommunications networks [1].

The SIP protocol provides a basic overload control mechanism through the 503 (Service Unavailable) response code. SIP servers that are unable to forward a request due to temporary overload can reject the request with a 503 response. The overloaded server can insert a Retry-After header into the 503 response, which defines the number of seconds during which this server is not available

---

* This work was supported in part by the Russian Foundation for Basic Research (grants 12-07-00108 and 13-07-00665).

S. Balandin et al. (Eds.): NEW2AN/ruSMART 2013, LNCS 8121, pp. 383–396, 2013.

for receiving any further requests from the upstream neighbor. A server that receives a 503 response from a downstream neighbor stops forwarding requests to this neighbor for the specified amount of time and starts again after this time is over. Without a Retry-After header, a 503 response only affects the current request and all other requests can still be forwarded to this downstream neighbor. A server that has received a 503 response can try to resend the request to an alternate server, if one is available. A server does not forward 503 responses toward the UA and converts them to 500 Server Internal Error responses instead. RFC 3261 also provides, in the case of overload the recipient to discard incoming messages without notifying the sender.

The rapid development of the market for services based on the SIP protocol and the growing user needs have revealed a number of shortcomings in the basic mechanism 503. In RFC 5390 [2] Rosenberg, one of the authors of the protocol, declared the mechanism's inconsistent in overcoming the problems related to network congestion and formulated the basic requirements for future overload control mechanisms. The IETF working group SOC (SIP Overload Control) proposed the Loss-based overload control scheme (LBOC). A mechanism based on this scheme should be used as the default one to deal with congestion in SIP-network.

Studies of servers' incoming flows based on measurements carried out on a real network show that the traffic to the server can be modelled by MMPP-flow (Markov Modulated Poisson Process) with a sufficient degree of accuracy [6], but not with Poisson process. In the paper we modelling SIP server with a queuing system with hysteretic overload control and MMPP input flow. First, we make an overview of overload control techniques, which are implemented on the server and the client side. Second, we construct a queuing system for SIP server modelling with bursty traffic and hysteretic overload control. And finally, we introduce two quick algorithm for finding the stationary probabilities and mean return time of the system.

2   Overview of the Overload Control Techniques

The basic idea of LBOC scheme is that the sending entity (SE) reduces the number of messages on the request of the receiving entity (RE) which will be send to RE by specified in the request amount of the total number of messages. The scheme based on the idea of feedback control loop between all neighboring SIP servers that directly exchange traffic. Each loop controls only two entities. The Actuator is located on the sending entity and throttles the traffic if necessary. The receiving entity has the Monitor which measures the current server load.

Hysteretic control technique is to use on the server side to determine the moments for sending messages with control information from SE to RE. The system during operation changes its states depending on the buffer occupancy n. Choose arbitrary numbers L and H such that $0 < L < H < R$, where R is the buffer capacity. When the system starts to work it is empty, $(n = 0)$, and as long as the total number of messages in the system remains below $H - 1$, system is considered to be in normal state, ($h = 0$). When total number of messages exceeds $H - 1$ for the first time, the system changes its state to overload, ($h = 1$), and RE informs SE that traffic load should be reduced: it stays in it as long as the number of messages remains between L and $R - 1$. Being in overload state, RE's system waits till the number of messages drops down below L after which it changes its state back to normal and informs SE about changes, or exceeds $R - 1$ after which it changes its state to blocking, ($h = 2$), and ask SE for temporary suspension of sending SIP requests. When the total number of messages drops down below $H + 1$, system's state changes back to overload, and RE informs SE that the process of sending of messages can be resumed with the current limitations.

On the client side LBOC scheme is implemented. The idea of the scheme presented in [3] is to sift the client's outgoing flow by the value indicated by the value of the 'oc' parameter. The parameter 'oc' is one of the four new parameters introduced by SOC group in [4], which are used for the exchange of control information between SE and RE. The value of parameter 'oc' defines what percentage of the total number of SIP requests are subject to reduction at the SE.

Let us consider the example of the implementation of the scheme. The client maintains two types of requests – the priority and non-priority. Prioritization of messages is done in accordance with local policies applicable to each SIP-server. In situations where the client has to sift the outgoing flow, it first reduces non-priority messages, and then if the buffer contains only priority messages and further reduction is still needed, the client reduces the priority messages.

Under overload condition, the client converts the value of the 'oc'=q parameter to a value that it applies to non-priority requests. Let N_1 denote the number of priority messages and N_2 denote the number of the non-priority messages in the client's buffer. The client should reduce the non-priority messages with probability $q_2 = \min\left\{1, q\frac{N_1+N_2}{N_2}\right\}$ and the priority messages with probability $q_1 = \frac{q(N_1+N_2)-q_2 N_2}{N_1}$ if necessary to get an overall reduction of the 'oc' value.

To affect the reduction rate with probability q_2 from the non-priority messages, the client draws a random number between 1 and 100 for the request picked from the first category. If the random number is less than or equal to converted value of the 'oc' parameter, the request is not forwarded; otherwise the request is forwarded. Recalculation of probabilities is performed periodically every 5-10 seconds by getting the value of the counters N_1 and N_2.

Besides the 'oc' parameter there is one more key parameter 'oc-validity' which establishes a time limit for which overload control is in effect. The problem of finding 'oc' and 'oc-validity' is still unsolved. One of the solution will be proposed below.

3   Queueing Model of SIP Server with Bursty Input Flow

We are modelling message processing by RE as a single-server queuing system shown in Fig. 1 with hysteretic overload control with two thresholds, L

and H, which is denoted by $MMPP|M|1|\langle L, H\rangle |R$ according to the Kendall classification.

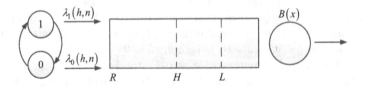

Fig. 1. $MMPP|M|1|\langle L, H\rangle |R$ queuing model

According to the hysteretic algorithm the server operates in three modes [5]: normal ($h = 0$), overload ($h = 1$), and blocking ($h = 2$), where h is the overload status. Let n denote the queue length, $n = 0, ..., R$.

We are modelling servers' input flow as MMPP process with infinitesimal generator Q and arrival rates $\lambda_l^h, l = 0, 1, h = 0, 1$:

$$Q = \begin{pmatrix} -\alpha & \alpha \\ \beta & -\beta \end{pmatrix}, \quad \Lambda_0 = \begin{pmatrix} \lambda_0^0 & 0 \\ 0 & \lambda_1^0 \end{pmatrix}, \quad \Lambda_1 = \begin{pmatrix} \lambda_0^1 & 0 \\ 0 & \lambda_1^1 \end{pmatrix}.$$

The input load function $\lambda(l, h, n)$ is shown in Fig. 2, where $q = 1 - p$ is the dropping probability and $\lambda_l^1 = p\lambda_l^0, l = 0, 1$.

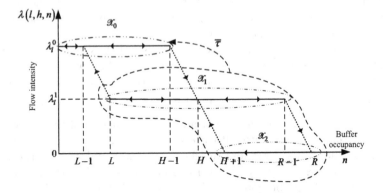

Fig. 2. Input load function

Customers arrive at the system and receive service in accordance with FIFO algorithm. The mean processing time is μ^{-1}, $B(x) = 1 - e^{-\mu x}, x > 0$. The state space of the system is defined by $\mathcal{X} = \mathcal{X}_0 \cup \mathcal{X}_1 \cup \mathcal{X}_2$, where $\mathcal{X}_0$ is the set of states of normal load, $\mathcal{X}_1$ is the set of overload states, and $\mathcal{X}_2$ is the set of discard states. These sets are given by following formulae:

$$\mathcal{X}_0 = \{(l, h, n) : l = 0, 1, h = 0, 0 \leq n \leq H - 1\},$$
$$\mathcal{X}_1 = \{(l, h, n) : l = 0, 1, h = 1, L \leq n \leq R - 1\},$$
$$\mathcal{X}_2 = \{(l, h, n) : l = 0, 1, h = 2, H + 1 \leq n \leq R\}.$$

The probability of the system being in the set of normal load states is denoted by $P(\mathcal{X}_0)$, the probability of being in the set of overload states by $P(\mathcal{X}_1)$, and the probability of being in the set of discard states by $P(\mathcal{X}_2)$. The key performance measures of the system are overload probability $P(\mathcal{X}_1)$, blocking probability $P(\mathcal{X}_2)$, and $\bar{\tau}$ as the mean return time from overload and discard states $\mathcal{X}_1 \cup \mathcal{X}_2$.

Stationary Performance Characteristics

The operation of the considered queueing system can be completely described by Markov process $\mathbf{X}(t) = (l(t); h(t); n(t))$ with three components: $l(t)$ — source's phase at time t, $h(t)$ — state of the system at time t, $n(t)$ — number of customers in the system at time t. Clearly $\mathbf{X}(t)$ is ergodic and thus stationary distribution exists.

Let $\boldsymbol{p}_n^h = (p_{n0}^h, p_{n1}^h)$ be row vector of size 1×2, whose first element p_{n0}^h is stationary probability of the fact that there are total of n customers in the system, system is the state h, $h = 0, 1, 2$ and source's phase is 0; the second element p_{n1}^h is stationary probability of the fact that there are n customers in the system, system is in the normal state and source's phase is 1.

Stationary distribution can be effectively computed using approach developed in the series of papers (see, e.g. [7], [8], [9]). In the Appendix one can find explanation of the approach and it is shown how it can be applied for the considered problem. Thus below, skipping all the details that follow the derivation of stationary distribution (see Appendix), we state the final computational algorithm 1.

Henceforth, for the sake of simplicity, notations U_i, W_i, V_i, M, I stand for the following matrices:

$$U_0 = \begin{pmatrix} \frac{\mu}{\lambda_0^0 + \mu + \alpha} & 0 \\ 0 & \frac{\mu}{\lambda_1^0 + \mu + \beta} \end{pmatrix}, V_0 = \begin{pmatrix} 0 & \frac{\alpha}{\lambda_0^0 + \mu + \alpha} \\ \frac{\beta}{\lambda_1^0 + \mu + \beta} & 0 \end{pmatrix}, W_0 = \begin{pmatrix} \frac{\lambda_0^0}{\lambda_0^0 + \mu + \alpha} & 0 \\ 0 & \frac{\lambda_1^0}{\lambda_1^0 + \mu + \beta} \end{pmatrix},$$

$$U_1 = \begin{pmatrix} \frac{\mu}{\lambda_0^1 + \mu + \alpha} & 0 \\ 0 & \frac{\mu}{\lambda_1^1 + \mu + \beta} \end{pmatrix}, W_1 = \begin{pmatrix} \frac{\lambda_0^1}{\lambda_0^0 + \mu + \alpha} & 0 \\ 0 & \frac{\lambda_1^1}{\lambda_1^1 + \mu + \beta} \end{pmatrix}, V_1 = \begin{pmatrix} 0 & \frac{\alpha}{\lambda_0^0 + \mu + \alpha} \\ \frac{\beta}{\lambda_1^1 + \mu + \beta} & 0 \end{pmatrix},$$

$$U_2 = \begin{pmatrix} \frac{\mu}{\mu + \alpha} & 0 \\ 0 & \frac{\mu}{\mu + \beta} \end{pmatrix}, V_2 = \begin{pmatrix} 0 & \frac{\alpha}{\mu + \alpha} \\ \frac{\beta}{\mu + \beta} & 0 \end{pmatrix}, M = \begin{pmatrix} \mu & 0 \\ 0 & \mu \end{pmatrix}, I = \begin{pmatrix} 1 & 0 \\ 0 & 1 \end{pmatrix}.$$

Let us dwell in more detail on the problem of finding a mean return time from the set of overload and discard states to the set of normal load states.

Let m_{nj}, $n = \overline{L, R-1}$ denote the mean return time to reach the moment when the number of customers in the system hits $L - 1$ for the first time, given that at some moment there were n customers in the system and the overload status was $h = 1$, and source state was equalled to $j = 0, 1$. Let m_{nj}^*, $n = \overline{H+1, R}$ be the mean return time to reach the moment when the number of customers in the system hits $L-1$ for the first time, given that at some moment there were n customers in the system and the overload status was $h = 2$, and source state was equalled to $j = 0, 1$.

Algorithm 1. Algorithm for computation of steady state probabilities

Initialize Λ_0, M, Q, Λ_1, A_i, G_i, D, B_i, D_i;

    $F_0 := \Lambda_0 + M - Q$;

    $F_1 := \Lambda_1 + M - Q$;

    $K_0 := I$;

  for $i = 1$ to $H - 2$ **do**

    if $i = L - 1$ **then**

      $K_i := K_{i-1}\Lambda_0(F_0 - \Lambda_0[A_L + G_L D])^{-1}$;

    else

      $X_i := \Lambda_0(F_0 - \Lambda_0 A_{i+1})^{-1}$;

      $K_i := K_{i-1}X_i$;

    end if

  end for

$K_{H-1} := K_{H-2}\Lambda_0 F_0^{-1}$;

$K_H := K_{H-1}\Lambda_0[F_1 - MB_{H-1} - \Lambda_1(A_{H+1} + G_{H+1} rm D_H)]^{-1}$;

for $i = H + 1$ to $R - 2$ **do**

  $Y_i := \Lambda_1(F_1 - \Lambda_1 A_{i+1})^{-1}$;

  $K_i := K_{i-1}Y_i$;

end for

$K_{R-1} := K_{R-2}\Lambda_1 F_1^{-1}$;

$K_R := K_{R-1}\Lambda_1(M - Q)^{-1}$;

$Z := M(M - Q)^{-1}$;

$N_H := K_H$;

for $i = H - 1$ to $L + 1$ **do**

  $Y_i := M(F_1 - MB_{i-1})^{-1}$;

  $N_i := N_{i+1}Y_i$;

end for

$N_L := N_{L+1}MF_1^{-1}$;

$F_2 := \sum_{n=0}^{R-1} K_n + \sum_{n=L}^{H-1} N_n + K_R\left(I + \sum_{n=H+1}^{R-1} Z^{R-n}\right)$;

Solve $\boldsymbol{p}_0^0(\Lambda_0(I - A_1) - Q) = \boldsymbol{0}$, $\boldsymbol{p}_0^0 F_2 \boldsymbol{1} = 1$;

for $i = 1$ to $H - 1$ **do**

  $\boldsymbol{p}_i^0 := \boldsymbol{p}_0^0 K_i$

end for

for $i = H$ to $R - 1$ **do**

  $\boldsymbol{p}_i^1 := \boldsymbol{p}_0^0 K_i$

end for

$\boldsymbol{p}_R^2 := \boldsymbol{p}_0^0 K_R$;

for $i = H + 1$ to $R - 1$ **do**

  $\boldsymbol{p}_i^2 := \boldsymbol{p}_0^0 K_R Z^{R-i}$;

end for

for $i = H - 1$ to L **do**

  $\boldsymbol{p}_i^1 := \boldsymbol{p}_0^0 N_i$;

end for

We denote the vectors of mean return times by $\boldsymbol{m}_n^T = (m_{n0}, m_{n1})$ and $\boldsymbol{m}^{*T}_n = (m^*_{n0}, m^*_{n1})$. Conditioning on the first step and using the following notations $S_1 = \mu^{-1}U_1$, $S_2 = \mu^{-1}U_2$, and the law of total expectation one can arrive at the following relations:

$$\boldsymbol{m}_L = S_1\boldsymbol{1} + V_1\boldsymbol{m}_L + W_1\boldsymbol{m}_{L+1},$$

$$\boldsymbol{m}_n = S_1\boldsymbol{1} + U_1\boldsymbol{m}_{n-1} + V_1\boldsymbol{m}_n + W_1\boldsymbol{m}_{n+1}, \ n = \overline{L+1, R-2},$$

$$\boldsymbol{m}_{R-1} = S_1\boldsymbol{1} + U_1\boldsymbol{m}_{R-2} + V_1\boldsymbol{m}_{R-1} + W_1\boldsymbol{m}^*_R,$$

$$\boldsymbol{m}^*_n = S_2\boldsymbol{1} + U_2\boldsymbol{m}^*_{n-1} + V_2\boldsymbol{m}^*_n, \ n = \overline{R, H+2},$$

$$\boldsymbol{m}^*_{H+1} = S_2\boldsymbol{1} + U_2\boldsymbol{m}_H + V_2\boldsymbol{m}^*_{H+1}.$$

It can be seen that the system can be solved recursively. Algorithm 2 gives steps for the computation of mean return times $\boldsymbol{m}_n$ and $\boldsymbol{m}^*_n$.

Above we have stated the algorithm for the computation of $2D - H - L$ different values of mean return time. For practical purposes it important to specify which of the computed values one should use as the value for the mean return time from the set of overload and discard states to the set of normal states. Here we propose to use the value

$$\overline{\tau} = \frac{p^1_{H0}m_{H0} + p^1_{H1}m_{H1}}{p^1_{H0} + p^1_{H1}},$$

basing on the fact that if we start to observe system when it is normal mode, the value of $\overline{\tau}$ will represent mean time it takes the system to get back to normal operation once it becomes overloaded. Naturally if one starts to observe the system when it is already in overload (discard) state, one should substitute the value of m_{nj} (m^*_{nj}) instead of the value of m_{Hj} in the expression for $\overline{\tau}$.

In the next section we proceed to the numerical example.

4   Numerical Example

In this section, the illustrative numerical example for solving the optimization problem from [5] is presented. Let τ be the mean control cycle time, which can be obtained in the following form:

$$\tau = \tau_0 + \overline{\tau}.$$

Given that the average number of transitions from the set and to the set should be equal in equilibrium, the value of τ_0 can be calculated using the following formula:

$$\tau_0 = \overline{\tau} \cdot \frac{P(\mathcal{X}_0)}{P(\mathcal{X}_1 \bigcup \mathcal{X}_2)}.$$

Algorithm 2. Algorithm for computations of mean return times

Initialize V_1, U_1, W_1, V_2, U_2, S_1, S_2; $F := (I - V_2)^{-1} S_2$; $T := (I - V_2)^{-1} U_2$;
 $x_{R-1} := (I - V_1)^{-1} [S_1 + W_1 \sum_{i=0}^{R-H-1} T^i F] 1$;
 $y_{R-1} := (I - V_1)^{-1} U_1$;
 $z_{R-1} := (I - V_1)^{-1} W_1 T^{R-H}$;
for $n = R - 2$ to $H + 1$ **do**
 $x_n := (I - V_1 - W_1 y_{n+1})^{-1} [S_1 1 + W_1 x_{n+1}]$;
 $y_n := (I - V_1 - W_1 y_{n+1})^{-1} U_1$;
 $z_n := (I - V_1 - W_1 y_{n+1})^{-1} W_1 z_{n+1}$;
end for
$d_H := (I - V_1 - W_1 y_{H+1} - W_1 z_{H+1})^{-1} [S_1 1 + W_1 x_{H+1}]$;
$e_H := (I - V_1 - W_1 y_{H+1} - W_1 z_{H+1})^{-1} U_1$;
for $n = H - 1$ to $L + 1$ **do**
 $d_n := (I - V_1 - W_1 e_{n+1})^{-1} [S_1 1 + W_1 d_{n+1}]$;
 $e_n := (I - V_1 - W_1 e_{n+1})^{-1} U_1$;
end for
$m_L := (I - V_1 - W_1 e_{L+1})^{-1} [S_1 1 + W_1 d_{L+1}]$;
for $n = L + 1$ to H **do**
 $m_n = d_n + e_n m_{n-1}$;
end for
for $n = H + 1$ to $R - 1$ **do**
 $m_n := x_n + y_n m_{n-1} + z_n m_H$;
end for
$m^*_{H+1} := F1 + T m_H$;
for $n = H + 2$ to R **do**
 $m^*_n := F1 + T m^*_{n-1}$;
end for

Note that the value of $P(\mathcal{X}_0)$ and $P(\mathcal{X}_1 \bigcup \mathcal{X}_2)$ can be calculated from the considered queue as follows:

$$P(\mathcal{X}_i) = \sum_{(l,h,n) \in \mathcal{X}_i} p^h_{nl},$$

$$P\left(\mathcal{X}_1 \bigcup \mathcal{X}_2\right) = 1 - P(\mathcal{X}_0).$$

Clearly that the best mode of operation of the server is a situation in which the server operates most of the time in the set of normal load states, and that can be achieved by minimizing the mean return time. The value of 'oc-value' parameter should be equalled to the value of control cycle time to stabilize the the process of sending and receiving messages. The value of control cycle time is supposed to be large enough to avoid possible oscillations. Thus the problem is stated as follows: minimise the mean return time needed with respect to the choice of the two thresholds, L and H, such that the requirements $R1$–$R3$ are satisfied

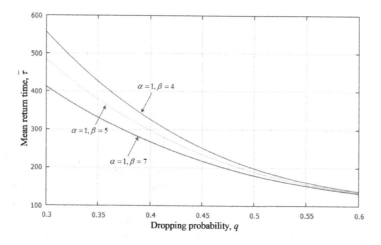

Fig. 3. The dependence of the mean return time τ on dropping probability q

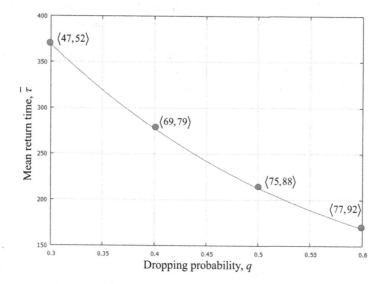

Fig. 4. The optimal value of the thresholds L and H ($\alpha = 1,\ \beta = 4$)

$$\overline{\tau}(L, H) \rightarrow \min;$$
$$R1: \quad P(\mathcal{X}_1) \leq \gamma_1;$$
$$R2: \quad P(\mathcal{X}_2) \leq \gamma_2;$$
$$R3: \quad \tau \geq \gamma_3.$$

The optimization problem for which the mean return time, $\overline{\tau}$, is minimized. The solution to the problem of the choice of L and H for a given dropping probability $q \in \{0, 3; 0, 4; 0, 5; 0, 6\}$, mean service time $\mu^{-1} = 5$ ms, and signalling load $\rho = 1, 2$ can now be sought. In Fig. 3 the changes of the mean return time from the dropping probability q for different value of pairs (α, β) are shown.

In Fig. 4 the solution of optimization problem for the following value of boundary conditions $\gamma_1 = 0.793, \gamma_2 = 0.0001, \gamma_3 = 400$ is shown.

5   Conclusion

In this paper, we developed a queueing system $MMPP|M|1| \langle L, H \rangle |R$ for modelling a SIP server hysteretic overload control with bursty traffic. The considered queueing system is a modification of the queueing system from [5], which provide a possibility to obtain more accurate value of the performance characteristic of a SIP server. New approach is proposed which allows fast computation of the joint stationary distribution of the underlying three dimensional Markov process. The key performance parameter of the queue –the mean delay time, was determined as a function of input load, ρ, dropping probability, q, and two thresholds L and H and a numerical example was presented. The optimization problem was formulated and solved. Solution of the problem allows us to formulate recommendations on the choice of control parameters. Clearly, the considered problem was only one of the possible formulations. Our further research will be devoted to the verification of obtained results by comparing them with simulation results based on real time traffic and to analysis of more general queueing systems suitable for SIP server overload control modelling.

References

1. Rosenberg, J., Schulzrinne, H., Camarillo, G., et al.: SIP: Session Initiation Protocol. RFC3261 (2002)
2. Rosenberg, J.: Requirements for Management of Overload in the Session Initiation Protocol. RFC5390 (2008)
3. Hilt, V., Noel, E., Shen, C., Abdelal, A.: Design Considerations for Session Initiation Protocol (SIP) Overload Control. RFC6357 (2011)
4. Gurbani, V., Hilt, V., Schulzrinne, H.: Session Initiation Protocol (SIP) Overload Control. draft-ietf-soc-overload-control-08 (2012)
5. Abaev, P., Gaidamaka, Y., Samouylov, K.E.: Modeling of Hysteretic Signaling Load Control in Next Generation Networks. In: Andreev, S., Balandin, S., Koucheryavy, Y. (eds.) NEW2AN/ruSMART 2012. LNCS, vol. 7469, pp. 440–452. Springer, Heidelberg (2012)
6. Bali, S., Victor, S.: Frost An Algorithm for Fitting MMPP to IP Traffic Traces. IEEE Communication Letters 11(2), 207–209 (2007)
7. Abaev, P., Gaidamaka, Y., Pechinkin, A., Razumchik, R., Shorgin, S.: Simulation of overload control in SIP server networks. In: Proc. of the 26th European Conference on Modelling and Simulation, ECMS 2012, pp. 533–539 (2012)
8. Abaev, P., Pechinkin, A., Razumchik, R.: On analytical model for optimal sip server hop-by-hop overload control. In: Proc. of the 4th International Congress on Ultra Modern Telecommunications and Control Systems, ICUMT 2012, pp. 303–308 (2012)

9. Abaev, P., Pechinkin, A., Razumchik, R.: Analysis of queueing system with constant service time for SIP server hop-by-hop overload control. In: Dudin, A., Klimenok, V., Tsarenkov, G., Dudin, S. (eds.) BWWQT 2013. CCIS, vol. 356, pp. 1–10. Springer, Heidelberg (2013)
10. Bocharov, P.P., DÁpice, C., Pechinkin, A.V., Salerno, S.: Queueing theory. Series "Modern Probability and Statistics". VSP Publishing, Utrecht (2003)
11. Latouche, G., Ramaswami, V.: Introduction to matrix analytic methods in stochastic modeling. SIAM, Philadelphia (1999)

Appendix

In order to find stationary state probabilities using approach developed in the series of papers (see, e.g. [7], [8], [9]) some auxiliary variables need to be introduced. Henceforth we make use of matrices U_i, W_i, V_i, M, introduced in Section 3 of the paper.

Let at some moment of time there be n, $n = \overline{H+1, R-1}$ customers in the system and the system is in "overload" mode. Denote by A_n matrix of size 2×2. The $(i,j)^{th}$ entry of A_n is the probability that at the moment of time when the total number of customers in the system equals $(n-1)$ for the first time, the source's phase will be j, given that at initial moment of time it was i, and until that moment the total number of customers in the system remained below R. In other words, following, e.g. [11], (i,j), $i = 0, 1$, $j = 0, 1$, entries of A_n represent taboo probabilities, i.e.

$$[A_n]_{(i,j)} = \mathbf{P}\{\mathbf{X}(\tau) = (j, 1, n-1); \mathbf{X}(t) \notin \mathcal{X}_2, t \in (0, \tau) | \mathbf{X}(0) = (i, 1, n)\},$$

where $\tau = \inf\{t > 0 : n(t) = R - 1\}$. Following the first step analysis one can observe that matrix A_{R-1} satisfies the equation $A_{R-1} = U_1 + V_1 A_{R-1}$ wherefrom it follows that $A_{R-1} = (I - V_1)^{-1} U_1$.

For other matrices A_n, $n = \overline{R-2, H+1}$, one can verify, using first step analysis, that it holds

$$A_n = U_1 + V_1 A_n + W_1 A_{n+1} A_n.$$

Now let at some moment of time there be n, $n = \overline{H+1, R-1}$ customers in the system and the system is in "overload" mode. Denote by G_n matrix of size 2×2. The $(i,j)^{th}$ entry of G_n is the probability that at the moment of time when the total number of customers in the system equals R for the first time, source's phase will be j, given that at initial moment of time it was i, and until that moment the total number of customers in the system remained above $(n-1)$. As in the previous case, entries $[G_n]_{(i,j)}$, $i = 0, 1$, $j = 0, 1$, can be seen as taboo probabilities

$$[G_n]_{(i,j)} = \mathbf{P}\{\mathbf{X}(\tau) = (j, 2, R); \mathbf{X}(t) \notin (0, 1, n-1) \cup$$
$$(1, 1, n-1), t \in (0, \tau) | \mathbf{X}(0) = (i, 1, n)\},$$

where $\tau = \inf\{t > 0 : n(t) = R\}$. As in the case with matrix A_{R-1} one can see, that matrix G_{R-1} satisfies the equation $G_{R-1} = W_1 + V_1 G_{R-1}$, i.e. $G_{R-1} =$

$(I - V_1)^{-1}W_1$. Other matrices G_n, $n = \overline{R-2, H+1}$ are computed from the following recurrence relations:

$$G_n = W_1(G_{n+1} + A_{n+1}G_n) + V_1G_n.$$

Assume that at some moment of time there be R customers in the system. Clearly the system is in "blocking" mode. Denote by D_n matrix of size 2×2. The $(i,j)^{th}$ entry of D_n is the probability that at the moment of time when the total number of customers in the system equals n, $n = \overline{H, R-1}$ for the first time, the sources phase will be j given that at initial moment of time it was i. Due to the fact that in "blocking" mode system does not accept newly arriving customers, matrix D_{R-1} satisfies relation $D_{R-1} = U_2 + V_2D_{R-1}$. Thus $D_{R-1} = (I - V_2)^{-1}U_2$. Noticing that probability of moving from state n to $n-1$ for $n = \overline{H+1, R-2}$ are completely described by matrix D_{R-1}, we have that $D_n = (D_{R-1})^{R-n}$, $n = \overline{R-2, H}$.

Now let at some moment of time there be n, $n = \overline{L, H-1}$ customers in the system and the system is in "normal" mode. Denote by A_n matrix of size 2×2. The $(i,j)^{th}$ entry of A_n is the probability that at the moment of time when the total number of customers in the system equals $(n-1)$ for the first time, the source's phase is j, given that at initial moment of time it was i and until that moment the total number of customers in the system remained below H. Using similar probabilistic arguments as before we find that $A_{H-1} = U_0 + V_0A_{H-1}$, wherefrom $A_{H-1} = (I - V_0)^{-1}U_0$. It can be easily checked that other matrices A_n, $n = \overline{L, H-2}$ are computed from the equation

$$A_n = U_0 + V_0A_n + W_0A_{n+1}A_n.$$

Assume that at some moment of time there are n, $n = \overline{L, H-1}$ customers in the system and the system is in "normal" mode. Denote by G_n matrix of size 2×2. The $(i,j)^{th}$ entry of G_n is the probability that at the moment of time when the total number of customers in the system equals H for the first time, the source's phase will be j, given that at initial moment of time it was i and until that moment the total number of customers in the system remained above $(n-1)$. Matrix G_{H-1} is determined from equation $G_{H-1} = W_0 + V_0G_{H-1}$ and for other matrices G_n, $n = \overline{L, H-2}$ it holds that

$$G_n = W_0(G_{n+1} + A_{n+1}G_n) + V_0G_n.$$

If at some moment of time there are n, $n = \overline{L, H-1}$ customers in the system and the system is in "overload" mode, denote B_n — matrix of size 2×2. The $(i,j)^{th}$ entry of B_n is the probability that at the moment of time when the total number of customers in the system equals $(n+1)$ for the first time, the source's phase will be j, given that at initial moment of time it was i and until that moment the total number of customers in the system remained above $L-1$. For matrix B_L is holds $B_L = W_1 + V_1B_L$ and other matrices B_n, $n = \overline{L+1, H-1}$, are computed from relations

$$B_n = W_1 + V_1B_n + U_1B_{n-1}B_n.$$

Now let us consider the following case. Let at some moment of time there be n, $n = \overline{L, H-1}$ customers in the system and the system is in "overload" mode. Denote by D_n matrix of size 2×2. The $(i, j)^{th}$ entry of D_n is the probability that at the moment of time when the total number of customers in the system equals $(L-1)$ for the first time, the source's phase will be j given that at initial moment of time it was i, and until that moment the total number of customers in the system remained below $n+1$. Fist step analysis shows that matrix D_L is computed from equation $D_L = U_1 + V_1 D_L$ and other matrices D_n, $n = \overline{L+1, H-1}$ from equations

$$D_n = U_1(D_{n-1} + B_{n-1}D_n) + V_1 D_n.$$

Now if at some moment of time there are H customers in the system then the probability that at the moment of time when the total number of customers in the system equals $(L-1)$ for the first time, the sources phase will be j, given that at initial moment of time it was i is given by matrix D of size 2×2 which can be determined from the equation

$$D = U_1 D_{H-1} + U_1 B_{H-1} D + W_1(A_{H+1} + G_{H+1}D_H) + V_1 D.$$

Finally we introduce the last sequence of matrices. Let at some moment of time there be n, $n = \overline{1, L-1}$ customers in the system and the system is in "normal" mode. Denote by A_n matrix of size 2×2. The $(i, j)^{th}$ entry of A_n is the probability that at the moment of time when the total number of customers in the system equals $(n-1)$ for the first time, the source's phase will be j, given that at initial moment of time it was i. Matrix A_{L-1} can be found from equation $A_{L-1} = U_0 + W_0(A_L + G_L\Delta)A_{L-1} + V_0 A_{L-1}$ and other matrices A_n, $n = \overline{1, L-2}$ are computed from equations

$$A_n = U_0 + W_0 A_{n+1} A_n + V_0 A_n.$$

After introducing all preliminary notations and auxiliary variables we can write out the system of equilibrium equations for $\boldsymbol{p}_n^h$. Following the elimination method which can be found e.g. in [10], one can verify that the following relations hold:

$$\boldsymbol{p}_0^0(\Lambda_0(I - A_1) - Q) = \boldsymbol{0}^T, n = 0, \tag{1}$$
$$\boldsymbol{p}_n^0(\Lambda_0 + M - Q) = \boldsymbol{p}_{n-1}^0\Lambda_0 + \boldsymbol{p}_n^0\Lambda_0 A_{n+1}, n = \overline{1, H-2},\ n \neq L-1, \tag{2}$$
$$\boldsymbol{p}_{L-1}^0(\Lambda_0 + M - Q) = \boldsymbol{p}_{L-2}^0\Lambda_0 + \boldsymbol{p}_{L-1}^0\Lambda_0[A_L + G_L D], n = L-1, \tag{3}$$
$$\boldsymbol{p}_{H-1}^0(\Lambda_0 + M - Q) = \boldsymbol{p}_{H-2}^0\Lambda_0, n = H-1, \tag{4}$$
$$\boldsymbol{p}_H^1(\Lambda_1 + M - Q) = \boldsymbol{p}_{H-1}^0\Lambda_0 + \boldsymbol{p}_H^1(MB_{H-1} + \Lambda_1(A_{H+1} + G_{H+1}D_H)), \tag{5}$$
$$\boldsymbol{p}_n^1(\Lambda_1 + M - Q) = \boldsymbol{p}_{n-1}^1\Lambda_1 + \boldsymbol{p}_n^1\Lambda_1 A_{n+1}, n = \overline{H+1, R-2}, \tag{6}$$
$$\boldsymbol{p}_{R-1}^1(\Lambda_1 + M - Q) = \boldsymbol{p}_{R-2}^1\Lambda_1, n = R-1, \tag{7}$$
$$\boldsymbol{p}_R^2(M - Q) = \boldsymbol{p}_{R-1}^1\Lambda_1, \tag{8}$$
$$\boldsymbol{p}_n^2(M - Q) = \boldsymbol{p}_{n+1}^2 M, n = \overline{H+1, R-1}, \tag{9}$$
$$\boldsymbol{p}_n^1(\Lambda_1 + M - Q) = \boldsymbol{p}_{n+1}^1 M + \boldsymbol{p}_n^1 MB_{n-1}, n = \overline{L+1, H-1}, \tag{10}$$
$$\boldsymbol{p}_L^1(\Lambda_1 + M - Q) = \boldsymbol{p}_{L+1}^1 M. \tag{11}$$

The normalization condition is

$$p_0^0 1 + \sum_{n=1}^{H-1} p_n^0 1 + \sum_{n=L}^{R-1} p_n^1 1 + \sum_{n=H+1}^{R} p_n^2 1 = 1. \tag{12}$$

Due to the lack of space we have presented here the results only for the case of two-state MMPP but the proposed method and corresponding algorithm are also true for n-state MMPP model $(n \geq 2)$ with minor changes in the matrices U_i, W_i, V_i, M, I, Λ_i and Q.

An Efficient Propagation Method for Emergency Messages in Urban VANETs

Kyuchang Lee[1], Keuchul Cho[2], Junhyung Kim[2], Youngmi Baek[2], Jeongbae Yun[2], Gihyuk Seong[2], and Kijun Han[2]

[1] Graduate School of Department of Mobile Telecommunications Engineering,
Kyungpook National University 1370, Sankyuk-dong, Buk-gu, Daegu, 702-701, Korea
kclee@netopia.knu.ac.kr
[2] Graduate School of Computer Science and Engineering,
Kyungpook National University, 1370, Sankyuk-dong, Buk-gu, Daegu, Daegu, 702-701, Korea
{k5435n,jhkim,maya,jbyun,khseong}@netopia.knu.ac.kr
[3] School of Computer Science and Engineering,
Kyungpook National University, 1370, Sankyuk-dong, Buk-gu, Daegu, 702-701, Daegu, Korea
kjhan@knu.ac.kr

Abstract. If an accident vehicle propagates emergency messages to other vehicles close to it, the other drivers may realize and avoid the accident spot. In this letter, we propose a broadcast scheme to propagate emergency messages fast in urban VANETs (Vehicular Ad-Hoc Networks) with the help of GPS (Global Position System). In our scheme, a transmitting vehicle chooses the farthest node as the next relay vehicle to propagate emergency messages. And, we suggest an algorithm for intersection recognition and SCF (Store-Carry-Forward) task by taking advantage of periodic hello packets to reduce the propagation time and enhance the delivery ratio.

Keywords: VANETs, emergency message, broadcast storm problems, intersection algorithm.

1 Introduction

VANETs are a recently proposed smart transportation network intended for road safety and commercial application [1][3]. Especially, with the emergency messages, the drivers can be aware of the car accidents happened in front of the vehicle even if the line of sight is bad. Then, the drivers can change their road lanes or something else to avoid hitting the abnormal cars. Or, they can change their route to the destination in time and thus avoid getting into a traffic jam.

The design of reliable and efficient broadcast protocols is a key enabler for the successful deployment of vehicular communication services. In order to design broadcast protocols for urban environment, there are some problems to be considered. First, the broadcast storm problem [5] may occur during message communication. Second, the inter-vehicle communication at an intersection is unreliable. Finally, the communication could be easily blocked by the buildings around the corner when the transmission starts.

S. Balandin et al. (Eds.): NEW2AN/ruSMART 2013, LNCS 8121, pp. 397–406, 2013.
© Springer-Verlag Berlin Heidelberg 2013

In this letter, we propose a broadcast scheme to propagate emergency messages in urban VANETs with the help of GPS navigation system. Most broadcast protocols proposed for VANETs choose relay vehicles to propagate emergency messages at the receiver nodes. But, in our scheme, the disseminator selects relay vehicles to reduce transmitting time and avoid unnecessary retransmissions. And, we suggest an algorithm for intersection recognition and efficient SCF task.

The rest of the letter is organized as follows: in Section 3, describes our scheme in detail. In Section 3, we evaluate the performance of the proposed protocol. Finally, we conclude in Section 4.

2   Proposed Scheme

We assume that vehicles can communicate with other vehicles with the same range wireless device through a single channel, and each vehicle knows its location by using GPS device.

All vehicles in network periodically transmit hello messages to their neighborhood in their communication range. The hello messages delivery information such as vehicle ID, location (coordinates), absence or presence on the intersection. The emergency message IDs are specified in hello packets. Hence, each vehicle can define neighbors and their locations.

2.1   Basic Concept

If a vehicle gets into an accident, the accident vehicle tries to propagate emergency message to other vehicles by using broadcasting mechanism. And, the others will again broadcast to their neighbor repeatedly. But, it is very inefficient and may cause broadcast storm problems if they all participate in propagation.

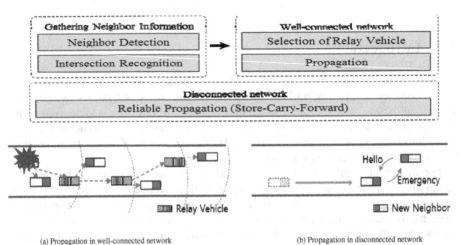

(a) Propagation in well-connected network   (b) Propagation in disconnected network

Fig. 1. Basic concept of propagating emergency messages in our scheme

In our scheme, a transmitting vehicle chooses the farthest neighboring node as the next RV which will have priority for message propagation if the network is well connected as depicted in Fig. 1(a). If the network is disconnected, it stores and carries the emergency message until it finds a new neighbor node within its transmission range to forward the messages as depicted in Fig. 1(b). This mechanism is called SCF (Store-Carry-Forward).

2.2 Selection of Relay Vehicle

In our scheme, the current relay vehicle chooses the farthest neighboring nodes within its transmission range as next relay vehicles.

The number of the next relay vehicles can be one or more depending on the location of the current relay vehicle RV_i. If it hears an emergency message moving around a 4-way intersection, three vehicles moving in other ways will be selected as the next relay vehicles. For example, when the current RV_i receives the emergency messages from previous sender RV_{i-1}, it chooses three farthest vehicles moving in directions of reference point (DP_1, DP_2 and DP_3) as the next relay vehiclesRV_{i+1} as depicted in Fig. 2(a).

On the other hand, if the current relay vehicle receives the emergency messages moving in a non-intersection, only one vehicle moving in the reverse way will be selected as the next relay vehicle. For example, when the relay vehicle receives an emergency message from previous sender, it chooses the farthest vehicle moving in the opposite direction (DP_2) as the next relay vehicle as illustrated in Fig. 2(b).

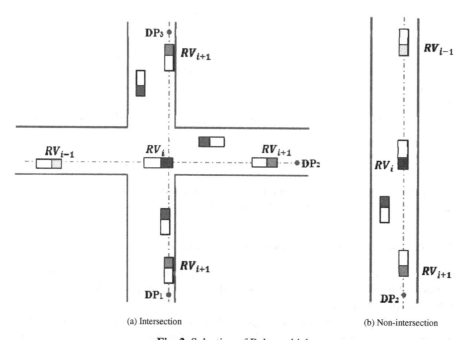

(a) Intersection   (b) Non-intersection

Fig. 2. Selection of Relay vehicles

When the current relay vehicle receives an emergency message, it chooses the next relay vehicles by first computing the coordinates of 3 directional points (DP), which are located on 90 (DP$_1$), 180 (DP$_2$) and 270 (DP$_3$) degrees from the straight-line connecting the previous relay vehicle RV$_{i-1}$ and the current relay vehicle RV$_i$ on the transmission circle illustrated in Fig. 3.

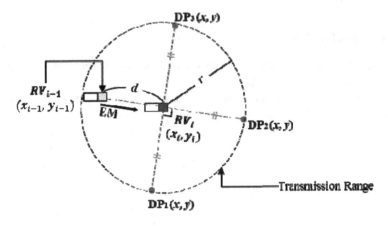

Fig. 3. Choice of the directional points

The coordinates of three DPs (x, y) are given by

$$X=\{(r+d)/d\}*(x_{i-1}-x_i)*\cos\theta-\{(r+d)/d\}*(y_{i-1}-y_i)*\sin\theta$$

$$Y=\{(r+d)/d\}*(x_{i-1}-x_i)*\sin\theta+\{(r+d)/d\}*(y_{i-1}-y_i)*\cos\theta \quad\quad (1)$$

where r is transmission radius, d is distance between the previous relay vehicle RV$_{i-1}$ and the current relay vehicle, RV$_i$ and   is 90, 180, 270 degrees for DP1, DP2, DP3, respectively. Finally, the current relay vehicle selects the closest vehicles from the directional points within the transmission range as the next relay vehicles. This can be rewritten as follows.

$$RV_{i+1} = \text{neighbor} \mid \min\{\text{distance}(\forall \text{neighbor}, DP_{\{1,2 \text{ or } 3\}})\} \quad\quad (2)$$

If some neighbors of current vehicle are on intersection, one of them is chosen as Intersection Relay Vehicle (IRV) independently of RV.

2.3   Reliable Propagation of Emergency Message

When the current relay vehicle broadcasts the emergency message, all neighboring vehicles within the transmission range hears the message and they should determine whether or not to rebroadcast the message.

Upon receiving an emergency message, each vehicle first checks if it is designated as the relay vehicle by decoding the address fields of the emergency message. If so, it should select the next relay vehicle to rebroadcast the messages and propagates the emergency message immediately. This process is repeated until the propagator cannot find the next RV.

If the vehicle is not designated as the relay vehicle, it just keeps hearing any emergency message from the selected relay vehicle during some time. This is to make the vehicle act as a candidate relay vehicle in case that there is communication error of emergency message. For this, it starts a timer (called CANDI_TIMER) to examine whether or not propagation of emergency message is successful. The timer stops when it hears the message successfully. When the timer is expired (this means there is communication error), it propagates emergency message in place of the erroneous relay vehicle.   Fig. 4 illustrates the state transition diagram for reliable propagation of emergency message at each vehicle.

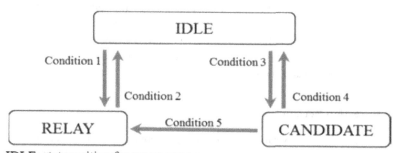

IDLE: state waiting for emergency message
RELAY: state propagating emergency message
CANDIDATE: state checking whether there is communication error

Condition 1	When receiving EM and being designated as RV
Condition 2	When selecting the next RV and propagating EM
Condition 3	When receiving EM but being not designated as RV
Condition 4	When receiving EM
Condition 5	When the CANDI_TIMER is expired

※ EM: Emergency Message

Fig. 4. State transition diagram of a vehicle

The expiration time of the timer is given depending on the distance to RV as follows:

$$\text{Expiration time} = \text{RANK} * \text{SLOT\_TIME} \qquad (3)$$

where SLOT_TIME is a predefined constant time unit, and RANK means the order of the distance to RV.   RANK value of the nearest vehicle to RV is given 1, and RANK values of the others are assigned 2,3,4,…, depending on their distances to RV.

Additionally, a mentioned IRV has some little wait time (e.g. half *SLOT_TIME*). However, the timer of IRV is not canceled even though IRV receives packet of RV.

When the timer is expired, IRV chooses RVs by using mentioned mechanism, and then rebroadcast to its neighbor.

2.4   SCF Mechanism for Disconnected Network

When a vehicle does not have any neighbor to forward message on receiving an emergency message (i.e., network is disconnected), it stores and carries the emergency message until it meets new neighbors to forward the message (called SCF scheme).

When it finds new neighbors to forward the emergency message, it starts a timer (called SCF_TIMER). This timer makes only one SCF vehicle to forward the emergency message at a time by suppressing redundant broadcast of emergency messages from two or more SCF nodes.

When an SCF vehicle carrying the emergency message hears a hello message, it should determines if or not to propagate the emergency message. The expiration time of the SCF_TIMER is given depending on the distance to the new neighbor. So, the closest neighboring vehicle will have the shortest waiting time.

When the timer is expired, the SCF node designates the new neighbor as the next relay vehicle and propagates the emergency message.

2.5   Intersection Recognition

The vehicles usually have various directional neighbors on the intersection. We suggest that each vehicle can identify being on intersection by taking advantage of their neighbor location information without RSU (Road Side Unit). Each vehicle recognizes that it is currently located at an intersection if the distance to the orthogonal point is less than the road width divided a tuning factor for adjusting accuracy (denoted by α). We can write as follows:

$$\text{Located at intersection if} \quad \text{DIST} < (\text{Road}_{\text{WIDTH}} / \alpha) \tag{4}$$

A larger value of α gives more accurate recognition of intersection, but reduces the probability of intersection discovery.

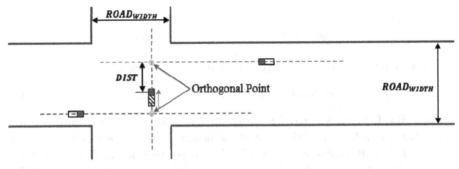

Fig. 5. Intersection Recognition

3   Performance Evaluation

In the simulations, we assume a map size of 2,000m x 2,000m such as Manhattan in New York, which is comprised of the blocks with a size of 200m x 200m and road width of 10m. Each road has four lanes and two directions. It is assumed that each vehicle is equipped with wireless communication device, and separated by equal space. Each vehicle is assumed to have beacon interval of 500ms. The velocity of each vehicle is set to 45 or 60km/s randomly. The transmission range is set to 150m.

We evaluate performance of our scheme with IRA (Intersection Recognition Algorithm) and RSU, and compare them with UV-CAST and the simple flooding. The simple flooding is a repeated broadcasting mechanism without using SCF and suppression of redundancy.

Fig. 6 shows that the propagation speed of our scheme is considerably faster than the UV-CAST in both low and high densities, where the low or high density means 300 or 1,000 vehicles, respectively, over the total map (2000*2000 m). The propagation speed here means the speed to broadcast the emergency message over the whole network. In low density, each vehicle has few neighbors, which results in disconnected network. On the other hand, in high density, each vehicle has one or more neighbors.

We can also see that the UV-CAST spends much more time on selecting of relay nodes because of long waiting time at non-intersection and restricted SCF vehicles. The repeated simple flooding fails to propagate emergency messages over the whole network in low density environment since it does not use SCF mechanism. However, our scheme offers a shorter propagation time than the existing schemes in both density environments because the relay vehicles do not need long waiting time regardless of location and all vehicles can work as SCF vehicles. The chart shows that the IRA scheme spends the almost the same processing time as the RSU.

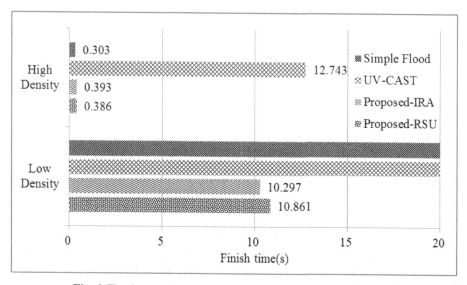

Fig. 6. The time until emergency message is delivered to all vehicles

Fig. 7 depicts the average number of propagations which indicates the amount of potential broadcast storm. In this graph, the simple flooding scheme is not included because it has vast values (over 30,000). The graph shows that the number of propagations continually increases in UV-CAST because SCF vehicles propagate message as often as it meets new neighbors. But, our scheme offers a less propagations by suppressing unnecessary rebroadcasting although it initially requires more propagations than the UV-CAST because of the faster propagation speed.

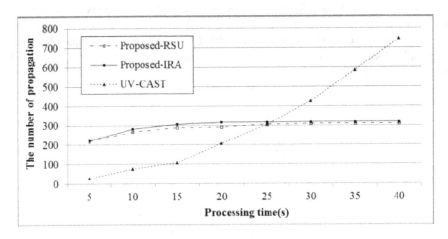

(a) at low density

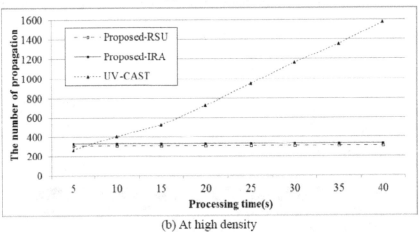

(b) At high density

Fig. 7. The number of propagations

In Fig. 8, we can see that our SCF scheme provides a higher delivery ratio than the UV-CAST because all nodes operate as SCF vehicles in our scheme. So, our scheme guarantees reliable communication not only in high density but also low density environment.

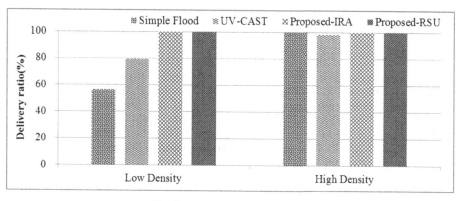

Fig. 8. The average delivery ratio

4   Conclusion

In this letter, we suggest a fast and reliable broadcast scheme for propagating emergency messages in urban environment for VANETs. In our scheme, a propagator chooses the farthest neighboring nodes messages, using directional points, as next relay vehicles to rebroadcast emergency messages. In disconnect network, the propagator stores and carries the emergency message until it meets new neighbors to forward the message. We used a contention mechanism using SCF_TIMER to make only one SCF vehicle to forward the emergency message at a time by suppressing redundant broadcast of emergency messages from two or more SCF nodes. In addition, we provide an algorithm for intersection recognition in environment without RSU.

The simulation results show that our scheme can reduce propagation time as well as unnecessary packet delivery and can improve propagation ratio than the UV-CAST.

Acknowledgements. This work was supported by the IT R&D program of MOTIE/KEIT. [10041145, Self-Organized Software platform(SoSp) for Welfare Devices]. This research was supported by Basic Science Research Program through the National Research Foundation of Korea (NRF) funded by the Ministry of Education, Science and Technology (2011-0029034).

References

1. Lee, K.C., Lee, U., Gerla, M.: Survey of Routing Protocols in Vehicular Ad Hoc Networks. In: Advances in Vehicular Ad-Hoc Networks, pp. 149–170 (2010)
2. Viriyasitavat, W., Tonguz, O.K.: UV-CAST: An Urban Vehicular Broadcast Protocol. IEEE Communications Magazine, 116–124 (November 2011)
3. Zhao, J., Cao, G.: VADD: Vehicle-assisted Data Delivery in Vehicular Ad Hoc Networks. In: 25th IEEE International Conference on Computer Communications, INFOCOMM 2006, pp. 1–12 (April 2006)

4. Lee, J.-F., Wang, C.-S., Chunag, M.-C.: Fast and Reliable Emergency Message Dissemination Mechanism in Vehicular Ad Hoc Networks. In: WCNC 2010. IEEE Communications Society (2010)
5. Wisitpongphan, N., Tonguz, O.K.: Broadcast Storm Mitigation Techniques in Vehicular Ad Hoc Networks. IEEE Wireless Communications, 84–94 (December 2007)
6. Yi, C.-W., Chuang, Y.-T., Yeh, H.-H., Tseng, Y.-C., Liu, P.-C.: Streetcast: An Urban Broadcast Protocol for Vehicular Ad-Hoc Networks. In: 2010 IEEE 71st Vehicular Technology Conference, VTC 2010-Spring, pp. 1–5 (May 2010)
7. Martinex, F.J., Cano, J.C., Calafate, C.T., Manzoni, P.: CityMob:A mobility model pattern generator for VANETs. In: IEEE International Conference on Communications Workshop, ICC, pp. 370–374 (May 2008)

A Cross Layer Balanced Routing Protocol for Differentiated Traffics over Mobile Ad Hoc Networks

Mariem Thaalbi[1,2], Nabil Tabbane[1], Tarek Bejaoui[1], and Ahmed Meddahi[2]

[1] Mediatron Lab, Higher Communication School of Tunis, University of Carthage. Tunisia
[2] Institut Mines Telecom/TELECOM Lille 1. France
{Thaalbi.Mariem,Nabil.tabbane}@supcom.rnu.tn,
tarek.bejaoui@issatm.rnu.tn, ahmed.meddahi@telecom-lille1.eu

Abstract. We propose a cross layer approach to achieve greater routing performance for applications with real time constraints in Mobile Ad Hoc Networks. Interactions between MAC, Network and Application layers are fully exploited to get accurate information about the end-to-end path quality, and the applications' characteristics. The improvements provided by our scheme come from considering a service class differentiation, a balanced routing protocol and a path quality cost function as well. It aims to enhance the routing performance for real time applications to meet the QoS requirements defined by the ITU-G1010 recommendation. The simulation results and analysis show that our contribution achieves a good performance and capacity gain.

Keywords: MANET, Link Quality, Path Quality, Traffic Class Differentiation, Cross Layer Design, Balanced Routing.

1 Introduction

Since real time applications present one of the main critical services in wireless and mobile networks, many protocol designs have been proposed to enhance the application QoS: *Quality of Service* and the QoE: *Quality of Experience* in such networks. Some existing contributions [1-6] deploy prediction methods to get information about the link quality and route the data packets through the best link. Other contributions [7-13] use cross layer approach to exploit the essential information derived from lower layers. In this paper, we aim to help the routing process to select stable and quality aware routes based on the application's requirements. Our contribution combines the cross layer and the prediction approaches to enhance the routing of real time application in MANETs. We define an enhanced cross layer design that takes into account the network and applications features. Prediction methods in MAC layer is deployed to get link quality information. Cross layer design between L3: Layer 3 and L2: Layer 2 is defined to get information about the whole path quality. In order to get information about the application characteristics, a cross layer design between L7 and L3 is defined. A new software entity is defined to manage the L7-L3 cross layer interaction. In order to define a routing protocol that fits the requirements of different application a per traffic balanced routing protocol is defined. This paper is organized as follows. We outline some cross layer approaches and prediction methods in section 2. Our cross

S. Balandin et al. (Eds.): NEW2AN/ruSMART 2013, LNCS 8121, pp. 407–419, 2013.

layer design approach is introduced in Section 3. Experimental results are reported in section 4, and finally we conclude and give some perspectives in Section 5.

2   Related Works

In this section, we first investigate some methods used to predict the network performance criteria like delay and packet loss ratio, and then we focus on the cross layer concept.

2.1   Prediction Methods Investigation

To get information about the link quality, prediction methods can be deployed. Accurate delay estimation in mobile and wireless networks stills a challenging issue. Some Research activities focused on this issue to improve the used methods and techniques for better efficiency and accuracy. According to [1], the delay at each wireless node is composed of input queuing delay, processing delay, output queuing delay, transmission delay, propagation delay, and retransmission delay. To implement the packet delay measurement at a node in 802.11 networks, authors in [1] propose to record the time when a packet enters the node (t_1) and the time when the data packet is acknowledged (t_5) after being relayed. The packet delay is calculated by $t_5 - t_1$. Authors in [2] use Chaos neural networks to predict delay. The neural network can model unknown system with a given precision while keeping the computation cost minimized. To predict end to end delay, the authors process first to the delay measurement using the ping tool to collect RTT: *Round Trip Time* traces. The critical issue of neural network model is to determine the structure of the network. In MANETs, the structure of the network is dynamic since nodes can join or leave the network in dynamic way. To predict the link quality, the authors in [3] measure the link quality and then perform a prediction algorithm. To measure the link quality, a link layer assessment request is sent periodically. The delay is computed as the RTT of the request and its reply. The network node keeps a track of the past measurements and then predicts future link quality based on these measurements using the WLSR: *Weighted Least Square Regression* algorithm. The WLSR prediction algorithm applies weights to the measurements. It takes as input a window of measurements of a given QoS metric and predicts the value of the metric.

In Mobile Ad Hoc Networks, the packet loss is due to the following most dominant factors [1]: the buffer overflow, the transmission loss and the link breakages. The authors in [4] predict the network loss based on a hierarchal model where the short term dynamics of losses is driven by 2-state Markov chains while long-term network losses are modeled by the HMM: *Hidden Markov Model*. Given a fixed window of several time units, the short term loss rate is the fraction between the number of loss packets in this window and the number of packets transmitted in it. The hidden state models longer-term events that change end-to-end loss statistics, e.g., router congestion, routing convergence, wireless signal fading. To achieve this, the authors constrain hidden state transitions to happen at large timescales [4]. Authors of [5] use a measurement method to predict the packet delivery of an IEEE 802.11n channel. According to [5], the wireless packet delivery can be accurately predicted using the CSI: *Channel State Information* measurements. The CSI provides more information than the RSSI: Received

Signal Strength Indicator since it describes the Signal Noise Rate (SNR) and the path phase [5]. A IEEE 802.11n node can probe the receiver to gather a CSI or use channel reciprocity to learn CSI from the received packets. An analytical method was proposed in [6] to predict the packet loss ratio. The authors focus on the GE: Gilbert-Elliot channels which refer to the wireless channel having two states, good and bad.

In wireless networks, prediction methods are efficient techniques to get information about the network capacity and availability [1-6]. Delay and packet loss are two key QoS parameters for real time applications [1, 2]. Conventional MAC layer functions can only predict the link characteristics. In our paper, a cross layer design is deployed to get information about the predicted end-to-end path quality and exploits this information to improve the behavior of upper layer protocols.

2.2 Cross Layer Designs Investigation

Routing in dynamic networks stills a challenging issue. To achieve greater performance, some works exploit the cross layer interactions as well as possible. [7] for example, describes the different cross layer designs based on the information flow: upward design, downward design, and back-and-forth. In [8] the authors demonstrate that changing a lower layer protocol will generate a significant impact on the upper layer protocol's performance and eventually affect the overall performance of the network.

The authors of [9] and [10] exploit the broadcasting nature of the wireless medium to capture packets that are not intended to the node. The MAC layer will overhear the medium and forwards all relevant information to the application layer. This layer will extract the useful information from the incoming packets to update its overlay table. Due to the cross layer design, overheard packets from neighboring nodes can be used for information update. Moreover the authors of [9] exploit the cross layer between Application and MAC layers to propagate P2P information over the network.

The authors of [11] describe the LEMO: *Less Remaining Hop More Opportunity* algorithm that gives more priority to the closest packets to their destinations. It calculates the ratio based on the remaining hop number. The information of total hop numbers or remaining hop numbers can be known from the network layer. A cross layer interaction between MAC and Network layers is then defined to communicate the required information.

In [12], the authors try to overcome the following situation when two Mobile Ad hoc Networks are overlapped while their P2P networks are disconnected in overlay layer. The authors propose then a cross layer approach to detect this situation and merge these P2P networks at overlay layer. To merge P2P networks over MANETs, authors of [12] defines two phases: at first, detection of the merging of P2P networks at overlay layer and after merging the P2P networks.

Authors of [13] describe a cross layer design for LEO satellite Ad Hoc networks. They define three cross layer optimizations: The first one is a specific integrated MAC/PHY layer that aims to provide accurate information about the link quality. The second one controls the sliding window of TCP in transport protocol and the third one is a Balanced Routing protocol (BPR) which adopts DSR: *Dynamic Source Routing* protocol to LEO satellite networks. The balanced routing mechanism is based on the 1-hop information.

In [13], the priority information is provided by the applications themselves, while the wireless link quality is provided by the defined MAC/PHY layer. The routing protocol uses links with better quality based on the wireless link quality information. The TCP layer adjusts the congestion window size according to the MAC/PHY layer information and the application priority.

In our paper we deploy the cross layer design in order to enhance the behavior of the routing protocol in dynamic networks such MANETs for different traffic class. Our enhanced balanced routing protocol is suggested as an enhancement to the AOMDV: *Ad hoc On-demand multipath Distance Vector* protocol. To achieve traffic differentiation, a cross layer interaction between Application and routing layer is designed. To take into account the dynamicity of the network, the routing layer will exploit the MAC layer in order to get accurate information about the end to end path quality.

3   Proposed Cross Layer Routing Protocol : Description and Architecture

In our paper, the interactions between MAC, Routing and Application layers are fully exploited to enhance the routing performance of real time applications in wireless networks. We propose a cross layer architecture that involves two cross layer designs. The first design aims to exploit MAC layer in order to provide accurate information about the end-to-end path quality. The second one adapts the routing protocol to the wireless environment and the application requirements. Based on this design, the routing protocol will consider quality aware paths and balance the traffic in the network based on the application layer's information. We use the cross layer infrastructure presented in Figure 1. As shown in Figure1, three new optimization modules are introduced. The first module is "Exploiting MAC layer". It provides accurate information about the link and the path quality using prediction methods. Information about the whole path's quality will be stored in the routing table of the L3. The second module is the "EBRP: *Enhanced Balanced Routing Protocol*", a routing protocol designed to select the optimized paths during the routing process and to provide a fair allocation of traffic among different paths. The third module is a "XLME: *Cross Layer Management Entity*" used to classify the traffics according to their QoS requirements. This entity is based on communications through sockets. According to our approach, the path quality information is provided by the cross layer design between the routing layer and the MAC layer, while the application class information is provided by the cross layer interaction between the routing layer and the upper-layers through the XLME. The routing layer protocol EBRP uses paths quality to perform a context aware routing. It also provides a load balancing mechanism based on the application class.   To achieve a traffic differentiation routing, we define three service classes based on the applications requirements defined by the ITU-G1010 recommendation [15]:

- Class 1: Error tolerant and delay sensitive applications
- Class 2: Error and delay tolerant applications
- Class 3: Timely and non-error tolerant applications

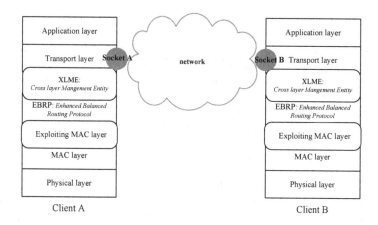

Fig. 1. The proposed Cross layer design and the end to end communication through sockets

The requirements of the application classes in terms of delay and packet loss are shown in Table 1:

Table 1. QoS classes

	Class 1	Class2	Class 3
Acceptable Packet loss (Pth)	3%	1%	0%
Acceptable delay (Dth)	150ms	400ms	4s

3.1   Exploiting MAC Layer

In this section, we propose a method to exploit MAC layer to get information about the whole path quality. We define a variable called "PPQ: *Predicted Path Quality*" that represents the whole path quality. *PPQ* will be used as an attribute to rank the paths to the same destination in the routing layer.

PPQ is computed using two metrics: 1) the loss probability P of the whole destination path, and 2) the path latency D. To select the best quality route that optimizes the PPQ, we use the SAW: *Simple Additive Weighting* method [15] as the multi-attribute decision making (MADM) method.

Based on SAW, PPQ is calculated as a weighted sum function of path loss ratio and the path latency. It is defined as the following:

$$PPQ = w_p \times \frac{P_{th}}{P} + w_d \times \frac{D_{th}}{D} \tag{1}$$

P and D are respectively the measured packet loss ratio and the end to end delay of the path from source to destination. P_{th} and D_{th} are the acceptable threshold values of packet loss and delay respectively defined on Table 1. w_p and w_d are the weights of packet loss and delay.

The weights value depends on the traffic class. More the parameter is critical for the traffic, more its weight is higher. For *class 1*, we assume that the *delay* is much more important than *packet loss*. For *class 2*, the weights of *delay* and *packet loss* are equal since the applications of this class are not critical in terms of delay and packet loss. For class3, we consider that the *packet loss* is much more important than the *delay* since no error is tolerated. To compute the w_p and w_d of each class, we use the AHP: *Analytic Hierarchy Process* method [16].

P represents the ratio of lost packets N_l from the total packets N as:

$$P = \frac{N_l}{N} \tag{2}$$

Each node computes periodically the packet loss ratio. In our contribution, we consider the NAV: *Network Allocation Vector* period as the periodic time to compute the packet loss ratio. Based on the assumption that Packet loss is a multiplicative metric, the whole packet loss ratio P from source *Src* to destination *Dest* is computed as the multiplication of the packet loss ratio of its composing links P_i according to (3):

$$P = 1 - \prod_{i \in route(Src, Dest)} (1 - P_i) \tag{3}$$

The *Hello* messages of the routing protocol are used to predict the link delay D_i. Since the delay is an additive metric [3], the end to end delay of a path D is computed as the sum of delays of its composing links according to (4):

$$D = \sum_{i \in route(Src, Dest)} D_i \tag{4}$$

In the conventional MAC layer, only links quality information can be provided. In order to provide the whole path information, a cross layer approach between MAC and routing layers is designed.

The link quality is predicted by the MAC layer by sensing the wireless media shared with its neighbors. Each node computes the link quality and stores this information in its neighbors' table. To store the link quality, two new fields are introduced in the neighbors' table of each node. One field stores the link delay and the other one stores the link packet loss ratio.

The whole path's quality will be acquired by exchanging information between MAC and Routing layers. To get information about the whole path's quality, a new field is added in the Layer 3's route reply packet.

3.2   Enhanced Balanced Routing Protocol (EBRP)

We propose a routing protocol called "EBRP: *Enhanced Balanced Routing Protocol*" to enhance the data routing performance according to the traffic class. EBRP is extended from the AOMDV protocol. AOMDV [18] is a reactive multi-path routing protocol. In this protocol, a source node can establish multiple loop free paths to a destination node in one route discovery. The source node selects the shortest path that minimizes the number of hops to forward the data packets.

Considering the AOMDV routing protocol, we introduce a new field called "PQ: *Path Quality*" in the RREP: *Route Reply* packet. The PQ field is divided into two subfields, *Loss* and *Delay* to carry the loss probability and the latency of the path. The RREP' structures of AOMDV and EBRP are shown in Figure2.

Type	ACK	Last hop	Hop Count	Destination IP address	Originator IP address	Lifetime

(a)   **AOMDV**

Type	ACK	Last hop	Hop Count	PQ		Destination IP address	Originator IP address	Lifetime
				Loss	Delay			

(b)   **EBRP**

Fig. 2. RREP format of (a) AOMDV and (b) EBRP

PQ will have "0" as default value for the Delay subfield, and "1" for the Loss subfield, and will be used to carry the path quality information between source and destination nodes.

Upon receiving a RREP, intermediate nodes update the PQ subfields by the appropriate values based on the information stored on its neighbors' table, and according to the metric type. The Loss subfield is a multiplicative metric; the intermediate nodes multiply the loss value in the RREP by the appropriate link loss. The Delay is a cumulative metric; the intermediate node adds the link's latency to the value of the delay subfield of the received RREP.

Upon receiving a RREP, the source node updates the PQ field and its routing table. A source node may receive multiple RREP from the same destination through different paths; it keeps at most the three best quality paths. To store the path's quality two new fields are added in the node's routing table.   The routing table fields of AOMDV and EBRP are shown in Figure 3.

EBRP uses a PPQ: *Predicted Path Quality* and a *Route Stability* metrics to sort the routes to the same destination. Also, it provides a load balancing mechanism based on the traffic class information provided from the upper layers.

In order to select the most stable route from the best quality ones, we add a "*Route Stability*" parameter *Stability* in the routing table, as shown in Figure 3, to indicate the route stability. The parameter *Stability* is considered as a counter which represents the durability of a specific path in the routing table.   Each process of routing table maintenance, we verify the perennity of each path present in this table and we increment the *Stability* parameter. With this parameter, the EBRP behaves differently from the conventional AOMDV routing protocol.

In the routing table, EBRP stores multiple routes to the same destination. These routes will be sorted according to their quality and then according to their stability. A node will store at most three different paths to the same destination as in AOMDV. The route with appropriate stability will be selected during the route discovery phase according to a load balancing approach that takes into account the traffic category. In order to achieve this load balancing, we propose a per traffic load balancing method.

"Per traffic load balancing" means that the node sends data packets that belong to the same traffic class over one path. Given two existing paths to the same destination, all packets from traffic class 1 will be sent over path 1, all packets from traffic class 2 go over path 2 and so on. In this way, we preserve packet sequence and synchronization.

Based on the traffic class information got from upper layers, the routing layer, first, computes the paths quality PPQ as defined in Eq.(1) based on the appropriate w_p and w_d and then selects the best path to route the traffic flow based on its stability. If two paths have the same PPQ and the same stability, the path that optimizes the traffic class requirements will be chosen for data routing. If the traffic belongs to class 1 or class 2, the path having the lower *Delay* will be chosen. If the traffic belongs to class3, the path offering the minimum *Loss* will be chosen.

destination	Sequence number	Advertised hop count	Route list		
			Next_hop1	Hop_count1	Timeout1
			Next_hop2	Hop_count2	Timeout2
			.		
			.		

(a)   **AOMDV**

destination	Sequence number	Advertised hop count	Route list					
			Next_hop1	Hop_count1	Loss1	Delay1	Stability 1	Timeout1
			Next_hop2	Hop_count2	Loss2	Delay2	Stability2	Timeout2
			.					
			.					

(b)   **EBRP**

Fig. 3. Routing table structure in (a) AOMDV and (b) EBRP

3.3   Application / Network Cross Layer Design

Communication's Socket

Every application protocol can be presented by a specific socket. In this contribution, we exploit the socket information to communicate the traffic characteristics from the Application layer to the Routing layer.

The socket is an inter-process communication point used to connect the service end points [17], as shown in Figure 1. Once the connection is established, both sides can send and receive data. A socket is characterized by its *domain*, its *type*, and its *address*. Two processes can communicate with each other only if their sockets have the same type and are in the same domain. The most widely used domains are: the Unix domain known as AF_UNIX and the Internet domain known as AF_INET. Each domain has its own address format. Considering the Internet domain' sockets, there are two widely used socket types: the *stream sockets* known as SOCK_STREAM and *datagram sockets* known as SOCK_DGRAM. Depending on the sockets' type a specific communication protocol will be chosen to transmit the data. Stream sockets use the reliable and connection oriented protocol TCP known as IPPROTO_TCP. Datagram sockets use the unreliable and connectionless transport protocol UDP known as IPPROTO_UDP. To communicate through sockets, the service creates first the socket by specifying its *domain*, its *type* and its *protocol*. This creation is performed by the *socket* function. If the socket is created successfully, the *socket* function returns an integer which identifies the socket known as the socket's *descriptor*. Once the socket is created, it must be assigned to an *address* through the *bind* function in order to communicate with the other network's entities. The *socket*, and *bind* functions are defined as the following:

---

int **socket** (int *domain*, int *type*, int *protocol*);

int **bind** (int *descriptor*, struct **sockaddr_in** **address*, int *address_length*);

Designed L3/L7 Cross Layer

In our contribution, we consider communications between the MANET nodes. We create a new socket's domain AF_AH to identify our ad hoc domain. To transmit the socket's data, we use the EBRP protocol designed by IPPROTO_EBRP. We use raw sockets known as SOCK_RAW. The SOCK_RAW is a type of sockets designed to send and receive IP packets without specifying a transport protocol.

The defined ad hoc domain AF_AH has a *sockaddr_ah* address structure defined as the following:

---

struct **sockaddr_ah** {

short   *family*;   /*AF_AH*/

u_short *port*;   /*associated port number*/

u_long *addr*;   /* IP address of the machine*/

}

The address structure of our defined domain (AF_AH) consists of the IP address of the host machine and the service port number. We exploit this address to handle routing functions since the socket port number provides information about the type of application; each service has its own port number such as 21 for FTP.

We define a new entity called "XLME: Cross Layer Management Entity" to manage and aggregate the cross layer interactions and to provide the required information to the routing layer, in order to perform the balanced routing. This software entity is also considered as L3-L7 communication interface, based on the socket concept. It extracts information about the traffic type from the socket. Based on this information, XLME associates the traffic to its specific class that will be communicated to the routing layer in order to perform the balancing routing mechanism. Therefore, our proposed XLME entity is located between the Routing and Transport layers as shown in Figure 1.

Three traffic classes, as defined on Table I, are considered to investigate the effectiveness of our protocol. In order to get information about the traffic type in the routing layer, the XLME entity will extract the port number of the socket using the "getsockport" function defined as the following:

```
u_short  getsockport(int descriptor, struct sockaddr  *address, int address_length);
```

The "getsockport" function takes as input the length and the value of the socket address in the Internet domain, and it gives the socket port number as an output. The port number is 2 Bytes unsigned integers.

4   Performance Evaluation

Through simulations, we compare our Cross layer routing algorithm with the deployed protocol AOMDV: Ad hoc On-demand multipath Distance Vector routing [18] for its ability to compute multiple paths. Simulations are performed using the Network Simulator NS2.35. Each data point in the curves is an average of ten simulation runs with different randomly generated value. We investigate upon *Class 1 traffic* services. In our future works, further simulations will be handled for remaining traffic classes. Simulation parameters are summarized in Table 2.

To simulate a MANET network as defined in the IETF Request for Comments RFC 2501, the nodes are randomly distributed. Also, source and destination nodes are randomly chosen.

Table 2. Simulation Parameters

Network size	20 nodes
Simulation area	4096m * 4096m
Simulation time	200 s
Transmission range	250m
Network load	{20%, 40%, 60%, 80%}
PHY/MAC technology	802.11b
Propagation model	Shadowing model
Mobility model	Random way Point (Pause time =0)
Mean node speed	2 m/ s
(source, destination) pairs	randomly
Class 1 traffic parameters (VoIP)	
w_d	0.6
w_p	0.4
Traffic model /rate	CBR /64 Kbit/s

In Figure 4.a, we report the number of traffic overhead (TOH) packets while varying the network load. The figure shows that the TOH increases with the network load. Also it shows better results for our protocol EBRP as compared to the AOMDV. When the network load is equal to 60%, EBRP reduces the overhead traffic by 43% as compared to AOMDV. In fact, our protocol EBRP takes into account the end-to-end path quality and it performs a routing based on path stability. As a result, the loss probability due to the link breakage or transmission loss is reduced. As a consequence, the number of sent RERR: *Route Error* packets is reduced. Figure 4.b shows the reliability of the two protocols by measuring the packet delivery ratio while varying the network load. EBRP outperforms AOMDV. When the network load varies from 20% and 40%, EBRP is slightly better than AOMDV. When the network load exceeds 40%, the outperformance of EBRP becomes clearer. With a network load equal to 60%, EBRP gives a 30% higher PDR than AOMDV. When the network load exceeds 40%, The PDR offered by AOMDV decreases dramatically; it moves from 83% to 51% when the network load varies from 40% to 80%. Results shown in Figure 4.c enlighten better average end-to-end delay to EBRP as compared to AOMDV. Based on the QoS thresholds defined in Table 1, EBRP gives acceptable delays until a network load equal to 60%. The AOMDV gives acceptable results when the network load is inferior to 40%. When the network load reaches 80%, the end-to-end delay offered by EBRP is enhanced by 50% as compared to AOMDV. Figure 4.d shows the offered throughput of EBRP and AOMDV. Our protocol EBRP exploits more efficiently the network resources since it offers better throughput than AOMDV. For example, when the traffic load reaches 80%, EBRP offers an average throughput equal to 47 Kb/s whereas it is equal to 25 Kb/s with AOMDV.

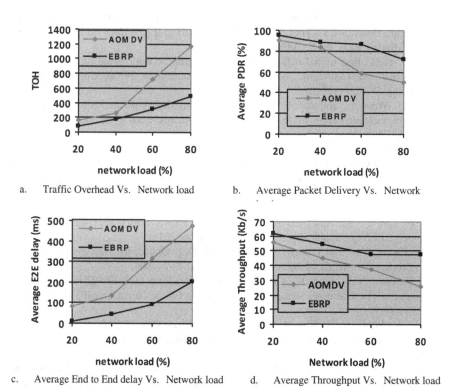

a.   Traffic Overhead Vs. Network load

b.   Average Packet Delivery Vs. Network

c.   Average End to End delay Vs. Network load

d.   Average Throughput Vs. Network load

Fig. 4. Performance evaluation for Class 1 traffic

5 Conclusion

In this paper, we present a cross layer approach to fully exploit the interactions between Application layer, Routing layer and MAC layer. A module to exploit MAC layer's information is defined to predict the link quality. A cross layer interaction between L3 and L2 is introduced to estimate the *end-to-end path quality* in term of *Packet loss* and *delay*. Based on this information, the routing protocol ranks the different paths leading to the same destination. A proposed selection mechanism extends this routing protocol to select the best path. In order to perform a load balancing routing, the traffic generated by the application layer is classified via a L7-L3 interaction.

The simulation results show that our approach optimizes *throughput, PDR, end-to-end delay* and *overhead traffic* to satisfy the real time applications' requirements. In terms of perspective, we plan to add other performance criteria to take into consideration other application constraints and network characteristics such as the node battery, the node speed and the link SNR for the route stability metric.

References

1. Shi, L., Fapojuwo, A., Viberg, N., Hoople, W., and Chan, N.: Methods for Calculating Bandwidth, Delay, and Packet Loss Metrics in Multi-Hop IEEE802.11 Ad Hoc Networks. In: VTC Spring 2008. IEEE (2008)
2. Sun, H., et al.: End to end delay prediction by neural network based on chaos theory. In: International Conference on Wireless Communications Networking and Mobile Computing, WiCOM (2010)
3. Duan, Q., Wang, L., Knutson, C.D., Zappala, D.: Link quality prediction for wireless devices with multiple radios. In: Proceedings of IEEE International Symposium on a World of Wireless Mobile and Multimedia Networks. IEEE (2008)
4. Silveira, F., de Souza e Silva, E.: Predicting packet loss statistics with hidden Markov models for FEC control. Computer Networks: The International Journal of Computer and Telecommunications Networking 56(2), 628–641 (2012)
5. Halperin, D., Hu, W., Shethy, A., Wetherall, D.: Predictable 802.11 Packet Delivery from Wireless Channel Measurements. In: SIGCOMM 2010, August 30-September 3 (2010)
6. Liu, F., Lin, C.: An Analytical Method to Predict Packet Losses over Bursty Wireless Channels. IEEE Communications Letters (99) (October 2011)
7. Stine, J.A.: Cross-Layer Design of MANETs: The Only Option. In: Military Communications Conference, MILCOM 2006, October 23-25, pp. 1–7 (2006)
8. Qin, Y., Gwee, C.L., Seah, W.: Cross layer interaction study on IEEE802.11e in wireless ad hoc networks. Communications and Networking in China, ChinaCom, 483–487, August 25-27 (2008)
9. Liu, C.-L., Wang, C.-Y., Wei, H.-Y.: Cross-Layer Mobile Chord P2P Protocol Design For VANET. International Journal of Ad Hoc and Ubiquitous Computing 6(3), 150–163 (2010)
10. Urresi, S., Canali, C., Renda, M.E., Santi, P.: Meshchord: Alocation-aware, cross-layer specialization of chord for wireless mesh networks. In: Proc. of PerCom, pp. 206–212 (2008)

11. Walia, M., Challa, R.: Performance analysis of cross-layer MAC and routing protocols in MANET. In: Second International Conference on Computer and Network Technology (ICCNT), pp. 53–59 (April 2010)
12. Shah, N., Qian, D.: Cross-Layer Design to Merge Structured P2P Networks over MANET. In: IEEE 16th International Conference on Parallel and Distributed Systems (ICPADS), December 8-10, pp. 851–856 (2010)
13. Chang, Z., Gaydadjiev, G.N.: Cross-layer designs architecture for LEO satellite ad hoc network. In: Harju, J., Heijenk, G., Langendörfer, P., Siris, V.A. (eds.) WWIC 2008. LNCS, vol. 5031, pp. 164–176. Springer, Heidelberg (2008)
14. ITU-T Recommendation G.1010: End-user multimedia QoS categories, http://www.itu-t.org
15. Savitha, K., Chandrasekar, C.: Vertical Handover decision schemes using SAW and WPM for Network selection in Heterogeneous Wireless Networks. Global Journal of Computer Science and Technology 11(9), Version 1.0, 19–24 (2011)
16. Saaty, T.L.: Decision making with the analytic hierarchy process. Int. J. Services Sciences 1(1), 83–98 (2008)
17. Douglas, E.: Internetworking with TCP/IP, 5th edn., vol. 1. Prentice Hall (2005)
18. Biradar, S.R., Koushik, M., Sarkar, S.: Performance Evaluation and Comparison of AODV and AOMDV (IJCSE) International Journal on Computer Science and Engineering 2(2), 373–377 (2010)

Modelling and Analysing a Dynamic Resource Allocation Scheme for M2M Traffic in LTE Networks[*]

Vladimir Y. Borodakiy[1], Ivan A. Buturlin[2],
Irina A. Gudkova[2], and Konstantin E. Samouylov[2]

[1] JSC "Concern "Sistemprom",
Nizhnaya Krasnoselskaya str. 13-1, 105066 Moscow, Russia
[2] Telecommunication Systems Department,
Peoples' Friendship University of Russia,
Ordzhonikidze str. 3, 115419 Moscow, Russia
bvu@systemprom.ru, ivan.buturlin@gmail.com,
{igudkova,ksam}@sci.pfu.edu.ru

Abstract. One of the main problems in LTE networks is the distribution of a limited number of radio resources among Human-to-Human (H2H) users as well as the increasing number of machine-type-communication (MTC) devices in machine-to-machine (M2M) communications. Different traffic types from user's equipment and MTC devices transmitted over the network suggests a dynamic resource allocation in order to provide a better quality of service (QoS). In this paper, we propose a dynamic resource allocation scheme for M2M traffic in LTE networks. The suggested method is based on fixed bandwidth intervals at which traffic from MTC devices is serviced according to the Processor Sharing (PS) discipline. By means of a Markov model, an estimation of the behaviour of LTE for H2H and M2M traffics characteristics is shown. We propose an analytical solution to calculate the model performance measures, such as blocking probabilities for H2H users.

Keywords: LTE, M2M, radio resource management, RRM, dynamic resource allocation, adaptive reservation, streaming traffic, elastic traffic, blocking probability.

1 Introduction

Modern information systems, such as Smart Grid networks and intelligent transportation systems include a large number of different technological sensors and controllers. These machine-type-communication (MTC) devices can transmit and receive data through wireless interfaces transmitting data independently and automatically. The increasing demand for various services without human intervention motivates

[*] The study was supported by The Ministry of education and science of Russia (projects 14.U02.21.1874, 8.7962.2013) and was partially supported by RFBR, research project No. 13-07-00953 a.

S. Balandin et al. (Eds.): NEW2AN/ruSMART 2013, LNCS 8121, pp. 420–426, 2013.

operators to introduce machine-to-machine (M2M) communications. According to Cisco VNI Global Mobile Data Traffic Forecast[1], the average M2M module will generate 330 megabytes of mobile data traffic per month in 2017, up from 64 megabytes per month in 2012. M2M will account for 5 percent of total mobile data traffic in 2017, compared to 3 percent at the end of 2012. 3GPP has been considering the support for increasing number of M2M communications in LTE networks [1] (3GPP TS 22.368, TR 22.888, TR 23.887). To ensure the proper QoS level for H2H users at the radio access network (evolved UMTS Terrestrial Radio Access Network, E-UTRAN), efficient radio resource management (RRM) techniques are implemented. According to 3GPP TS 36.300 [2], RRM includes:

- radio bearer control – configuration of radio resources associated with radio bearers during its establishment, maintenance and release;
- radio admission control – admission or rejection of the establishment requests for new radio bearers;
- connection mobility control – management of radio resources in connection with idle or connected mode mobility;
- dynamic resource allocation or packet scheduling – allocation and de-allocation of resources to user and control plane packets;
- inter-cell interference coordination – management of radio resources such that inter-cell interference is kept under control;
- load balancing – handling of uneven distribution of the traffic load over multiple cells.

This paper discusses the influence of M2M communications on dynamic resource allocation in LTE-Advanced network. For today, there is no working standard that defines the radio resources allocation strategy for M2M communications. There is a number of researches [4–6] and 3GPP recommendations that are helpful for operators to select a radio resource allocation scheme for transmitting data from MTC devices. Among all existing researches the common statement is that the data units from M2M devices have an extremely small size and the network operator shall be able to reduce the frequency of mobility management procedures per MTC device. That's why there is no use of a whole Physical Resource Block (PRB) allocation to transmit data from a single MTC device, as it is done by existing techniques for H2H users.

Allocation of one PRB to transmit data from MTC devices is described in [5]. In [4], a resource allocation scheme for M2M communications using virtual separated subcarriers is proposed. This paper discusses one possible dynamic resource allocation scheme, in which one PRB may be used to transmit data units from many MTC devices or to serve one H2H user. To serve the M2M communications the available PRBs are allocated at fixed intervals at which traffic from MTC devices is served according to the PS discipline. The number of allocated PRBs for H2H user depends on specified requirements for data rate of provided service.

---

[1] Cisco Visual Networking Index: Global Mobile Data Traffic Forecast Update, 2012–2017 (www.cisco.com/en/US/solutions/collateral/ns341/ns525/ns537/ns705/ns827/white_paper_c11-520862.html).

The remainder of this paper is organized as follows. In Section 2, we propose a model of a LTE-Advanced cell supporting H2H and M2M communications. In Section 3, the main performance measures of the proposed model via a numerical example is illustrated. Finally, we conclude the paper in Section 4.

2   Mathematical Model

Consider a single cell that receives call setup requests from H2H users and MTC devices. We make the simplifying assumption that all subscriber devices do not change their position relative to the cell for the period under review. Thus, the data rate will depend only on the number of allocated frequency resources – bandwidth units (b.u.). As the b.u. of the considered cell will be understood a minimum data rate of MTC device's data block.

Let's suppose the existence of C b.u., which corresponds to a predetermined number of PRBs. Requests from H2H users arrive according to an independent Poisson process with rate λ_{H2H} and require b_{H2H} b.u. The resource occupancy durations are exponential distributed with mean μ_{H2H}^{-1}. Note, we follow works of F.P. Kelly [7] and K.W. Ross [8] by assuming the Poisson arrival process.

Requests from M2M users arrive according to an independent Poisson process with rate λ_{M2M}. Establishing a connection for a MTC device means transmitting elastic data block, that's why the service of such requests usually performs according to PS discipline. The block size is a random value exponentially distributed with mean θ. All incoming data blocks from the MTC devices are characterized by minimum rate guarantee b_{M2M}. Let's assume S granted capacity intervals (g.c.i.) for servicing MTC devices requests and M be the maximum number of MTS devices that could be served at the g.c.i. subject to the requirements rate b_{M2M}.

This model represents a combination of models with unicast and elastic traffics (Fig. 1). We assume that H2H user's connections and transfer elastic data blocks from MTC devices operate independently of each other.

Let us introduce the following notations:

- $C_1 := M \cdot b_{M2M}$ – the number of b.u. for g.c.i.;
- $n := (n_{M2M}, n_{H2H})$ – the state of the system, where n_{M2M} is the number of M2M calls allocating bandwidth in reserved capacity C_{M2M} and n_{H2H} is the number of H2H calls allocating bandwidth in capacity C_{H2H};
- $\rho_{H2H} := \lambda_{H2H}/\mu_{H2H}$ – the traffic load due to H2H users;
- $\rho_{M2M} := \lambda_{M2M} \cdot \theta$ – the elastic traffic intensity from M2M calls, where θ mean a payload size per MTC devices;
- $\lfloor x \rfloor = \max\{y \in \mathbb{Z}: y \leq x\}$ – the floor function;
- $\lceil x \rceil = \min\{y \in \mathbb{Z}: y \geq x\}$ – the ceiling function;

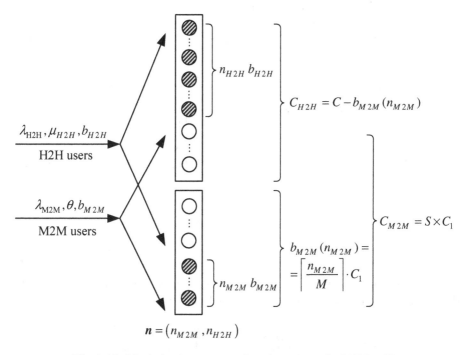

Fig. 1. Model of adaptive resource allocation scheme for M2M traffic

- $b_{\mathrm{M2M}}(n_{\mathrm{M2M}}) := \lceil n_{\mathrm{M2M}}/M \rceil \cdot C_1$ – is the number of bandwidth units assigned for n_{M2M} M2M users, where $\lceil n_{\mathrm{M2M}}/M \rceil$ means the number of g.c.i. assigned for n_{M2M} M2M users.

Given the above considerations, when a new call from H2H user arrives, two scenarios are possible.

1. The call from H2H user will be accepted with bandwidth b_{H2H} allocated in capacity $C_{\mathrm{H2H}} = C - C_{\mathrm{M2M}}$, which is possible if the call finds capacity C_{H2H} having at least b_{H2H} free.
2. The call will be blocked without any after-effect on the corresponding Poisson process arrival rate.

Similarly, when a new call from MTC device arrives, two scenarios are possible.

1. The call from MTC device will be accepted with guaranteed bit rate b_{M2M}, which is possible if the call finds reserved capacity $C_{M2M} = S \cdot C_1$ having at least b_{M2M} b.u. free.
2. The call will be blocked without any after-effect on the corresponding Poisson process arrival rate.

Then, the system state space is given by

$$\mathcal{X} := \left\{ n \geq 0 : \; n_{\text{H2H}} b_{\text{H2H}} \leq C - b_{\text{M2M}}(n_{\text{M2M}}), \; n_{\text{M2M}} b_{\text{M2M}} \leq C_{\text{M2M}} \right\}, \tag{1}$$

$$|\mathcal{X}| = \sum_{n_{\text{M2M}}=0}^{\lfloor C_{\text{M2M}}/b_{\text{M2M}} \rfloor} \left(\left\lfloor \frac{C - b_{\text{M2M}}(n_{\text{M2M}})}{b_{\text{H2H}}} \right\rfloor + 1 \right) = $$

$$\sum_{n_{\text{M2M}}=0}^{\lfloor C_{\text{M2M}}/b_{\text{M2M}} \rfloor} \left(\left\lfloor \frac{C - \lceil n_{\text{M2M}}/M \rceil \cdot C_1}{b_{\text{H2H}}} \right\rfloor + 1 \right). \tag{2}$$

Due to the fact that S g.c.i. are allocated for MTC devices, the system state space can be represented as follows

$$\mathcal{X} = \bigcup_{s=0}^{S} \mathcal{X}_s, \tag{3}$$

$$\mathcal{X}_s := \left\{ n \in \mathcal{X} : \; b_{\text{M2M}}(n_{\text{M2M}}) = s \cdot C_1 \right\}. \tag{4}$$

It could be obtained that the process representing the system states is a reversible Markov process with equilibrium product form distribution

$$p(n_{\text{M2M}}, n_{\text{H2H}}) = G^{-1}(\mathcal{X}) \cdot \left(\frac{\rho_{\text{M2M}}}{Mb_{\text{M2M}}} \right)^{n_{\text{M2M}}} \left(\prod_{i=1}^{n_{\text{M2M}}} \left\lceil \frac{i}{M} \right\rceil \right)^{-1} \cdot \frac{\rho_{\text{H2H}}^{n_{\text{H2H}}}}{n_{\text{H2H}}!}, \tag{5}$$

$$n = (n_{\text{M2M}}, n_{\text{H2H}}) \in \mathcal{X}$$

where $G(\mathcal{X}) = p^{-1}(0,0)$ is the normalization constant.

3   Performance Analysis

Let us introduce blocking sets $\mathscr{B}_{\text{M2M}}$ and $\mathscr{B}_{\text{H2H}}$ as follows:

$$\mathscr{B}_{\text{M2M}} := \left\{ n \in \mathcal{X} : \; n_{\text{H2H}} b_{\text{H2H}} > C - b_{\text{M2M}}(n_{\text{M2M}}+1) \lor (n_{\text{M2M}}+1)b_{\text{M2M}} > C_{\text{M2M}} \right\}, \tag{6}$$

$$\mathscr{B}_{\text{H2H}} := \left\{ n \in \mathcal{X} : \; (n_{\text{H2H}}+1)b_{\text{H2H}} > C - b_{\text{M2M}}(n_{\text{M2M}}) \right\}. \tag{7}$$

The performance metric of the model is blocking probabilities for H2H and M2M users.

Having found the probability distribution $p(n)$, $n \in \mathcal{X}$, the resource allocation scheme for M2M traffic, one may compute its performance measures, notable blocking probabilities B_{H2H} and B_{M2M}:

$$B_{M2M} := \sum_{n \in \mathscr{B}_{M2M}} p(n) =$$

$$\sum_{s=0}^{S-1} \sum_{n_{H2H}=\lfloor (C-(s+1)\cdot C_1)/b_{H2H} \rfloor +1}^{\lfloor (C-s\cdot C_1)/b_{H2H} \rfloor} p(s \cdot M, n_{H2H}) + \sum_{n_{H2H}=0}^{\lfloor C_{H2H}/b_{H2H} \rfloor} p(S \cdot M, n_{H2H}), \tag{8}$$

$$B_{H2H} := \sum_{n \in \mathscr{B}_{H2H}} p(n) =$$

$$p\left(0, \left\lfloor \frac{C}{b_{H2H}} \right\rfloor \right) + \sum_{s=1}^{S} \sum_{n_{M2M}=(s-1)\cdot M+1}^{s \cdot M} p\left(n_{M2M}, \left\lfloor \frac{C-s\cdot C_1}{b_{H2H}} \right\rfloor \right). \tag{9}$$

We present an example of a single cell supporting video services for H2H (2 Mbit/s) users and M2M communications to illustrate the performance measures defined above. Assume that data traffic from MTC devices characterized by a minimum bandwidth requirement of 110 Kbps and mean size of 100 Kb. We let 1 b. u. equal to 110 Kbps. Let us consider a cell with a total capacity of $C = 50$ Mbit/s = 57 b.u. Let the traffic load due to H2H users $\rho_{H2H} = 10$. By changing the offered load ρ_{M2M} from 0 to 25 and number of g.c.i S (max 50% capacity can be allocated for MTC devices) we compute the performance measures (8), (9).

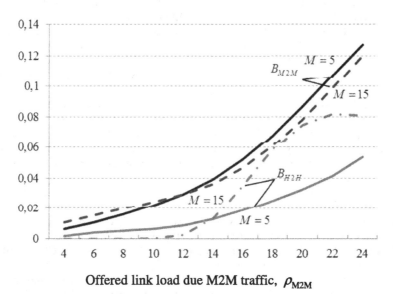

Offered link load due M2M traffic, ρ_{M2M}

Fig. 2. Performance measures

4   Conclusion

In this paper, we address a resource sharing problem for M2M traffic in LTE cell and give the analytical model with elastic traffic from MTC devices and minimum rate guarantees. The resource allocation scheme is based on fixed bandwidth intervals at which traffic from the MTC devices is served according to the PS discipline. We propose an analytical solution to calculate the model performance measures under the assumption of simplified physical model. An interesting task for future studies is the analysis of various admission control schemes for M2M communication and we plan to verify the model with simulations.

References

1. Stasiak, M., Glabowski, M., Wisniewski, A., Zwierzykowski, P.: Modelling and dimensioning of mobile wireless networks: from GSM to LTE, 340 p. Willey (2010)
2. 3GPP TS 36.300: Evolved Universal Terrestrial Radio Access (E-UTRA) and Evolved Universal Terrestrial Radio Access Network (E-UTRAN); Overall description; Stage 2 (Release 11) (2013)
3. Global Initiative for M2M Standardization,
   `http://www.3gpp.org/Global-Initiative-for-M2M`
4. Beale, M.: Future challenges in efficiently supporting M2M in the LTE standards. In: Proc. of the 10th Wireless Communications and Networking Conference WCNCW 2012, Paris, France, April 1-4, pp. 186–190. IEEE (2012)
5. Shin, S.Y., Triwicaksono, D.: Radio resource control scheme for machine-to-machine communication in LTE infrastructure. In: Proc. of the 3rd International Conference on ICT Convergence ICTC 2012, Jeju Island, Korea, October 15-17, pp. 1–6. IEEE (2012)
6. Zheng, K., Hu, F., Wang, W., Xiang, W., Dohler, M.: Radio resource allocation in LTE-advanced cellular networks with M2M communications. IEEE Communications Magazine 50(7), 184–192 (2012)
7. Kelly, F.P.: Reversibility and stochastic networks, 238 p. Cambridge University Press (2011)
8. Ross, K.W.: Multiservice loss models for broadband telecommunication networks. Springer, 343 p (1995)
9. Basharin, G.P., Samouylov, K.E., Yarkina, N.V., Gudkova, I.A.: A new stage in mathematical teletraffic theory. Automation and Remote Control 70(12), 1954–1964 (2009)
10. Gudkova, I.A., Samouylov, K.E.: Modelling a radio admission control scheme for video telephony service in wireless networks. In: Andreev, S., Balandin, S., Koucheryavy, Y. (eds.) NEW2AN/ruSMART 2012. LNCS, vol. 7469, pp. 208–215. Springer, Heidelberg (2012)
11. Samouylov, K.E., Gudkova, I.A., Maslovskaya, N.D.: A model for analysing impact of frequency reuse on inter-cell interference in LTE network. In: Proc. of the 4th International Congress on Ultra Modern Telecommunications and Control Systems ICUMT 2012, St. Petersburg, Russia, October 3-5, pp. 298–301. IEEE (2012)
12. Buturlin, I.A., Gaidamaka, Y.V., Samuylov, A.K.: Utility function maximization problems for two cross-layer optimization algorithms in OFDM wireless networks. In: Proc. of the 4th International Congress on Ultra Modern Telecommunications and Control Systems ICUMT 2012, St. Petersburg, Russia, October 3-5, pp. 63–65. IEEE (2012)

Random Access NDMA MAC Protocols
for Satellite Networks

José Vieira[1], Francisco Ganhão[1,2], Luis Bernardo[1], Rui Dinis[1,2],
Marko Beko[1,3], Rodolfo Oliveira[1], and Paulo F. Pinto[1]

[1] CTS, Uninova, Dep.º de Eng.ª Electrotécnica, Faculdade de Ciências e Tecnologia,
FCT, Universidade Nova de Lisboa, 2829-516 Caparica, Portugal
[2] Instituto de Telecomunicações, Lisboa, Portugal
[3] Universidade Lusófona de Humanidades e Tecnologias, Lisboa, Portugal

Abstract. Random Access (RA) approaches in satellite networks may
limit the achievable energy efficiency due to the random nature of channel
access, especially when hand-held terminals must operate with a very low
signal-to-noise ratio. This paper shows that Network Diversity Multiple
Access (NDMA) principles can be used to provide an energy efficient RA
scheme which satisfies some Quality of Service requirements. The paper
proposes performance models for the throughput and energy efficiency
considering a finite queue at each terminal and multiple packet retransmissions. Optimal parameters are calculated to maximize the energy efficiency while satisfying the throughput and error requirements, taking
into account the bit-rate constraint. The proposed system's performance
is evaluated for a Single-Carrier with Frequency Domain Equalization
(SC-FDE) scheme at the uplink. Results show that the proposed system
is energy efficient and can provide sustained bandwidth.

Keywords: Multipacket Detection, Network Diversity Multiple Access
(NDMA), Random Access, Satellite Network, Analytical Performance
Evaluation.

1   Introduction

Hand-held terminals in satellite networks have to be very energy efficient and the
network must be able to operate with a very low Signal-to-Noise ratio (SNR).
Random Access (RA) approaches in satellite networks usually limit the maximum supported Quality of Service (QoS). An example of a pure RA approach
is the Feedback Free-Network Diversity Multiple Access (FF-NDMA) [1] which
overcomes the transmission power limitations by sending multiple packet copies.
It estimates the number of transmitting users and defines the required number of
packet copies to successfully receive the packet at the satellite. However, most of
the Medium Access Control (MAC) protocols proposed for satellite networks that
satisfy QoS requirements [2], [3] either use Demand Assigned Multiple Access
(DAMA) or an hybrid mode with random access and reservations mechanisms.
Although these mechanisms introduce an initial overhead to reserve resources,
they allow the optimization of the subsequent channel accesses.

S. Balandin et al. (Eds.): NEW2AN/ruSMART 2013, LNCS 8121, pp. 427–438, 2013.

This paper proposes an hybrid-mode MAC protocol, Satellite Random NDMA (SR-NDMA), that combines RA with Scheduled Access (SA) transmissions. SR-NDMA is capable of satisfying the Packet Error Rate (PER) and maximum delay QoS constraints, and still optimize the energy consumption by balancing the SA and RA slots allocation. SR-NDMA is an evolution of Satellite NDMA (S-NDMA) [4], which only considered DAMA SA transmissions. S-NDMA schedules concurrent multiple transmissions from the hand-held Mobile Terminals (MTs) and applies a MultiPacket Reception (MPR) algorithm [5] at the satellite to separate them. Additional retransmissions are scheduled to enhance packet reception when difficult reception conditions persist, e.g. low SNR or high interference. SR-NDMA replaces the S-NDMA's initial reservation mechanism with a set of RA transmission slots, reducing the initial delay transmission at the cost of some energy efficiency. The developed work considers a Low Earth Orbit (LEO) satellite network, based on the Iridium satellite constellation, using Single Carrier-Frequency Domain Equalization (SC-FDE) scheme for the uplink transmission. Analytical performance models are proposed to compute the throughput and energy consumption, considering a constrained bandwidth and limited queue size constraints, while assuming an homogeneous Poisson load. This model is used to calculate the optimal energy transmission parameters that guarantees the QoS requirements for a known uniform average load.

The system's overview and the MAC protocol are presented in section 2. The system's performance and optimization are analyzed in sections 3 and 4. The performance is presented in section 5, and section 6 contains the conclusions.

2   System Characterization

This section characterizes the MT uplink transmission. MTs are low-resource battery-operated devices, send data to one satellite, and are grouped in sets; the satellite is a high resource device and runs a multi-packet detection algorithm with Hybrid-ARQ (H-ARQ) error control in real-time. The MPR capability of the receiver (the maximum number of packets that can be concurrently received) limits the maximum number of MTs in a set.

The uplink channel is composed of a continuous sequence of super-frames. Super-frames last for Round-Trip Time (RTT) seconds and are divided into an RA slot part and an SA slot part. If, for instance, two sets were defined in the system the structure would be (RA1, SA1, RA2, SA2). Each RA slot part has a fixed structure of N groups of n_0 slots each. The structure of the SA part is defined by the satellite in reaction to what happened (success, collision, etc.) during the previous super-frame. One downlink control channel is used to announce this information. Each time an MT has a data packet to transmit, it randomly chooses one group of n_0 RA slots and transmits the packet n_0 times, to increase the total packet energy and redundancy available at the receiver. If it has n packets to transmit it chooses n groups of n_0 slots. For each n_0 group the satellite detects which MTs transmitted (p) and decides how many SA slots ($n_1^{(p)}$) will be used in the next (1) super-frame to resolve the transmission.

A similar procedure happens now for the SA slots: the satellite decides how many slots should they now transmit in the following super-frame $(n_2^{(p)})$.

A sequence of consecutive RA and SA slots that deal with one packet's recovery from each MT is denoted as an epoch. The epoch's duration is limited to a maximum of R super-frames with SA slots ($R + 1$ in total). Data packets not received within an epoch are retransmitted in a future epoch as fresh start packets.

If there is only one set active, all slots of the super-frame belong to the set. If there are two, the slots are divided by 2. In general terms each set has B_{max} slots available per RTT. The satellite uses the energy optimization algorithm proposed in the following sections for the overall scheduling constrained to a maximum of B_{max} slots per set.

2.1   Medium Access Control Protocol

The SR-NDMA MAC protocol runs within an MT set. Let us assume that a set has J MTs, and T is the period equal to the highest RTT in a given MT set.

Random Access Mode. The number of MTs that transmit in each RA slot is a random variable P, that should satisfy $P \leq J$. The maximum throughput of the system is given by $\frac{NJ}{T}$, due to the multipacket reception (MPR) from J MTs. In order for the transmission queue to be stable, the throughput should be higher or equal to the total load, λJ, where λ denotes the average packet generation rate per MT.

When n_0 is set high enough, it is possible to successfully transmit packets using the initial RA transmissions (as in FF-NDMA). However, this also reduces the energy and bandwidth efficiency, which is maximized for $n_0 = n_i^{(p)} = 1$ (additional retransmissions are sent one at a time). Therefore, there is a trade-off of n_0 which also depends on value of T.

Scheduled Access Mode. For each unsuccessful reception of data packets using RA slots, additional SA slots are scheduled. From this point on, the protocol behaves like in S-NDMA [4] and uses a pure DAMA approach for further retransmissions: i.e., individual packets not received at super-frame l are scheduled for retransmission in super-frame $l + 1$, and can be distributed over up to R retransmission super-frames, after which the epoch ends. If all packets are correctly received before the Rth retransmission, the epoch also ends.

The number of allocated SA slots for a packet in the lth retransmission super-frame of an epoch is denoted as $n_l^{(P)}$, where $l \leq R$, and the vector with all $n_l^{(P)}$ values is denoted as $\mathbf{n}^{(P)} = \left[n_0, n_1^{(P)}, ..., n_R^{(P)} \right]$.

In [4] it was shown that the maximum energy efficiency is achieved when $n_l^{(P)} = 1$ for all P and l values, possibly leading to a very large R value when SNR is low. However, due to the high RTT for a satellite network, R cannot be too high. Given a maximum epoch time duration, τ_{max}, R's value should be below $\lfloor \tau_{max}/T \rfloor$.

SR-NDMA differs from S-NDMA because SR-NDMA defines a matrix $\mathbf{n} = [\mathbf{n}^{(1)}, ..., \mathbf{n}^{(J)}]$, where all possible values of $0 < P \leq J$ must be specified, whereas for a pure DAMA approach P is known *a priori*. SR-NDMA trades-off a lower delay for a worse bandwidth efficiency, since some initially allocated slots might be unused. The next section addresses the problem of defining the optimal values for $\mathbf{n}$ that minimize energy consumption when the bandwidth allocated to the set is B_{max} slots per RTT/super-frame.

The receiver adopted at the satellite is the linear version of the uncoded multipacket receiver in [5] for SC-FDE systems. This paper considers the uplink transmission of a satellite system with Single Carrier with Frequency-Domain Equalization (SC-FDE).

3   Analytical Model

This section proposes an analytical model for the throughput and energy consumption, which quantifies the influence of PER and the reception power at the satellite. It is assumed that each of the J MTs generates new packets according to an homogeneous Poisson process with rate λ, and that the satellite is able to discern up to J colliding packets. It is also assumed that no errors occur in the downlink control channels; a known uniform average bit energy to noise ratio (E_b/N_0) value is received at the satellite from each MT; and perfect time advance mechanisms are applied.

3.1   Scheduled Access Phase

After the initial RA transmission of a packet in n_0 RA slots, the satellite knows that p MTs transmitted, and it schedules further retransmissions of failed packets on SA slots minimizing energy consumption. This is exactly what happens in S-NDMA for the second and further retransmissions. Therefore, the SR-NDMA performance model is incremental and adds only the influence of the initial RA transmission to the S-NDMA model in [4].

It was shown in [4] that the performance of the scheduled access phase depends on p, on the slot allocation vector used, $\mathbf{n}^{(p)}$, and on E_b/N_0 value at the reception. The packet transmission behavior can be defined by a hidden Markov chain with a random state vector denoted by $\Psi^{(p,R)} = \{\psi_k^{(p,R)}, k = 0...R\}$, which defines the number of MTs whose packets were successfully received and stopped transmitting at the end of the retransmission super-frame $k = 0, ..., R$.

Considering only the retransmission super-frames up to $l \leq R$, the state space of $\Psi^{(p,l)}$ is denoted by the set $\Omega_p^{(l)}$. It contains all the vector elements $K^{(l)}$ (with dimension $l + 1$) that satisfy $\sum_{k=0}^{l} K_k^{(l)} = p$. Each state $\Psi^{(p,l)} = \{\psi_0^{(p,l)} = K_0^{(l)}, ..., \psi_l^{(p,l)} = K_l^{(l)}\}$ defines the set of transmission sequences where $K_0^{(l)}$ MTs stopped transmitting after the initial RA slots (super-frame 0), $K_1^{(l)}$ MTs stopped transmitting after the SA slots at super-frame 1, and so on until $K_l^{(l)}$ MTs transmitted in the SA slots at super-frame l.

The average PER at the $(l+1)$th super-frame with p MTs is denoted by $PER_P\left(\Psi^{(p,l)}\right)$, and for the proposed receiver it is calculated using [5], where H_k is the matrix with the channel response. This matrix has zero coefficients for the epoch's slots with idle transmissions, i.e. non-existing transmissions.

A packet is incorrectly received after l retransmission super-frames if it is transmitted in all slots and its reception still fails for the $l+1$ super-frame. The expected number of packets received with errors during an epoch until the lth retransmission super-frame [4] is $\mathbb{E}\left[err\left(\Psi^{(p,l)} = K^{(l)}\right)\right] = K_l^{(l)}PER_p\left(\Psi^{(p,l)}\right)$. Assuming that packet failures are independent, the packet error probability for an epoch $\Omega_P^{(l)}$ (with up to l retransmission super-frames) is given by

$$P_{err}\left(\Omega_p^{(l)}\right) = \sum_{K^{(l)} \in \Omega_p^{(l)}} \frac{1}{p} Pr\left\{\Psi^{(p,l)} = K^{(l)}\right\} \mathbb{E}\left[err\left(\Psi^{(p,l)} = K^{(l)}\right)\right]. \quad (1)$$

An upper bound for the average packet error probability during an epoch can be calculated using [4]

$$P_{err}\left(\Omega_p^{(R)}\right) \le PER_P\left(\Psi^{(p,R)} = [0,0,...,p]\right). \quad (2)$$

The expected number of used slots can be calculated using [4]

$$\mathbb{E}\left[tx\left(\Omega_p^{(R)}\right)\right] = \sum_{K^{(R)} \in \Omega_p^{(R)}} Pr\left\{\Psi^{(p,R)} = K^{(R)}\right\} tx\left(\Psi^{(p,R)} = K^{(R)}\right), \quad (3)$$

where $tx\left(\Psi^{(p,R)} = K^{(R)}\right)$ denotes the expected number of used slots for a given state $K^{(R)} \in \Omega_P^{(R)}$.

3.2   System's Steady State

Let q_m denote the number of packets in the MT's queue at the beginning of the mth super-frame. Let X_m^0 denote the number of packets really transmitted in the RA slots of the mth super-frame, and X_m^l the number of packets being retransmitted in the SA slots for the lth time, with $1 < l \le R$. Obviously, the number of packets really sent in RA slots is directly related with the number of packets in the MT's queue, and is limited by N. Therefore the probability mass function of X_m^0 is

$$Pr\left\{X_m^0 = x\right\} = \begin{cases} Pr\left\{q_m = x\right\}, x < N \\ Pr\left\{q_m \ge N\right\}, x = N \end{cases}. \quad (4)$$

The probability of an MT transmitting a packet in one of the RA slots at the mth super-frame, P_{tx}^m, is given by

$$P_{tx}^m = \frac{\mathbb{E}\left[X_m^0\right]}{N} = \frac{1}{N} \sum_{x=1}^{N} x Pr\left\{X_m^0 = x\right\}. \quad (5)$$

The sending of packets in subsequent retransmissions depends on the packets' success on previous transmissions. Therefore, the probability of v packets having success when x packets are transmitted, $Pr\{V_m^l = v \mid X_m^l = x\}$, for $0 \leq l \leq R$, is introduced. This probability depends on the number of MTs contending in each epoch. Assuming that epochs are independent and that errors are uncorrelated with q_m, the number of MTs contending in each epoch and the number of packets successfully received can be approximated with binomial distributions, where

$$Pr\{V_m^l = v \mid X_m^l = x\} = \sum_{p=0}^{J-1} f_b\left(J-1, p, P_{tx}^m\right) f_b\left(x, v, 1 - P_{err}\left(\Omega_{(p+1)}^{(l)}\right)\right).$$

(6)

$f_b(J, k, p) = \binom{J}{k} p^k (1-p)^{J-k}$ denotes the Binomial probability mass function. The PER can be calculated using (1), and depends on the $\mathbf{n}$ and E_b/N_0 values. With this, the conditional probability mass function of the number of retransmitted packets for the lth time in an epoch is given by

$$Pr\{X_m^{l+1} = x \mid X_m^l\} = \sum_{i=x}^{N} Pr\{X_m^l = i\} Pr\{V_m^l = i - x \mid X_m^l = i\}, \quad (7)$$

which recursively defines the probability mass function, where the expected value is $E\left[X_m^l\right] = \sum_{x=0}^{N} x\, Pr\{X_m^l = x\}$. The expected number of successfully received packets at retransmission l, $E\left[V_m^l\right]$, is simply the number of packets that were transmitted during super-frame $l+1$, i.e., $E\left[V_m^l\right] = E\left[X_m^l\right] - E\left[X_m^{l+1}\right]$.

The expected number of unsuccessfully received packets during retransmission super-frame R can be calculated using (7) with $l = R + 1$, thus the expected number of successful transmissions is given by $E\left[V_m^R\right]$. Remember that a packet is incorrectly received at the end of the epoch when the reception fails at the Rth super-frame. The probability of having f failed packets, $Pr\{\phi_m^R = f\}$, is given by

$$Pr\{\phi_m^R = f\} = \sum_{x=f}^{N} \left(Pr\{X_m^R = x\} Pr\{V_m^R = x - f \mid X_m^R = x\}\right). \quad (8)$$

When a packet arrives at an MT's queue, it waits until its first transmission, defining the beginning of a new epoch (for it). Different epochs are separated by multiples of T seconds. Therefore, there will be concurrent epochs running. Assuming that packet arrivals are defined by a Poisson process with an average load λ, the number of packets arriving in T seconds is defined by $Pr\{A = k\} = (\lambda T)^k e^{-\lambda T}/k!$.

RA slots are repeated every T seconds, and have a service capacity of N packets per super-frame. Failed epoch transmissions are returned to the queue.

Assuming that the queue has a maximum capacity of Q_{max} packets, the q_m distribution can be defined by the following Discrete Time Markov Chain (DTMC):

$$Pr\{q_{m+1} = Q\} = \sum_{f=0}^{N} (Pr\{\phi_m^R = f\} Pr\{q_m < N\} Pr\{A = Q - f\} +$$

$$\sum_{i=0}^{min(Q_{max}-N,Q-f)} Pr\{q_m = N + i\} Pr\{A = Q - f - i\}) \qquad (9)$$

for $Q < Q_{max}$. When $Q = Q_{max}$,

$$Pr\{q_{m+1} = Q_{max}\} = 1 - \sum_{i=0}^{Q_{max}-1} Pr\{q_{m+1} = i\} . \qquad (10)$$

When the RA slot utilization is less than one ($\rho_R < 1$), it can be shown that this DTMC converges to a steady state distribution, denoted by ξ, which can be calculated using an iterative numerical method [6]. In the following equations the m index is dropped and the queue distribution refers to ξ. The RA slots utilization is defined by

$$\rho_R = \frac{\lambda T}{N} . \qquad (11)$$

3.3   Throughput and Channel Utilization Analysis

The throughput is calculated by taking into account the number of successfully received packets during a RTT (T), for all retransmission stages. Considering all J MTs, it is given by

$$S = \frac{J}{T} \sum_{l=0}^{R} \mathbb{E}[V^l] = \frac{J}{T} (\mathbb{E}[X^0] - \mathbb{E}[X^{R+1}]) . \qquad (12)$$

The measured throughput results from lost packets due to channel errors and dropped packets due to queue overflow (Q_{max} is the queue size). From S, it is possible to obtain the total packet loss ratio, ε, given by

$$\varepsilon = 1 - \frac{S}{\lambda J} . \qquad (13)$$

A relevant parameter is the channel utilization ratio, ρ_b, defined as the ratio of the average number of used slots over the number of allocated slots (B_{max}) per RTT, which should be always below 1. ρ_b can be calculated using (14), which counts the total allocated RA slots and the average number of scheduled SA slots per RTT. The number of used SA slots for a given number of transmitted packets at the lth super-frame and P MTs is defined by the configuration parameters $n_l^{(P)}$, and its average value is influenced by the probability of having P transmitting MTs per epoch.

$$\rho_b = \frac{1}{B_{max}} \left(N n_0 + \sum_{l=1}^{R} \mathbb{E}[X^l] \sum_{p=0}^{J-1} f_b(J - 1, p, P_{tx}) n_l^{(p+1)} \right) . \qquad (14)$$

3.4   Energy Analysis

The energy per useful packet, denoted by $EPUP$, measures the necessary average energy to successfully receive a packet at the satellite. The expression in (15) depends on the average number of slots that the MT uses during an epoch, the transmitted packet's success probability, and a packet's transmission energy (E_p) - the last one is related to the MT's transmission power [4]. A Bayesian approach is used to account the influence of the number of transmitting MTs per epoch.

$$EPUP = \sum_{p=0}^{J-1} f_b\left(J-1, p, P_{tx}\right) \frac{\mathbb{E}\left[tx\left(\Omega_{p+1}^{(R)}\right)\right] E_p}{1 - P_{err}\left(\Omega_{p+1}^{(R)}\right)} . \tag{15}$$

4   Optimization

The objective is to minimize the EPUP and to support the total MTs' load λJ using only B_{max} slots per RTT such that a PER threshold (PER_{max}) is satisfied for a given R value. SR-NDMA's performance is influenced by the average E_b/N_0 measured at the satellite, the parameters $\mathbf{n}$, N, Q_{max}, and the network load. This section illustrates how such optimization problem can be solved.

The value of N is greater or equal to $\lceil \lambda T \rceil$ (ceiling operation), to allow $\rho_R < 1$. A higher value of N reduces packet drops but leads to some energy inefficiency since the number of SA slots is reduced.

When $R = 0$, all slots must be allocated as RA, and n_0 must be set to the highest value of $N\zeta_R^{J1}$ that matches the available B_{max}, to minimize the EPUP, satisfying $\varepsilon < PER_{max}$. For $R > 0$, for a given N such that $\rho_R < 1$, it is necessary to calculate the optimal matrix $\mathbf{n}^\star$ and $E_b^\star/N_0$. The optimization problem can be defined as:

$$\text{Minimize:}\quad EPUP\left(\mathbf{n}^\star, E_b^\star/N_0\right)$$
$$\text{Subject to:}\quad \rho_b < 1 \quad \text{and} \quad \varepsilon < PER_{max} . \tag{16}$$

The proposed heuristic is to initially calculate the minimum number of total slots (RA and SA), $\zeta_R^{(p)}$, that guarantees a PER value below the PER_{max} bound for a range of E_b/N_0 values and for all $1 \le p \le J$ using (2). Next, similarly to the S-NDMA optimization proposed in [4], for each p value, the state space defined by all combinations of E_b/N_0, RA and SA slots with R retransmissions and $\zeta_R^{(p)}$ slots is searched using an alternating minimization approach, to look for the $\mathbf{n}^{p\star}$ and E_b/N_0 configuration that minimizes the EPUP (15), satisfying the bandwidth and QoS restrictions.

$\mathbf{n}^{p\star}$ cannot be directly used, because the optimal value of n_0 depends on p, $n_0(p)$. Therefore, sub-optimal matrices were devised, where the n_0 value for all p is replaced by one of the $n_0(p)$ of the optimal matrix. A second optimization

---

[1] Being ζ_R^J the total number of RA and SA slots.

iteration is done for each sub-matrix by reconfiguring the SA slots to maintain the $\zeta_R^{(p)}$ value, and searching neighbour values of the optimal $E_b^\star/N_0$ for the pth probability mass function distribution. The $n^\star$ and $E_b^\star/N_0$ values are defined by the sub-optimal matrix that leads to the minimum EPUP value for a given load. The optimal matrix depends on the B_{max} value, so different matrices should be calculated for different values of B_{max}.

The optimization process can be run offline, to produce a set of configuration tables, indexed by B_{max}, D_{max}, J and the total load, which can be used by the satellite to configure in real time the RA and SA slots configuration and the transmission parameters for each individual MT.

5   Performance Analysis

The system performance is analysed in this section, considering the PER, throughput and EPUP. A LEO satellite constellation is modelled, with circular orbits at an altitude of 781 Km (like in Iridium), and a RTT of 23.8 ms was considered for the farthest MT. The severely time dispersive channel of [4] was considered, with rich multipath propagation and uncorrelated Rayleigh fading for each path and MT (similar results were obtained for other fading models). To cope with channel correlation for different retransmissions, the Shifted Packet technique from [7] was considered, where each transmitted block has a different cyclic shift. MTs transmit uncoded data blocks with $N = 256$ symbols selected from a QPSK constellation with Gray mapping with an $814.9\mu s$ transmission time, which is spread with a factor of 128. Therefore the RTT is defined as $T = 32$ slots, making $B_{max}=32$. As an example, the paper considers $PER_{max} \leq 1\%$ for $R = 2$, $Q_{max}=25$ packets and an MT set composed by $J = 5$ MTs.

5.1 $n^{p\star}$ Calculation

The minimum number of packet transmissions $\zeta_R^{(p)}$, that satisfy the PER_{max} threshold should be computed first. Figure 1(a) illustrates the calculation of $\zeta_R^{(p)}$, considering PER_{max}. It shows that the PER condition can be satisfied for all values of E_b/N_0, requiring the allocation of more slots for lower E_b/N_0 or higher p values. The number of slots allocated per MT decreases for higher p making SR-NDMA more bandwidth efficient.

In a second step, the optimal $n^{p\star}$ is calculated for the different p values. Figure 1(b) depicts the optimal $n_0(p)$ values, showing that it has a significant variation with p for higher E_b/N_0 values, where the packet copies are mainly used to separate the packets concurrently transmitted. Near and below 0dB, $n_0(p)$ has a small variation with p since the copies are mainly used to reduce the PER. From this set of values it is possible to calculate the sub-optimal matrices and select the combination of E_b/N_0 and $n^\star$ that minimize the EPUP and satisfy the QoS constraints for B_{max}.

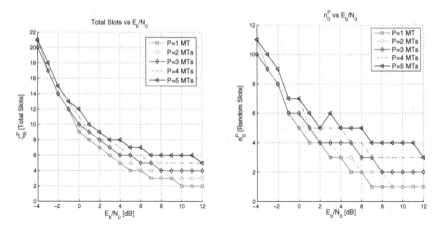

(a) ζ_R^P minimum that satisfies PER $\leq 1\%$ (b) Optimal $n_0(p)$ values over minimum over E_b/N_0 E_b/N_0

Fig. 1. Optimal parameters

5.2 B_{max} Parameter

The number of slots allocated to a set of MTs, B_{max}, depends on the number of MTs connected to the satellite. For a small number, a single set of MTs is used; otherwise, slots have to be divided by different sets. The best achievable energy efficiency is when a single set of MTs use all slots, corresponding to $B_{max} = 32$ slots. Figure 2(a) depicts the variation of ρ_b over B_{max} for different values of E_b/N_0 for the optimal configuration that satisfies the requirements. It shows that an higher channel utilization is required for smaller E_b/N_0 values, i.e., more packet copies are required to satisfy the same PER_{max} requirement. Therefore, the minimum transmission power (associated to lower E_b/N_0 values) can only be used when B_{max} is high. Otherwise, to make a better utilization of the available bandwidth, the energy has to be raised. The ideal value of E_b/N_0 should be the one that is closest to the limit of the band utilization, but does not surpasses it, allowing a better energy efficiency (i.e. lower EPUP).

5.3 N and n_0 Parameter

N is the number of RA groups per super-frame/RTT. The figures in this section illustrate how parameter N influences the throughput, when B_{max} is fixed. The E_b/N_0 in use is variable: for the various λJ and N, the lowest E_b/N_0 that verifies the optimization constraints is used. $n_0(p)$ is chosen according to the E_b/N_0 value for 5 MTs (represented in Figure 1(b)). As expected, Figure 2(b) shows that N limits the maximum throughput to NJ/T. The system is saturated for a total load above this limit, leading to discarded packets above PER_{max}, and packet transmissions over multiple epochs.

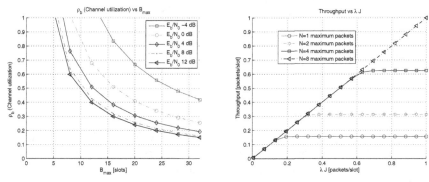

(a) ρ_b over B_{max} and E_b/N_0, for $N=1$,   (b) Throughput over the total load λJ
$\lambda J = 0.50$ packets/slot, optimal $n_0(p)$ for
3 MTs

Fig. 2. ρ_b and throughput tendencies when the E_b/N_0 and N varies respectively

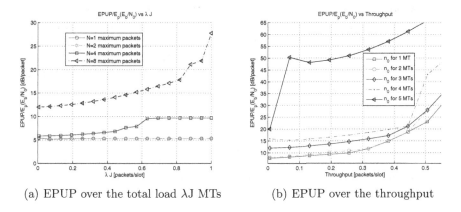

(a) EPUP over the total load λJ MTs   (b) EPUP over the throughput

Fig. 3. Different number of N maximum allowed packets, $B_{max} = 32$ slots, optimal $n_0(p)$ for 3 MTs

Figure 3(a) depicts $(EPUP/E_p)(E_b/N_0)$, which considers the energy at the receiver, removing the dependence on all propagation issues [4]. This figure shows that when N is increased until $N = 4$ the EPUP does not change significantly. However, for $N = 8$ it increases continuously, because the slots left for SA retransmissions $(B_{max} - n_0 N)$ start to be too short, requiring higher E_b/N_0 values to satisfy the $\rho_b < 1$ condition. N should balance the required network capacity and the $EPUP$ increase, being mainly conditioned by the B_{max} value.

Figure 3(b) depicts the EPUP measured for different values of throughput for $B_{max} = 16$ slots and $N = 4$ packets/RTT. Given that all setups satisfy the PER_{max} requirement and use the same N, they carry the same maximum throughput. However, a higher n_0 value increases the use of RA slots, thus reducing the available number of SA slots, leading to a lesser energy efficiency. The figure exposes a configuration trade-off for SR-NDMA: n_0 can be used to maximize the energy efficiency (setting $n_0(p)$ for $p = 1$ MT), but this might lead to an higher delay, since the error would be higher for the first transmission.

6 Conclusion

This paper presented SR-NDMA, an energy efficient RA protocol designed for scenarios with a high RTT, such as a satellite network. SR-NDMA combines an initial random access with additional scheduled accesses for packet transmissions, avoiding the requirement for an initial reservation phase that exists for pure DAMA protocols, such as S-NDMA. The results for the SR-NDMA system configured with optimal parameters show that SR-NDMA is capable of satisfying the application's quality of service requirements and adapt to the available bandwidth, minimizing the energy consumption. Therefore, it can be a valid option for the satellite component of future integrated telecommunication networks.

Acknowledgments. This work was supported by the FCT/MEC projects CTS PEst-OE/EEI/UI0066/ 2011; IT PEst-OE/EEI/LA0008/ 2013; MP-Sat PTDC/EEA-TEL/099074/2008; OPPORTUNISTIC-CR PTDC/EEA-TEL/115981/ 2009; ADCOD PTDC/EEA-TEL/099973/2008 and Femtocells PTDC/EEA-TEL/120666/2010; Ciência 2008 Post-Doctoral Research grant as well as grant SFRH/BD/66105/2009.

References

1. Madueño, M., Vidal, J.: Joint PHY-MAC layer design of the broadcast protocol in ad-hoc networks. Journal on Selected Areas in Comm. (2005)
2. Peyravi, H.: Medium Access Control Protocols Performance in Satellite Communications. IEEE Comm. Mag. 37, 62–71 (1999)
3. De Gaudenzi, R., Herrero, O.: Advances in Random Access Protocols for Satellite Networks. In: IEEE IWSSC 2009, pp. 331–336 (2009)
4. Ganhão, F., Bernardo, L., Dinis, R., Barros, G., Santos, E., Furtado, A., Oliveira, R., Pinto, P.: Energy-Efficient QoS Provisioning in Demand Assigned Satellite NDMA Schemes. In: IEEE ICCCN 2012 (2012)
5. Ganhão, F., Dinis, R., Bernardo, L., Oliveira, R.: Analytical BER and PER Performance of Frequency-Domain Diversity Combining, Multipacket Detection and Hybrid Schemes. IEEE Trans. on Comm. 60, 2353–2362 (2009)
6. Bolch, G., Greiner, S., deMeer, H., Tivedi, K.: Queueing Networks and Markov Chains: Modeling and Performance Evaluation with Computer Science Applications. John Wiley & Sons (1998)
7. Dinis, R., Montezuma, P., Bernardo, L., Oliveira, R., Pereira, M., Pinto, P.: Frequency-domain multipacket detection: a high throughput technique for SC-FDE systems. IEEE Trans. on Wireless Comm. 8, 3798–3807 (2009)

Author Index